PLANETARY INTERIORS

PLANETARY
INTERIORS

William B. Hubbard

Professor of Planetary Sciences
University of Arizona

VNR VAN NOSTRAND REINHOLD COMPANY

Copyright © 1984 by Van Nostrand Reinhold Company Inc.

Library of Congress Catalog Card Number: 83-23296
ISBN: 0-442-23704-9

Manufactured in the United States of America

Published by Van Nostrand Reinhold Company Inc.
135 West 50th Street
New York, New York 10020

Van Nostrand Reinhold Company Limited
Molly Millars Lane
Wokingham, Berkshire RG11 2PY, England

Van Nostrand Reinhold
480 Latrobe Street
Melbourne, Victoria 3000, Australia

Macmillan of Canada
Division of Gage Publishing Limited
164 Commander Boulevard
Agincourt, Ontario M1S 3C7, Canada

14 13 12 11 10 9 8 7 6 5 4 3 2 1

Library of Congress Cataloging in Publication Data
Hubbard, William B.
   Planetary interiors.

   Includes index.
   1. Planets—Internal structure.   I. Title.
QB603.I53H83   1984   559.9′2   83-23296
ISBN 0-442-23704-9

Preface

Space exploration has brought rapid development in the study of the solar system; the reference books of this new era are only just beginning to be written. The field has moved so fast and come so far during the past few years that a single author has difficulty in writing a current, comprehensive book. Nevertheless, new books on planetary science are not lacking. Quite a few books on various specialized topics in planetary science are now available, but most of these are actually collections of papers written by various experts. This is the way that books tend to be written in a fast-moving area of science. Many of these books are very well conceived and edited; for example, the space science series published by the University of Arizona Press.

The flow of new data on the planets may slow for a while, because new U.S. deep-space missions, which typically have a long lead time, will be few in number for some time to come. So perhaps the time has come for a more leisurely assessment of what we have learned about the interiors of the planets during this period of high scientific activity, and, most important, to explain it to interested individuals who may not specialize in space science, or who are just beginning to study the subject. For this latter task, a compendium of research papers and review papers, however expertly written and edited, is not really very appropriate.

There is a broad spectrum of possible planetary science books. This ranges from the entirely nontechnical, descriptive general text intended for liberal-arts majors up through a monograph on, say, celestial mechanics. This book is aimed at the middle of the spectrum. It is intended for those who have some knowledge of the various disciplines which form the basis for the study of planetary interiors, such as physics, chemistry, geology, geophysics, astronomy, and their various subspecialties. I have tried to maintain a level of presentation more or less appropriate for an individual just finishing a bachelor's degree or just beginning graduate study, in one of these subjects. Most of the material is intended to be current as of about mid-1983.

I have tackled the job of presenting the subject of planetary interiors from a unified point of view. I hasten to add that, as will become evident in the following pages, there frequently is more than one point of view about important issues in the subject. I have tried to represent these as well as possible, but the reader who wants to go more deeply into the subject should consult the references given in the various chapters.

It is a pleasure to acknowledge the assistance of many colleagues during the writing of this volume. These include Bruce Bills, William Boynton, Michael Drake, Gail Georgenson, Richard Greenberg, Lonnie Hood, Eugene Levy, and Charles Sonett.

William B. Hubbard

Contents

PLANETARY INTERIORS

1
Introduction

The study of planetary interiors represents a subdiscipline of the relatively new field of planetary science. This new field has grown up on the boundary of astronomy and earth science, and derives many of its fundamental approaches from the two older fields. Similarly, the sub-area of planetary interiors is ultimately derived from stellar interior physics and geophysics. Like these two parent disciplines, the study of planetary interiors is primarily concerned with deducing the properties of matter under conditions where direct observation is difficult or impossible. It is a science of modeling, using mathematical constructs to synthesize and ultimately predict quantities which are accessible to observation. The intellectual fertility of the discipline depends upon the scope of the synthesis which it is able to achieve. In order to put the present level of achievement in planetary interiors research into perspective, let us consider two great advances in adjoining fields.

Prior to the 1960s, the study of stellar interiors was about where planetary interior work is today. It was possible to use known physical principles to construct static models of stars which were in reasonable agreement with observation. The bulk composition of stars could be deduced with a reasonable degree of certainty. What was missing was a grand, unified approach, which could begin with stellar material of a standard initial composition, allow it to form a star, follow the star through all its processes of evolution, and at every stage satisfactorily reproduce all the observable parameters of stars. The missing element was supplied in the 1960s by L. G. Henyey and collaborators, who showed how modern computers could be used to integrate the time-dependent equations of stellar evolution, and to follow stellar interiors in detail through eons of changes. For the first time, it became possible to begin with a very general set of initial models and mathematically evolve them to reproduce the observed stellar main sequence (see Chapter 2). Moreover, the technique made it possible to study advanced evolutionary phases of stars, and also to mathematically simulate the observed properties of highly evolved stellar assemblages. Except for a number of troublesome problems which still remain, the major aspects of the evolution of spherically symmetric, individual stars now seem to be well understood.

Why can't we use the same approach to understand planets? The basic reason is that planets are far more complicated than stars. Chemistry exists in a planetary interior, unlike a stellar interior, where temperatures are too high. Because of the existence of chemistry, reactions can occur which may or may not proceed fully to equilibrium. Important processes which affect the evolution of planetary interiors depend to an extraordinary extent on the precise temperature range, yet the temperature in turn depends upon still other processes; and few of these processes are well understood in terms of physical models.

Moreover, the present state of planets is in general sensitive to the initial state, unlike the case for stars, where we usually start with a standard composition, and the object soon "forgets" the initial configuration. There is no "standard" composition for planets, although we now suspect that there are rules which govern their initial compositions (about which we will have more to say later).

A corresponding great synthesis in geophysics has doubtless been the development of plate tectonics (a brief description is given in Chapter 7). With this concept, we are able to relate the major observed geological features of the earth's surface to fundamental processes in the earth's interior. We now understand, to a large extent, why the surface of the earth looks the way it does. This synthesis has required the introduction of additional new concepts such as the lithosphere, the asthenosphere, and solid-state convection. We are now hoping that these concepts will prove useful in understanding the interiors of other planets, although the "breakthrough" nature of plate tectonics has not yet really extended beyond the earth.

But if the field of planetary interiors is still looking for its first major breakthrough, it is not without solid achievements and promising leads. Unlike a stellar interior, a planetary interior can at least be landed upon, and we can directly sample its surface. Geological characteristics of the surface may bear traces of long-extinct interior processes. The external gravitational and magnetic fields of a planet can be minutely studied for further clues about the planet's interior. Such investigations have largely been made possible by sophisticated unmanned spacecraft launched by the U.S. and the U.S.S.R. during the last two decades, which have now reached all the planets in the solar system except the outermost three. We now have available a broad and rapidly expanding data base for the basic parameters of the planets and their satellites. Table 1-1 presents some of the most fundamental of these parameters; additional data are presented in appropriate chapters later in the book.

Planetary scientists have a general hunch about the way a unifying picture of planetary interiors will come to be created. We must start at the same point as the stellar interiors specialist, with a diffused mass of primordial material of standard composition. Out of this material we make not only a star, but also planets. The difference is that something happens to the composition of the material as a small fraction of it condenses to form planets. If we can correctly deduce what happened during the critical initial stage, it should be possible in principle to account for every aspect of planetary interiors observed today.

Thus, if there is a crucial chapter in this book, it is Chapter 2. This chapter presents the deduced primordial composition of the solar system, which comprises the set of basic building blocks for modern theories about planetary interiors. To be accepted, a synthesis of planetary interiors must show how this composition, acted upon by accepted physical processes, leads to the present bulk compositions of all planetary bodies. We shall not present such a synthesis in this book, for (a) the book is not about cosmogony, and (b) none yet exists. We will, however, mention some useful general concepts about relative volatility which seem to have some predictive power.

Chapter 3 discusses thermodynamic behavior of the principal constituents of planetary interiors in the relevant temperature and pressure regimes. Chapters 4, 5, and 6 discuss the principal tools used as diagnostics of planetary interiors: gravitational fields, heat flow, and magnetic fields.

This book is not intended as a text on the earth's interior, to which we therefore devote relatively little space. However, the earth's interior serves as a starting point for generalizations to the interiors of

Table 1-1. Physical and Orbital Parameters of Planets and their Satellites*

Note that for the four Jovian planets, a is the equatorial radius at one bar pressure; for Mars and the earth, a is the mean equatorial surface radius; and for all other planets and satellites, a is the mean surface radius, except that the mean semiaxes are given for highly nonspherical satellites. R is the semimajor axis of planetary orbits, and the mean distance from the primary in the case of satellites. Remaining entries are the total mass M, orbital eccentricity e (in the case of Io and Europa, the forced eccentricity is given), orbital period P ($= 2\pi/n$, where n is the mean motion), and rotational period P_{rot}. The mass of the earth is 5.976×10^{27} g.

OBJECT	a (km)	R (km)	M (earth masses)	e	P (days)	P_{rot} (days)
Mercury (no satellites)	2439	5.79×10^7	0.055	0.206	87.96935	58.6461
Venus (no satellites)	6051.5	1.082×10^8	0.815	0.007	224.70	243.01
Earth	6378.16	1.496×10^8	1.000	0.017	365.26	0.99726
Moon	1737.53	3.844×10^5	0.0123	0.055	27.322	27.322
Mars	3394	2.279×10^8	0.108	0.093	686.98	1.026
Phobos	$27 \times 22 \times 19$	9378	1.6×10^{-9}	0.015	0.318910	0.318910
Deimos	$15 \times 12 \times 11$	23,459	3.3×10^{-10}	0.00052	1.262441	1.262441
Jupiter	71,492	7.78×10^8	317.735	0.0484	4333	0.41354
Adrastea	20	1.28×10^5		0	0.295	
Metis	20	1.28×10^5		0	0.295	
Amalthea	$135 \times 85 \times 75$	1.81×10^5		0.003	0.489	0.489
Thebe	40	2.21×10^5		0	0.675	
Io	1815	4.22×10^5	0.0149	0.0041	1.769	1.769
Europa	1569	6.71×10^5	0.00815	0.0101	3.551	3.551
Ganymede	2631	1.070×10^6	0.0249	0.0015	7.155	7.155
Callisto	2400	1.880×10^6	0.0180	0.01	16.69	16.69
Leda	5	1.111×10^7		0.146	240	
Himalia	90	1.147×10^7		0.158	251	
Lepithia	10	1.171×10^7		0.130	260	
Elara	40	1.174×10^7		0.207	260	
Ananke	10	2.070×10^7		0.17	617	

Carme	15	2.235×10^7		0.21	692
Pasiphae	20	2.33×10^7		0.38	735
Sinope	15	2.37×10^7		0.28	758	0.44403
Saturn	60,330	1.427×10^9	95.147	0.0557	10,759
1980S28	20 × 10	137,670		0.002	0.6019
1980S27	70 × 50 × 40	139,350		0.003	0.6130
1980S26	55 × 45 × 35	141,700		0.004	0.6285
1980S3	70 × 60 × 50	151,422		0.009	0.6943
1980S1	110 × 100 × 80	151,472		0.007	0.6947
Mimas	196	185,540	7.6×10^{-6}	0.020	0.9424
Mimas'	?	186,000			
Enceladus	250	238,040	1×10^{-5}	0.0044	1.3702
Tethys	530	294,670	1.3×10^{-4}	0.000	1.8878
1980S13	17 × 14 × 13	294,670			1.8878
1980S25	17 × 11 × 11	294,670			1.8878
Dione	560	377,420	1.76×10^{-4}	0.002	2.7369
1980S6	18 × 16 × 15	378,060		0.005	2.7391
Rhea	765	527,100	4.17×10^{-4}	0.001	4.5171
Titan	2575	1.222×10^6	0.0225	0.029	15.945
Hyperion	205 × 130 × 110	1.481×10^6		0.104	21.277
Iapetus	730	3.56×10^6	3.15×10^{-4}	0.028	79.331
Phoebe	110	1.30×10^7		0.163	550.45
Uranus	25,650	2.870×10^9	14.54	0.0472	30,685	0.68
Miranda	160	130,000	4×10^{-6}	0.017	1.4135
Ariel	470	192,000	9×10^{-5}	0.0028	2.520
Umbriel	320	267,000	3×10^{-5}	0.0035	4.144
Titania	520	438,000	1×10^{-4}	0.0024	8.706
Oberon	460	586,000	9×10^{-5}	0.0007	13.46
Neptune	24,820	4.497×10^9	17.23	0.00858	60,189	0.57?
1981N1	100	75,000			
Triton	2000	353,400	0.02	0.000	5.877
Nereid	300	5.56×10^6		0.749	359.881
Pluto	1500–2000	5.900×10^9	0.0022	0.2502	90,465	6.3874
Charon	750–1000	19,700	0.0003	0	6.3874

Table 1-1 *continued*

OBJECT	a (km)	R (km)	M (earth masses)	e	P (days)	P_{rot} (days)

* Unless otherwise stated, data are from Allen, C. W. *Astrophysical Quantities*, London: Athlone Press, 1973.

Mercury: radius from Howard, T. *et al.* Mercury: results on mass, radius, ionosphere, and atmosphere from Mariner 10 dual frequency radio signal. *Science* **185**: 179–180 (1974). Mass from Esposito, P. B., Anderson, J. D., and Ng, A. T. Y. Experimental determination of Mercury's mass and oblateness. *Space Res.* **17**: 639–644 (1977). Rotation period from Klaasen, K. P. Mercury's rotation axis and period, *Icarus* **28**: 469–478 (1976).

Venus: radius from Pettengill, G. *et al.* Pioneer Venus radar results: altimetry and surface properties. *JGR* **85**: 8261–8270 (1980). Mass from Goettel, K. A., Shields, J. A., and Decker, D. A. Density constraints on the composition of Venus. *Proc. Lunar Sci. Conf.* **12B**:1507–1516 (1981). Rotation period from Shapiro, I. I., Campbell, D. B., and DeCampli, W. M. Nonresonance rotation of Venus? *Astrophys. J.* **230**: L123–L126 (1979).

Mars satellites: data from Veverka, J., and Burns, J. A. The moons of Mars. *Ann. Rev. Earth and Planet. Sci.* **8**: 527–558 (1980).

Jupiter: one-bar radius derived from data of Lindal, G. F. *et al.* The atmosphere of Jupiter: an analysis of the Voyager radio occultation measurements. *J. Geophys. Res.* **86**: 8721–8727 (1981). Mass from Null, G. W. Gravity field of Jupiter and its satellites from Pioneer 10 and Pioneer 11 tracking data. *Astron. J.* **81**: 1153–1161 (1976).

Jupiter satellites: data from Morrison, D. Introduction to the satellites of Jupiter, in *Satellites of Jupiter* (D. Morrison, ed.). Tucson: University of Arizona Press, 1982.

Saturn: one-bar radius is the value adopted by the Voyager project. Mass from Null, G. W., Lau, E. L., Biller, E. D., and Anderson, J. D. Saturn gravity results obtained from Pioneer 11 tracking data and Earth-based Saturn satellite data. *Astron. J.* **86**: 456–468 (1981). Rotation period from Desch, M. D., and Kaiser, M. L. Voyager measurement of the rotation period of Saturn's magnetic field. *Geophys. Res. Let.* **8**: 253–256 (1981).

Saturn satellites: note that the small coorbital satellites 1980S1 and 1980S3 periodically exchange orbits. Data on satellites are from several Voyager 1 papers in *Science* **212** (1981): Stone, E. C., and Miner, E. D. Voyager 1 encounter with the Saturnian system, pp. 159–163; Smith, B. A. *et al.* Encounter with Saturn: Voyager 1 imaging science results, pp. 163–191; Synnott, S. P. *et al.* Orbits of the small satellites of Saturn, pp. 191–192; Tyler, G. L. *et al.* Radio science investigations of the Saturn system with Voyager 1: preliminary results, pp. 201–206. Additional results are from Voyager 2 papers in *Science* **215** (1982): Stone, E. C., and Miner, E. D. Voyager 2 encounter with the Saturnian system, pp. 499–504; Smith, B. A. *et al.* A new look at the Saturn system: the Voyager 2 images, pp. 504–537; Tyler, G. L. *et al.* Radio science with Voyager 2 at Saturn: atmosphere and ionosphere and the masses of Mimas, Tethys, and Iapetus, pp. 553–558.

Uranus: data from Elliot, J. L. Rings of Uranus: a review of occultation results, in *Uranus and the Outer Planets* (G. E. Hunt, ed.). Cambridge: Cambridge University Press, 1982.

Uranus satellites: from Cruikshank, D. The satellites of Uranus, in *Uranus and the Outer Planets* (G. E. Hunt, ed.). Cambridge: Cambridge University Press, 1982. Radii are computed from an assumed geometric albedo of 0.5. Masses are computed from these radii and an assumed density of 1.3 g/cm³.

Neptune: radius from Hubbard, W. B. *et al.* Results from observations of the 15 June 1983 occultation by the Neptune system. To be published (1984).

Neptune satellites: 1981N1 was reported by Reitsema, H. J. *et al.* Occultation by a possible third satellite of Neptune. *Science* **215**: 289–291 (1982). Data given are based upon a presumed prograde, circular, equatorial orbit. This object has not yet been confirmed by further observations.

Pluto and Charon: from Harrington, R. S., and Christy, J. W., The satellite of Pluto. III. *Astron. J.* **86**: 442–443 (1981).

other planets. Chapter 7 is written to set the perspective for this limited purpose.

Chapter 8 contains a discussion of all of the planetary bodies whose interiors have been studied to date. The reader may find hints that a general conceptual framework exists for each planet, and even one for all of the planets. However, we will find jarring inconsistencies as well. Such problems are the starting point for future progress.

2
Chemical Composition and Structure
of the Sun

Our understanding of the structure, composition, and evolution of the planets owes much to earlier work on the structure of the sun. The sun is important for this purpose for a number of reasons:

(1) The composition of the sun provides an important clue to the primordial composition of the material out of which the planets formed.
(2) Energy produced by the sun strongly influences the surface temperatures of the planets and thus indirectly affects their interior evolution.
(3) Many of the same theoretical techniques developed for the study of stellar structure are also applicable to the study of planetary structure.

SOLAR AGE AND COMPOSITION

Abundances of elements in the solar atmosphere can be determined by spectroscopic techniques. These measurements are supplemented by determinations of the composition of material ejected from the sun, and by compositional measurements of meteorites. It is now believed that, with the exception of a few elements such as Li, Be, B, and the hydrogen isotope H^2, the composition of the present solar atmosphere is essentially equal to the composition of the material out of which the solar system formed. This conclusion follows from an evaluation of the processes of solar evolution which might alter the primordial composition of the sun.

It is also possible to deduce the *primordial* composition of the sun by analyzing primitive meteorites known as carbonaceous chondrites. The relative abundances of elements in these objects are, except for extremely volatile elements such as hydrogen, helium, and those mentioned above, strikingly similar to abundances in the solar atmosphere. The oldest known carbonaceous chondrites were formed about 4.6×10^9 years ago, and their compositions are believed to reflect the primordial composition of the nebula from which the sun and planets formed. We will take their age to define the age of the solar system.[1] Other nonastronomical information about the composition of the sun comes from measurements of abundances in the solar wind.

Table 2-1 presents results from two compilations[2,3,4] of abundances of elements in the sun. These compilations are represented in two ways: by number of atoms of the element relative to the number of hydrogen atoms, and by mass fraction (which is useful in construction of planetary interior models). The two compilations are not equivalent. Cameron's (1973) abundances are a reconstruction of the *primordial* composition of the sun, and make use of data from meteorites and other sources. Pagel's (1976) data are obtained solely from spectroscopic data on the composition of the *present* solar atmosphere. Note particularly the discrepancies for Li and B. Elements with atomic number $Z > 40$ (after Zr) have not been included because of their low abundances, although Th and U are included because of their importance in planetary heat production.

We have not included error bars for these determinations because it is difficult to assign true uncertainties. However, an idea of the errors can be obtained by comparing Table 2-1 with Table 2-2. The latter is based upon the most recent available compilation of elemental abundances derived from carbonaceous chondrites and astronomical and spacecraft data.[5] In Table 2-2, all abundances are given for the *primordial* solar composition, except that both present and primordial values are given for uranium and thorium, whose abundances change with time because of radioactive decay.

The term "primordial composition of the solar system" requires careful definition. Observations of chemical abundances in the atmospheres of other stars have shown that the abundances given in Table 2-1 and 2-2 are typical of many other objects in the sun's galaxy and other galaxies. However, this is true only to first approximation.

Hydrogen was formed during the initial stages of the universe at

Table 2-1. Solar Abundances*

ELEMENT	CAMERON'S DATA		PAGEL'S DATA	
	NUMBER/H	MASS FRACTION	NUMBER/H	MASS FRACTION
H**	1.	0.770	1.	0.784
He	0.069	0.212	0.0630958	0.196
Li	1.557E-09	8.252E-09	1.E-11	5.400E-11
Be	2.547E-11	1.753E-10	1.585E-11	1.111E-10
B	1.101E-08	9.087E-08	2.512E-10	2.112E-09
C	3.711E-04	3.403E-03	4.571E-04	4.271E-03
N	1.176E-04	1.258E-03	8.511E-05	9.275E-04
O	6.761E-04	8.259E-03	7.586E-04	9.442E-03
F	7.704E-08	1.118E-06	3.981E-08	5.884E-07
Ne	1.082E-04	1.667E-03	6.310E-05	9.905E-04
Na	1.887E-06	3.313E-05	1.950E-06	3.487E-05
Mg	3.336E-05	6.193E-04	3.981E-05	7.528E-04
Al	2.673E-06	5.508E-05	2.512E-06	5.273E-05
Si	3.145E-05	6.745E-04	3.981E-05	8.697E-04
P	3.019E-07	7.141E-06	3.162E-07	7.620E-06
S	1.572E-05	3.850E-04	1.585E-05	3.953E-04
Cl	1.792E-07	4.853E-06	2.512E-07	6.928E-06
Ar	3.686E-06	1.124E-04	2.512E-06	7.805E-05
K	1.321E-07	3.944E-06	1.413E-07	4.296E-06
Ca	2.267E-06	6.940E-05	1.995E-06	6.221E-05
Sc	1.101E-09	3.779E-08	1.E-09	3.497E-08
Ti	8.726E-08	3.192E-06	6.310E-08	2.351E-06
V	8.239E-09	3.205E-07	1.E-08	3.963E-07
Cr	3.994E-07	1.586E-05	5.012E-07	2.027E-05
Mn	2.925E-07	1.227E-05	2.512E-07	1.074E-05
Fe***	2.610E-05	1.113E-03	3.162E-05	1.374E-03
Co	6.950E-08	3.128E-06	1.E-07	4.586E-06
Ni	1.509E-06	6.7682-05	1.995E-06	9.112E-05
Cu	1.698E-08	8.241E-07	1.259E-08	6.223E-07
Zn	3.912E-08	1.953E-06	2.512E-08	1.278E-06
Ga	1.509E-09	8.037E-08	6.310E-10	3.422E-08
Ge	3.616E-09	2.005E-07	2.512E-09	1.417E-07
As	2.075E-10	1.188E-08		
Se	2.113E-09	1.274E-07		
Br	4.245E-10	2.591E-08		
Kr	1.471E-09	9.419E-08		
Rb	1.849E-10	1.207E-08	3.981E-10	2.647E-08
Sr	8.459E-10	5.661E-08	7.943E-10	5.415E-08
Y	1.509E-10	1.025E-08	1.259E-10	8.708E-09
Zr	8.805E-10	6.134E-08	6.310E-10	4.478E-08
Th	1.824E-12	3.232E-10	1.585E-12	2.861E-10
U	8.239E-13	1.498E-10		

* In this table and in Table 2-2, computer exponential notation is used for convenience. Thus 1.0E5 means 1.0×10^5, etc. Note that the number of significant figures given does not necessarily reflect the accuracy of the determination.
** The primordial number abundance of the hydrogen isotope H^2 relative to H^1 is[4] $(2.25 \pm 0.25) \times 10^{-5}$.
*** Pagel's iron abundance is to be preferred.

Table 2-2. Abundances from C1 Chondrites and Astronomical Data*[5]

ELEMENT	NUMBER/H	MASS FRACTION
H	1	0.744
He	0.080	0.237
Li	2.194E-09	1.124E-08
Be	2.868E-11	1.908E-10
B	8.824E-10	7.041E-09
C	4.412E-04	3.912E-03
N	9.081E-05	9.389E-04
O	7.390E-04	8.725E-03
F	3.099E-08	4.346E-07
Ne	1.283E-04	1.910E-03
Na	2.096E-06	3.556E-05
Mg	3.934E-05	7.058E-04
Al	3.121E-06	6.217E-05
Si	3.676E-05	7.621E-04
P	3.824E-07	8.742E-06
S	1.892E-05	4.477E-04
Cl	1.926E-07	5.042E-06
Ar	3.824E-06	1.127E-04
K	1.386E-07	4.000E-06
Ca	2.245E-06	6.643E-05
Sc	1.243E-09	4.124E-08
Ti	8.820E-08	3.119E-06
V	1.084E-08	4.075E-07
Cr	4.919E-07	1.888E-05
Mn	3.496E-07	1.418E-05
Fe	3.310E-05	1.364E-03
Co	8.272E-08	3.599E-06
Ni	1.815E-06	7.866E-05
Cu	1.890E-08	8.863E-07
Zn	4.623E-08	2.231E-06
Ga	1.390E-09	7.152E-08
Ge	4.338E-09	2.322E-07
As	2.496E-10	1.381E-08
Se	2.282E-09	1.329E-07
Br	4.338E-10	2.559E-08
Kr	1.667E-09	1.031E-07
Rb	2.607E-10	1.645E-08
Sr	8.751E-10	5.660E-08
Y	1.706E-10	1.120E-08
Zr	3.934E-10	2.649E-08
Th (present)	1.232E-12	2.110E-10
Th (primordial)	1.544E-12	2.645E-10
U (Present)	3.311E-13	5.816E-11
U (primordial)	8.761E-13	1.539E-10

* As in Table 2-1, the number of significant figures given does not necessarily reflect the accuracy of the determination.

the time of the "big bang." Helium was also formed at this time, with an abundance relative to hydrogen on the same order as the values given in Tables 2-1 and 2-2. However, heavier elements were not formed in any appreciable abundance during the big bang. The present solar abundances of these elements must have been established in large part by nuclear reactions in generations of massive stars, (i.e., with masses on the order of ten solar masses and more), prior to the sun's formation. These stars became unstable in later stages of their evolution, returning their matter, now enriched in high-Z elements, to the interstellar medium. As one might expect, this process of Z-enrichment of the interstellar medium is a patchy one: loci of intense formation and destruction of massive stars are loci of Z-enrichment. The enrichment process may be a self-accelerating one, for there is evidence that the explosion of massive objects, along with formation of interstellar dust from their debris, is conducive to the formation of new stars. Thus there is considerable variation in the abundances of high-Z elements in the galaxy.

We are not even assured that abundances were uniform in the solar system at the time that the sun and planets were forming, and some evidence exists that they were not. Nevertheless, the composition of the sun, the largest aggregate in the solar system, is relevant to any attempt to synthesize the compositions of the planets. We turn now to a discussion of the relationship between the sun's atmospheric composition and its interior structure and previous history.

INTERIOR STRUCTURE OF THE SUN

The large mass of the sun in comparison with the planets leads to important qualitative differences in its interior structure. However, the interior structure is governed by essentially the same set of equations as for planets. First, we have the equation of hydrostatic equilibrium:

$$\nabla P = -\mathbf{g}\rho, \tag{2-1}$$

where P is the pressure, $\mathbf{g}$ is the local gravitational acceleration, and ρ is the mass density. Next, we have the equation which relates the temperature profile to the heat flux vector **H**:

$$\mathbf{H} = -K \nabla T. \tag{2-2}$$

This equation must be replaced by the appropriate convective heat transport equation in convectively unstable regions, as described in Chapter 5. If the object is in thermal equilibrium (not a bad approximation for the present-day sun), then the heat flux at a specified radius r is related to energy sources by

$$4\pi r^2 H = \int_0^m \epsilon \, dm', \tag{2-3}$$

where $m(r)$ is the mass enclosed within a sphere of radius r and ϵ is the rate of energy generation per gram. At $r = a$ (the surface radius of the sun), we have $4\pi r^2 H = L_i$, the intrinsic luminosity of the sun. At any radius r, evidently $g = Gm(r)/r^2$, where G is the gravitational constant.

To close the set of equations specified above, it is necessary to add constituent relations for the effective thermal conductivity K, and an equation of state in the form $P = P(\rho, T)$. The principal difference between solar and planetary structure arises because the solar equation of state is, to an excellent approximation, given by the ideal gas relation

$$P = R_g \rho T / \mu, \tag{2-4}$$

where R_g is the gas constant and μ is the mean molecular weight, which is close to unity in the sun. Unlike the case in planetary interiors, then, the temperature profile in the sun is the controlling factor in the object's density distribution. Because the temperature distribution is in turn governed by the effective conductivity (Eq. 2-2), the sun's overall structure is determined to zeroth order by the conductivity of its interior. The high temperatures in the solar interior imply that photon transport is the most effective heat conduction mechanism. Calculations of photon opacities under solar interior conditions have shown that the relatively small fraction of atoms with $Z > 2$ control the photon transport, because many of these atoms are still not fully ionized in the solar interior and thus have a large cross-section for absorbing photons via photoionization. If the $Z > 2$ atoms (called "metals" in astrophysical parlance) were absent, the opacity would be controlled by the absorption of photons by free electrons in the vicinity of ions (inverse *bremsstrahlung*). Given that the sun obeys the ideal gas equation of state, it is easy to estimate a typical temperature value in its interior. From dimensional considerations, the pressure in the interior of a spherical object in hydrostatic equilibrium is given by

$$P \sim GM^2/a^4, \qquad (2\text{-}5)$$

where M is the total mass and a is the surface radius. For the sun, with $M = 1.9 \times 10^{33}$ g and $a = 7 \times 10^{10}$ cm, one finds an interior pressure $P \sim 10^{16}$ dyne/cm^2. Now the mean density of the sun is of order unity, and so Eq. 2-4 gives for a typical interior temperature $T \sim 10^8$ °K. This number is somewhat of an overestimate; because densities increase very rapidly toward the center of the sun, reaching values on the order of 10^2 g/cm^3, the actual temperatures near the sun's center are nearly a factor of ten lower.

As with any problem involving diffusion, evolution of the sun is an irreversible process, and it is not possible to uniquely reconstruct the sun's past history from a knowledge of its present configuration. The earliest stages of the sun's evolution are the subject of controversy, and are closely bound up with problems of the origin of the solar system. However, the present state of the sun does not depend sensitively upon assumed starting conditions, because the sun is now in a nearly steady-state configuration, such that the energy released by fusion of hydrogen at the center of the sun is radiated into space at an approximately equal rate. In this case, the time derivative terms which would appear in the general expression for the heat flux (see Chapter 5) are essentially zero.

Evolutionary models for the sun are computed under the assumption that the initial configuration is an extended, low-density, spherically symmetric cloud having a composition approximately equal to Cameron's abundances given in Table 2-1. This configuration is initially at too low a temperature to produce any energy by fusion processes, so the only source of heating is radioactive decay and changes in gravitational binding energy. Calculations by the Japanese astrophysicist C. Hayashi and collaborators have shown that the initial configuration becomes unstable when temperatures at the center become high enough to dissociate hydrogen (a few thousand degrees K) and a rapid, hydrodynamic collapse ensues. This collapse results in an object having a substantially higher luminosity and larger radius than the present sun. The luminosity is high enough that the condition for convective instability (see Chapter 5) is satisfied everywhere in the protosun. Although the luminosity is higher than the present sun's, the object is much more extended, and therefore the central density is much lower. The central temperature is still not high enough for fusion to proceed. Once the initial collapse phase is concluded, the sun is once again in a state of hydrostatic equilibrium, and evolves slowly through a sequence of hydrostatic equilibrium configurations. As discussed in Chapter 5, loss of heat from an object composed of an ideal gas leads to contraction of the object and increasing temperatures in the center. Thus the primordial sun approaches a state where central temperatures are high enough to permit hydrogen fusion to begin.

It is customary in astrophysics to represent the evolution of a star in terms of two quantities which are at least observable in principle. These quantities are the star's internal power L_i and its effective temperature T_e. These two quantities are related to the star's radius a by

$$L_i = 4\pi a^2 \sigma T_e^4, \tag{2-6}$$

where σ is the Stefan-Boltzmann constant, equal to 5.6703×10^{-5} erg/(s $\cdot$ cm^2 $\cdot$ °K^4). Figure 2-1 shows a schematic evolutionary track for the protosun as it approaches its present configuration. According to Hayashi's theory, the sun evolves through a sequence of fully convective configurations (the so-called *Hayashi track* shown in Fig. 2-1), gradually decreasing its value of L_i but increasing its central density and temperature. Initiation of hydrogen fusion occurs near the end of the "Hayashi phase," and the sun then enters a phase with the central core stable against convection and energy carried by photon conduction. It is believed that the present abundances of lithium and boron, and the essential absence of deuterium in the sun, are all the result of thermonuclear reactions which destroyed these nuclei during the "Hayashi" phase at temperatures lower than the

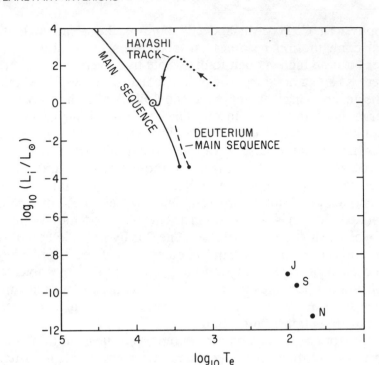

Fig. 2-1. Intrinsic luminosity (L_i) relative to solar luminosity ($L_\odot$) as a function of effective temperature. The main hydrogen-burning sequence is shown, as well as the lower end of the deuterium-burning sequence. J, S, and N are observed points for Jupiter, Saturn, and Neptune.

temperature required for the main hydrogen reaction to proceed. If the entire sun was involved in convection at this time, virtually all of its mass would have been purged of these isotopes.[6,7,8]

Once thermonuclear reactions in hydrogen are established at the core of the sun, further evolution of the object in the (L_i, T_e) plane is arrested because of the great abundance of hydrogen fuel. The proton-proton chain is the main reaction in hydrogen which is thought to be responsible for the sun's present luminosity. This reaction proceeds basically as follows:

$$H^1 + H^1 \rightarrow H^2 + e^+;$$
$$H^2 + H^1 \rightarrow He^3;$$
$$He^3 + He^3 \rightarrow He^4 + 2\ H^2. \tag{2-7}$$

The end result of this sequence of reactions is the conversion of six protons (H^1) to one helium nucleus (He^4) plus two protons, and the excess binding energy of the helium nucleus with respect to the protons is released.

The great abundance of hydrogen in stars causes virtually all stars to reach a temporary stationary or near-stationary phase in their evolution when they begin the process of energy release by hydrogen fusion. The locus of points on the (L_i, T_e) plane defined by this phase, as a function of stellar mass, is known as the *main sequence,* and is readily observable because of the tendency of stars to congregate along this sequence.

As the sun approaches the main sequence and begins hydrogen fusion, the convective portion of the solar interior shrinks to a small fraction of the total mass and moves toward the surface of the sun. At present the solar convective zone is estimated to comprise the outermost few percent of the sun's mass, and does not involve any of the inner core where nuclear fusion reactions are occurring. The composition of the solar atmosphere should therefore not have been altered by the conversion of hydrogen to helium in the core.

About half the hydrogen in the core has now been converted to helium (Fig. 2-2). According to theoretical estimates, the main sequence phase of the sun has so far lasted about 4.6×10^9 years, and the sun will continue in the main sequence phase for approximately

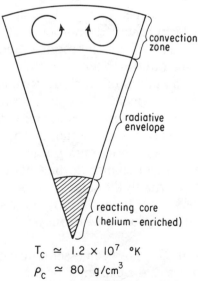

$$T_c \simeq 1.2 \times 10^7 \ °K$$
$$\rho_c \simeq 80 \ g/cm^3$$

Fig. 2-2. Interior structure of the sun.

the same amount of time, until the central hydrogen has been exhausted. We take the main-sequence lifetime of the sun to be essentially equal to the age of the solar system, and thus to the time which has elapsed since the planets formed.

There is relatively little doubt that hydrogen fusion is the dominant source of energy in main-sequence stars. Comparisons of distributions of actual stars on the (L_i, T_e) plane with theoretical computations from stellar evolution codes have strikingly confirmed the concept. However, one major discrepancy between theory and observation still exists. It is theoretically possible to make a direct observation of the hydrogen fusion region deep within the solar interior, because certain branches of the proton-proton reaction lead to the emission of neutrinos (not shown in Eq. 2-7). Neutrinos are uncharged particles with zero rest mass, to which the solar interior is essentially transparent. If neutrinos from the interior of the sun could be measured with a suitable detector on the earth, fundamental concepts about the origin of the solar luminosity could be verified. To date, the solar neutrino flux predicted by standard solar models has not been observed.[1] Neutrino detectors are based upon the principle that a neutrino can occasionally cause a reaction in a Cl^{37} nucleus. The probability of such a reaction is so low that solar neutrino fluxes are expressed in terms of the Solar Neutrino Unit (SNU), which is 10^{-36} captures per second per Cl^{37} nucleus at the earth. No solar neutrino has so far been observed; the observational upper limit is less than 1 SNU, whereas typical solar models predict a detection rate at least an order of magnitude higher. The most popular explanation of the discrepancy is that the sun's luminosity and interior structure may be variable on geologically short time scales with a very low amplitude. The nuclear reaction rates at the center of the sun are highly temperature-sensitive, and a small, temporary decrease in the central temperature could lead to a temporary cessation of neutrino production without a major change in the solar luminosity, because of the great thermal inertia of the solar interior.

From the point of view of the influence of the sun on the study of planetary interiors, our major conclusions are then the following:

(1) The solar luminosity has been essentially constant over all but the initial phases of the existence of the planetary system.
(2) Abundances in the solar atmosphere are, with a few important

exceptions noted above, representative of the overall composition of the material out of which the sun and the planets formed.

CONDENSATION OF SOLAR COMPOSITION MATERIAL

As we shall see, the inferred bulk compositions of planets are not in precise correspondence with the bulk composition of the sun. There is of course no reason to expect them to be in correspondence. One can imagine several ways that initial solar abundances could be modified as substances are incorporated in planets. The simplest mechanism for modifying abundances becomes apparent when we note that the sun is gaseous while the planets are mainly in condensed states of matter. If we take an initial gas phase at sufficiently high temperatures, so that all components are gaseous (as is the case in the solar atmosphere, for example), and then gradually reduce the temperature of this gas, keeping the pressure constant, various condensed phases will appear in certain temperature ranges. If one reduces the temperature all the way down to 0 °K, even hydrogen and helium condense, and the condensed phase again has solar composition. But no part of the solar system is, or ever was, at sufficiently low temperature to condense all components. Therefore existing condensates in the solar system must differ from the composition of the solar atmosphere.

Naturally, the solar system has not evolved in strict thermodynamic equilibrium. It has not been enclosed in a box and given indefinite amounts of time to come to equilibrium. Therefore any calculation of the composition of a condensate in equilibrium with a vapor may not be relevant to the actual processes of formation of planets. Nevertheless, the predictions of equilibrium condensation theory provide a convenient framework for interpretation of the observed compositions of planetary bodies. The theory proceeds as follows. We consider the solid (or liquid) phases which are in equilibrium with a vapor at a given temperature and pressure. The overall composition is taken to be solar, and the pressure is taken to be very low, say 10^{-3} bar, as might be appropriate for a cloud of interstellar matter which has accumulated to form a solar system. There will probably be temperature and density gradients in this cloud because of central compression and heating, and therefore the composition and amount of the condensate will vary within the cloud. At a certain instant, the uncondensed component is removed by some unspecified mechanism (perhaps an intense primordial solar wind), and the condensed component goes on to form planets. The regions of the primordial cloud which were relatively hot (presumably those regions close to the sun) will leave behind smaller amounts of low-volatility condensate, and those regions which were relatively cool will leave behind larger amounts of both low- and high-volatility condensate. This general picture undoubtedly has a substantial correspondence to reality, because the bulk compositions that we will infer for the planets seem to fit, to some extent, within the framework. Many planetary scientists now believe that equilibrium condensation theory has major deficiencies, but the theory continues to serve as a basic starting point for attempts to synthesize compositions of planetary objects.

We will now summarize the principal condensates which appear when a high-temperature solar-composition gas is cooled at a constant pressure of 10^{-3} bar, and is allowed to always remain in strict thermodynamic equilibrium with the condensates.[9] The first condensates appear in the temperature range of 1760–1500 °K, and are minerals which are rich in the elements Ca and Al. These include corundum (Al_2O_3), perovskite ($CaTiO_3$), melilite ($Ca_2Al_2SiO_7$-$Ca_2MgSi_2O_7$), and spinel ($MgAl_2O_4$)—the so-called refractory compounds. The first really abundant condensate, iron, appears at 1473 °K, alloyed with nickel. The major silicates, which are composed of the abundant elements Si, Mg, and O, appear at 1444 °K (forsterite, Mg_2SiO_4) and 1349 °K (enstatite, $MgSiO_3$). At 700 °K, sulfur leaves the gas phase (as H_2S) and reacts with iron to form FeS (troilite). Oxygen reacts with iron to form Fe_3O_4 (magnetite) at 405 °K, although most of the oxygen is still present in the gas as H_2O. The abundant elements O, N, and C condense at about 200 °K (H_2O), 120 °K (NH_3), and 80 °K (CH_4) respectively.[10] Later in the book we will consider whether models of planetary and satellite interiors provide evidence for the existence of such condensation sequences in the early solar system.

REFERENCES

1. Wood, J. A. *The Solar System*. Englewood Cliffs, N.J.: Prentice-Hall, 1979.
2. Cameron, A. G. W. Abundances of elements in the solar system. *Space Sci. Rev.* **15:** 121–146 (1973).
3. Pagel, B. E. J. Solar abundances: a new table (October 1976). In *Physics and Chemistry of the Earth* **11:** 79–80 (L. H. Ahrens, ed.). Oxford: Pergamon, 1977.
4. Bruston, P. *et al.* Physical and chemical fractionation of deuterium in the interstellar medium. *Astrophys. J.* **243:** 161–169 (1981).
5. Anders, E. and Ebihara, M. Solar-system abundances of the elements. *Geochim. Cosmochim. Acta* **46:** 2363–2380 (1982).
6. Graboske, H. C., Jr. and Grossman, A. S. Evolution of low-mass stars. IV. Effects of multilevel atomic partition functions for the ideal-gas region. *Astrophys. J.* **170:** 363–370 (1971).
7. Grossman, A. S. and Graboske, H. C., Jr. Evolution of low-mass stars. V. Minimum mass for the deuterium main sequence. *Astrophys. J.* **180:** 195–198 (1973).
8. Hayashi, C. Evolution of protostars. *Ann. Rev. Astron. and Astrophys.* **4:** 171–192 (1966).
9. Grossman, L. Condensation in the primitive solar nebula. *Geochim. Cosmochim. Acta* **36:** 597–619 (1972).
10. Lewis, J. S. Low temperature condensation from the solar nebula. *Icarus* **16:** 241–252 (1972).

3
Constituent Relations

The constituent relations for the sun are relatively simple. Temperatures within the sun are much higher than an atomic unit of temperature (which is defined following these paragraphs), and to an excellent approximation, the material can be considered a fully ionized, ideal gas. But planets and satellites are composed of condensed materials, whose thermodynamic variables are far more complicated functions of density, temperature, and composition. It is primarily for this reason that the chemistry of planetary interiors is far more complex than that of stellar interiors.

It is not feasible to carry out a full *a priori* calculation of the thermodynamics of planetary material at arbitrary pressures and for arbitrary composition. The problem is far too complex. Instead, we rely on simplified models and experimental data. The elements at the beginning of the periodic table, hydrogen and helium, have simple electronic structures amenable to theoretical treatment in certain approximations. At the same time, because of their great compressibility, it is difficult to experimentally study these elements at pressures greater than a megabar. Because hydrogen and helium are found in great abundance and at high pressures in the Jovian planets, theoretical constituent relations play an especially important role in models of these planets. The smaller planetary bodies, such as the terrestrial planets and the satellites, are substantially depleted in elements of low atomic number, and for these bodies we rely primarily on experimental thermodynamic data.

We now proceed to a discussion of some of the principal sources of constituent relations used in the construction of models of planetary interiors.

THEORETICAL METHODS FOR CALCULATING
THERMODYNAMICS AT ZERO TEMPERATURE

It will be useful in the following discussion to express all quantities in terms of atomic units. The atomic unit of length is the Bohr radius, $a_0 = 0.529 \times 10^{-8}$ cm, and thus the atomic unit of volume is a_0^3. The atomic unit of charge is the electron charge, $e = 4.80 \times 10^{-10}$ electrostatic units of charge. Thus the atomic unit of energy is $e^2/a_0 = 4.36 \times 10^{-11}$ erg, and the atomic unit of pressure is $e^2/a_0^4 = 2.94 \times 10^{14}$ dyne/cm^2 = 294 Mbar. The pressure unit is particularly significant, because it indicates the range of pressures needed to effect significant modification of the energy levels of the valence electrons in the atoms of a substance. The atomic unit of temperature is also significant: it is $e^2/a_0k = 3.16 \times 10^5$ °K (k = Boltzmann's constant). For temperatures well below this value, electrons are essentially in their ground energy state. Finally, the unit of mass is the electron mass, $m_e = 1.90 \times 10^{-28}$ g.

There is a natural hierarchy of energy levels in any material. The "coarsest" energy levels are those corresponding to different states of electron excitation. They are, not surprisingly, separated by energy intervals on the order of one atomic unit of energy. Next, one has, for a given electron state, different energy levels corresponding to vibration of the nuclei about their equilibrium positions. These levels are separated by intervals on the order of $\sqrt{(m_e/M)}$, where M is the nuclear mass, and thus they are at least a factor of 40 closer together than the electronic levels. The "finest" energy levels are those corresponding to translation and rotation of the nuclei, which are separated by energies on the order of m_e/M.

The interior structure of planets can be treated to first approximation as if the material were at zero temperature. Temperatures in planetary interiors are typically on the order of 10^4 °K or lower, and are therefore much lower than the temperatures required to significantly populate higher electron energy levels. On the other hand, such temperatures are high enough to populate a large number of higher energy levels of the nuclei, corresponding to their vibrational and translational states in the material. However, although the population of these states is crucially important to the thermal properties of planetary interiors, it is not a major factor in determining the pressure-density relation, which in turn governs the density stratification in a planetary interior through the equation of hydrostatic equilibrium.

According to the standard thermodynamic relationship, infinitesimal changes in the internal energy (E), volume (V), and entropy (S) are related by

$$dE = TdS - PdV, \qquad (3\text{-}1)$$

where T is the temperature, P is the pressure, and the number of particles in the sample is assumed to be constant. At $T = 0$, the internal energy E_0 is just the ground state energy of the system, and the zero-temperature pressure is then

$$P_0 = -dE_0/dV. \qquad (3\text{-}2)$$

Evidently, at zero temperature, $P_0 = P_0(V)$ only.

At very low pressures, $P_0(V)$ can be measured experimentally. It is also possible to measure E_0 at zero pressure, i.e., where $dE_0/dV = 0$. The value of E_0 at this point, relative to the energy of the same number of isolated molecules, is called the binding energy of the lattice. As pressure increases and volume decreases, E_0 must increase. In the limit of very high pressure, the form of E_0 can be predicted theoretically, using the following arguments. The result will then be used to calculate an asymptotic pressure-density relation valid for any material. While it is not quantitatively useful in modeling planetary interiors, this result is helpful in assessing more complicated theoretical models which are applied at lower pressures.

The quantum-mechanical energy operator of an assemblage of N_e electrons and N_e/Z nuclei of atomic number Z can be written, schematically,

$$H = \sum_i \mathbf{p}_i^2/2 + \sum_j (m_e/M)\mathbf{p}_j^2/2 + \Phi, \qquad (3\text{-}3)$$

where the index i runs over all electrons, j runs over all ions, $\mathbf{p}_i$ and $\mathbf{p}_j$ are the momentum operators of the ith electron and jth ion, respectively, and the function Φ symbolizes the potential energy operator, which depends on all of the particle coordinates $\{\mathbf{r}_i, \mathbf{r}_j\}$, in the form

$$\Phi = \sum_{i<i'} |\mathbf{r}_i - \mathbf{r}_{i'}|^{-1} - \sum_{i,j} Z_j |\mathbf{r}_i - \mathbf{r}_j|^{-1}$$
$$+ \sum_{j<j'} Z_j Z_{j'} |\mathbf{r}_j - \mathbf{r}_{j'}|^{-1}. \tag{3-4}$$

The ground state energy of the system is found by calculating the expectation value of H in the ground state of the system. In the range of temperatures of interest for planetary interiors, the electron state can be taken to be the ground state. At the same time, the more massive ions move essentially classically, because their translational and vibrational levels are essentially a continuum. The low-mass electron clouds follow the ions around, adjusting their instantaneous state to the lowest-energy one consistent with a given ion configuration. To obtain an estimate of the ground state energy, we first neglect the term in the ion kinetic energies because it is of order $m_e/M \lesssim 1/1800$ relative to the electron kinetic energies. Consider first the limit of extremely high pressure. For volumes per atom (expressed in atomic units) sufficiently less than one, essentially all atomic structure is obliterated. The order of magnitude of the first term can be estimated because at high pressures the electron kinetic energies are very high compared with typical values for the interaction energy (we will verify this *a posteriori* in a moment), and so the electrons can be taken to be free particles, in plane wave states. In a plane wave state, the expectation value for the kinetic energy of an electron is given by

$$\langle p^2/2 \rangle = k^2/2, \tag{3-5}$$

where k is the electron's wave number in atomic units. Now, in the lowest energy state, k will be minimized, but it cannot be zero in general because electrons are fermions and cannot populate any given quantum state with more than one electron. Because of the electrons' spin of ½, there are two possible spin states associated with each k-state. Thus each k-state is twofold degenerate. Each pair of electrons then populates a given k-state up to some maximum value. Since k has dimensions of 1/length, where the length parameter is on the order of the average distance between electrons, the maximum value of k must scale as $V^{-1/3}$, and so the average electron kinetic energy must scale as $V^{-2/3}$. Let us define a dimensionless parameter which measures the mean distance between electrons:

$$r_e = \left(\frac{3}{4\pi}\right)^{1/3} (V/N_e)^{1/3}, \tag{3-6}$$

where V is expressed in atomic units as before. Thus the average electron kinetic energy scales as r_e^{-2}. An exact calculation for plane wave states gives

$$\langle K \rangle = 1.105/r_e^2, \tag{3-7}$$

where $\langle K \rangle$ is the average kinetic energy per electron in atomic units.[1]

In the case of plane wave states, the electron charge density is everywhere uniform. The average potential energy per electron, $\langle P.E. \rangle$, must obviously scale as $1/r_e$ since the Coulomb potential scales as $1/r$. Thus $\langle K \rangle / \langle P.E. \rangle$ scales as r_e^{-1} and therefore becomes large as the system is compressed to very small values of r_e. Materials thus become more "ideal" in the limit of very large compressions. This result justifies the original assumption that at high pressures, the electrons were in plane wave states, interacting only slightly with the ions.

In order to determine how rapidly the plane-wave limit is approached, it is next necessary to obtain a quantitative estimate of the value of $\langle P.E. \rangle$ in this limit. Since the system is assumed to be in its ground state, the ions will be arrayed in some regular lattice. Consider an individual ion in this lattice, and construct a surface about the ion which encloses a volume containing the one ion and Z electrons, so that the enclosed volume is electrically neutral. Furthermore, require the surface to be non-overlapping with the corresponding surfaces of adjacent ions, and let the surface be identical from one ion to the next. Such a surface will in general have a polygonal shape. The volume enclosed by the surface is $4\pi Z r_e^3/3$, in atomic units. We now make the *Wigner-Seitz* approximation, in which we replace this polygonal surface with a sphere of identical volume, centered on the ion. The radius of the Wigner-Seitz sphere is $r_s = Z^{1/3} r_e$.

We further assume that adjacent Wigner-Seitz cells do not interact, so that the electrostatic potential energy of the ion lattice is just the sum of the electrostatic energies of each independent Wigner-Seitz sphere. The calculation is vey straightforward, since one must merely calculate the electrostatic energy of a sphere containing a point positive charge of value Z at the center, and a compensating negative charge distributed uniformly through the sphere. To obtain $\langle P.E. \rangle$, the average electrostatic potential energy per electron, divide the cell energy by Z, which gives the result

$$\langle P.E. \rangle = -0.9 \, Z^{2/3}/r_e, \tag{3-8}$$

expressed in atomic units. The Wigner-Seitz approximation is actually a rather good one; the exact electrostatic lattice energy for a face-centered cubic lattice of ions of charge Z is given by an expression identical to Eq. 3-8 but with the coefficient -0.9 replaced by -0.896.

Still making the approximation of plane wave electron states, one more quantity must be calculated in order to include all terms for the energy which scale as r_e^{-1}. This quantity is known as the *exchange energy* and arises because an electron in a given spin state interacts with other electrons not only through the direct, classical electrostatic interaction, but also through a purely quantum interaction which results from the antisymmetric nature of the spatial wave function for electrons which are in the same spin state. Like the classical potential energy, the exchange energy scales as r_e^{-1}. An exact calculation for plane wave states gives

$$\langle X \rangle = -0.458/r_e, \tag{3-9}$$

where $\langle X \rangle$ is the average exchange energy.[2]

We are now in a position to derive the $T = 0$ pressure-density relation for strongly compressed matter, valid in the limit $r_e \to 0$. The total energy per electron is given, to lowest order, by

$$E_0 = \langle K \rangle + \langle P.E. \rangle + \langle X \rangle. \tag{3-10}$$

Now the volume per electron is $4\pi r_e^3/3$, so the pressure, using Eq. (3-2), is

$$P_0 = 0.176 \, r_e^{-5} \, [1 - (0.407 \, Z^{2/3} + 0.207) \, r_e], \tag{3-11}$$

expressed in atomic units. Although Eq. 3-11 appears to be a sort of expansion in powers of r_e, note that no such expansion actually exists. Among the subsequent terms in this expansion is a term called the correlation pressure, which depends on the logarithm of r_e. In what domain of pressure is Eq. 3-11 quantitatively useful? An estimate can be obtained by considering the pressure required to completely ionize an atom of an element of atomic number Z. Ionization can be considered complete, and the electrons in plane wave states, when the energy density of the material is much greater than the energy density of the innermost electrons in the uncompressed atom. The latter is estimated as follows. Electrons in the K-shell have energies which scale approximately as Z^2. The radius of the K-shell scales approximately as Z^{-1}, and hence the volume of the K-shell scales as Z^{-3}. Thus Eq. 3-11 becomes valid when[3]

$$P_0 \gg Z^5. \qquad (3-12)$$

The highest pressure in a planet in the solar system occurs at the center of Jupiter, and is on the order of 100 megabars, or about ⅓ of an atomic unit of pressure. Thus Eq. 3-11 is not directly applicable to the study of planetary interiors. However, in the case of hydrogen ($Z = 1$), we are not too far from its range of validity, and it is in fact marginally applicable to metallic hydrogen in Jupiter.

The principal utility of Eq. 3-11 is that it serves as a point of departure for more elaborate theories, and is in any case an asymptotic limit. Although one might in principle try to interpolate between experimental results at low pressures and the asymptotic limit, such an approach has to be used with care because of the possibility of discontinuities in the pressure-density relation.

If we try to proceed to calculate a theoretical pressure-density relation which will be valid at lower pressures than Eq. 3-11, it is necessary to take into account the departure of the electrons from plane wave states. Two different approximations can be made at this point. The first approach, which we will describe next, is more rigorous but also more limited than the second approach. It is based upon a systematic calculation of corrections to the free electron states which were initially assumed.

The energy operator in Eq. 3-3, neglecting the ion kinetic energy, depends on the electron momenta and on the electrostatic potential energy. If we neglect the potential energy, the eigenstates corresponding to the kinetic energy are just plane waves, and the average energy of the system is just given by Eq. 3-7. Now, by elementary quantum theory, the first-order correction to this average energy is given by the average value of the potential energy *evaluated in the zero-order quantum states,* which are of course plane waves, corresponding to uniform electron density. This first-order correction yields just the direct and exchange potential energies (Eqs. 3-8 and 3-9). Thus Eq. 3-10 and the pressure relation derived from it, Eq. 3-11, are rigorous solutions to first order in the potential energy. These results are accurate only to the extent that the potential energy is a small perturbation to the total energy.

If we next calculate a first-order correction to the free electron states, this improved approximation to the electron states can then be used to generate a second approximation to the energy, etc. Stopping at the linear correction to the initial electron states has a major advantage: it is still possible to express the average energy in terms of an expression resembling Eq. 3-3, which involves a sum over electron kinetic energies and a pairwise sum over an interaction potential Φ'. This new interaction potential, which is no longer the pure Coulomb potential of the bare ions but now includes the screening clouds of electrons which they carry along, is called a *pseudopotential.* Since the screening clouds are considered to be a small effect compared with the main Coulomb interaction, screening cloud-screening cloud interactions are neglected. If such interactions are included, in a higher-order theory, then it clearly makes a difference whether multiple screening clouds overlap and interact, and then the interaction energy can no longer be expressed as in Eq. 3-3, in terms of a pairwise sum over terms involving separations $\mathbf{r}_{ij}$.

A self-consistent calculation of the first-order correction to electron plane wave states proceeds as follows. Assume that the electrons respond in some (initially unknown) way to the presence of the ions. Let the average electron density be $n_e^0 = N_e/V$, and let the fluctuation of the electron density as a function of position $\mathbf{r}$ be $\delta n_e(\mathbf{r})$. The spirit of this approximation will be that $\delta n_e \ll n_e^0$. In the zeroth approximation, one takes only the first term in Eq. 3-3, corresponding to the electron kinetic energy. In the first (next) approximation, we include the potential energy term Φ, but with the proviso that this term include not only the potential exerted on the electrons by the bare ions but also the potential produced by the electrons' own response to the potential, i.e., a self-consistent average potential. Let the perturbation to the electrons' energy be written as

$$U(\mathbf{r}) = -\sum_j Z_j|\mathbf{r} - \mathbf{r}_j|^{-1} + \int d^3r' \, \delta n_e(\mathbf{r}')|\mathbf{r} - \mathbf{r}'|^{-1}, \quad (3\text{-}13)$$

where U is the potential felt by an electron at coordinate $\mathbf{r}$, and we assume that $\langle U \rangle = \langle P.E. \rangle \ll \langle K \rangle$. We then use this perturbing potential to derive a first-order correction to the electron wave functions in the form

$$\psi(\mathbf{r}) = \psi_0 + \psi_1, \quad (3\text{-}14)$$

where ψ is an electron wave function, ψ_0 is a plane wave state, and ψ_1 is the first-order correction to this wave function. Now note that $\psi\psi^*$ is proportional to the density distribution of an electron, so that if this quantity is summed over all electrons, we can calculate the quantity $n_e(\mathbf{r})$ and thus the quantity $\delta n_e(\mathbf{r})$. Surprisingly, although $\delta n_e(\mathbf{r})$ depends on itself through Eq. 3-13, the problem can be solved analytically provided that exchange interactions with the electron screening cloud $\delta n_e(\mathbf{r})$ are neglected. The result is most easily obtained in Fourier space, with the result

$$\delta n_{e\mathbf{k}} = Z\rho_{\mathbf{k}}(\epsilon_{\mathbf{k}} - 1)/\epsilon_k, \quad (3\text{-}15)$$

where $\delta n_{e\mathbf{k}}$ is the three-dimensional Fourier transform of $\delta n_e(\mathbf{r})$ and $\rho_{\mathbf{k}}$ is the corresponding Fourier transform of the ion charge distribution. This expression allows us to solve the Schrödinger equation to first order, for *any* arbitrary ion distribution. Furthermore, this calculation only has to be done once, because the solution is contained in the function ϵ_k, which is known as the *dielectric function* (although it is defined in Fourier space instead of real space like a normal dielectric function). The dielectric function depends on the electron spacing parameter r_e as well as on the Fourier wave number $\mathbf{k}$, and is given by[4]

$$\epsilon_k = 1 + 0.166r_ex^{-2}\{0.5 + [(1 - x^2)/(4x)]ln[(1 + x)/|1 - x|]\},$$

$$(3\text{-}16)$$

where $x = k/2k_F$, and k_F is the Fermi wave number, given by

$$k_F = (3\pi^2)^{1/3} (N_e/V)^{1/3}. \qquad (3\text{-}17)$$

Using these expressions, it is straightforward to add a correction term to the expression for the energy given in Eq. 3-10. This correction term represents the extra electrostatic interaction energy between the ions and the electron screening cloud density, $\delta n_e(\mathbf{r})$. Correspondingly, an extra term is added to Eq. 3-11 for the zero-temperature pressure. Figure 3-1 shows the effect of adding such screening corrections to the high-pressure equations of state of hydrogen and helium. As is apparent, the modifications to the pressure-density relations are surprisingly small, even for relatively large values of r_e. The reason for this behavior seems to be the following. At high pressures, where the inequality (Eq. 3-12) is satisfied, the effect of the screening clouds is negligible, and the electrostatic energy of the system is given by Eq. 3-8. As the pressure and density decline, each ion develops an enhancement of electron charge around itself, which tends to screen out interactions with more distant ions. However, the increased interaction energy between an ion and its screening cloud roughly compensates for the decrease in the energy of interaction with the more distant ions, so that to first approximation, the electrostatic energy of the ion lattice changes little. However, we may expect this "cancellation" effect not to persist in the vibrational properties of the lattice, which affect its finite-temperature thermodynamics. Electron screening generally changes lattice frequencies considerably from their unscreened values.

Theories which use expansions from plane wave states are only useful for low-Z elements, and so this approach has primarily been used for study of Jovian planet interiors, which contain a large number fraction of hydrogen. We will now consider an alternative purely theoretical approach, which, like the dielectric function theory approach, approaches the limit of Eq. 3-10 as the pressure goes to infinity, but is not limited to the assumption that $\delta n_e \ll n_e^0$. The price we pay for this increased generality is that the theory is not fundamentally justified in the regime where it differs significantly from dielectric function theory. Ultimate justification of its utility depends upon comparisons with experimental high-pressure data.

We begin by assuming that the electrons are in *local* plane wave states, such that the parameters of the states (wavelength, etc.) change very slowly from one locale in the atomic array to another. This approach only makes sense if the distance over which electron energies vary is large compared with a typical electron wavelength. We then break up the Wigner-Seitz sphere into very small subvolumes, within each of which the electrostatic potential energy is assumed to be constant. Under these assumptions, the total energy per electron is given by an expression similar to Eq. 3-10, except that the average electrostatic potential energy per electron is given by

$$\langle P.E. \rangle = -\phi(\mathbf{r}), \tag{3-18}$$

where ϕ is the electrostatic potential at $\mathbf{r}$ produced by all of the charges in the system. In this approach, which is known as *Thomas-Fermi-Dirac* theory, the electrons are considered to be a gas of free particles which responds to the potential produced by the ions and the self-consistent electron distribution exactly as if the distribution of ions were on a macroscopic instead of a microscopic scale. We are therefore making a quasiclassical approximation; quantum mechanics only enters into the theory by determining the occupation number of the plane wave states and the first-order exchange correction to the energy. The equilibrium distribution of electrons is determined by the requirement that the chemical potential of the electrons, μ, be uniform throughout the substance. First, consider an idealized case where the ions are replaced by a uniform positive background with the same average charge density. The average energy per electron is then given by Eq. 3-10 with the average potential energy term $\langle P.E. \rangle$ set equal to zero. Let the electron chemical potential for this case be μ_e; it is given by

$$\mu_e = d[N_e(\langle K \rangle + \langle X \rangle)]/dN_e$$
$$= 1.842 \, r_e^{-2} - 0.611 \, r_e^{-1}. \tag{3-19}$$

We now "unsmear" the ions again, so that the electrostatic potential ϕ varies as a function of potential. The total electron chemical potential is then given by

$$\mu = \mu_e - \phi = \text{const.} \tag{3-20}$$

This equation imposes a relationship between the electron density distribution $n_e(\mathbf{r}) = N_e(\mathbf{r})/V$ and the electrostatic potential ϕ. These quantities are also related by Poisson's equation

$$\nabla^2 \phi = 4\pi n_e(\mathbf{r}), \tag{3-21}$$

and so both are determined. The boundary condition which is imposed on $\phi(\mathbf{r})$ is that it approach the potential due to an unscreened ion in the vicinity of an ion, and the other boundary condition which completes the specification of the problem is the requirement that the average charge density due to electrons be equal and opposite to the average charge density due to the ions. It is customary to solve the problem in the Wigner-Seitz approximation, but like the dielectric function approach, the TFD approach can be generalized to calculate the energy of an arbitrary configuration of ions. It can be shown (although not easily) that in the limit $r_e \to 0$, the TFD pressure reduces to the result given by Eq. 3-11. This result is to be expected, because the electron distribution in any valid theory should become more nearly uniform in this limit. It is a peculiarity of unmodified TFD theory that the electron density always approaches infinity in the vicinity of the ions. This is a consequence of the breakdown of the quasiclassical approximation in the region where the potential varies rapidly over a distance on the order of an average DeBroglie wavelength. This infinite density does not cause divergences in any measurable quantity such as pressure or energy, and can in any case be removed by a proper application of quantum corrections.

TFD theory differs little from dielectric function theory for hydrogen, but as Z increases, the differences become much larger. Figure 3-1 shows a comparison of zero-temperature pressure-density curves for hydrogen, helium, and lithium, calculated using both theories. Figure 3-2 compares TFD results for iron with experimental data on the compression of iron up to four megabars.[5,6] In this pressure range, TFD theory is qualitatively but not quantitatively correct, whereas unmodified dielectric function theory would not be even qualitatively valid. The major problem with TFD theory is that it does not allow for the existence of atomic shell structure at moderate pressures. Thus it predicts a monotonic behavior of equations of state as a function of Z. TFD theory does not recognize the existence of the periodic table at low pressures. In principle, this deficiency can be

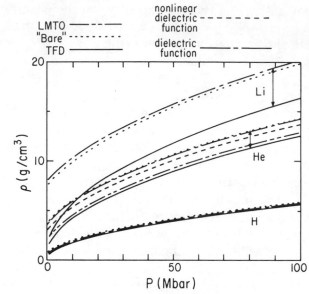

Fig. 3-1. Zero-temperature equations of state, in various approximations, for H, He, and Li.

remedied by solving in detail for the electronic wave functions at various degrees of compression, but the computational problems are formidable and relatively few results are available to date. Figure 3-1 also shows, for comparison, a zero-temperature equation of state for helium calculated using a combination of atomic orbitals, which may be the most accurate available equation of state for this element.[7]

THERMODYNAMICS AT FINITE TEMPERATURE

We turn now to a discussion of the thermal properties of compressed matter. As before, we begin with the limit of extremely high compression and then generalize to lower densities. First consider a lattice composed of a single species of ion of charge number Z, immersed in a uniform negative background of electrons in plane wave states. Consider some arbitrary displacement of the ions from their equilibrium positions, and let this displacement have amplitude ξ in atomic units. The increase in potential energy of the lattice scales as $Z^2\xi^2/r_e^3$, and so the frequencies of the lattice modes scale as

$$\omega \propto [Z^2 m_e/r_e^3 M]^{1/2}. \tag{3-22}$$

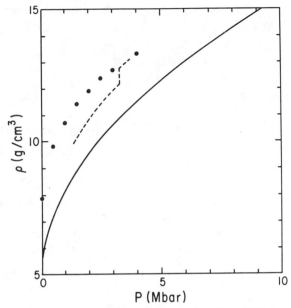

Fig. 3-2. Zero-temperature pressure-density relation for iron from TFD theory (solid curve). Dots are experimental shock Hugoniot points for iron (except for the $P = 0$ point, which refers to α-iron; see Fig. 3-16). Dashed curve is the seismic pressure-density relation for the earth's core: see Fig. 3-17.

Assume that the temperature is sufficiently high for all lattice modes to be highly excited. This means that T in atomic units is much higher than $\hbar\omega_m$, where ω_m is the highest normal frequency of the lattice and $\hbar$ is Planck's constant/2π. Provided that this is true, the thermal contribution to the internal energy per electron is given by

$$E_T = 3T/Z, \tag{3-23}$$

in atomic units. The total internal energy of the solid is obtained by adding this result to E_0. The total pressure is likewise obtained by adding a small thermal correction P_T to P_0, where

$$P_T = 3(N_e/V)\gamma T/Z, \tag{3-24}$$

and

$$\gamma = -[d \ln(\omega_m)/d \ln r_e]/3. \tag{3-25}$$

In this instance, $\gamma = 0.5$. It is straightforward to show that the slope of adiabats in the temperature-density plane is also given by γ:

$$\gamma = (\partial \ln T / \partial \ln \rho)_S. \qquad (3\text{-}26)$$

All of the above expressions are valid in the limit of ultra-high pressures as defined by the inequality Eq. 3-12. At lower pressures, the effects of electronic structure have to be taken into account. Dielectric function theory can be applied to hydrogen and helium to calculate the lattice modes at finite values of r_e. Thomas-Fermi theory (TFD theory without the exchange correction) has also been used to calculate ω_m and γ at lower pressures. Figure 3-3 shows results for γ as a function of pressure for three elements. Results for H were computed using dielectric function theory, while results for Fe and Si were computed using Thomas-Fermi theory.[8] Note that electron screening lowers the value of ω_m and increases γ (γ is known as the *Grüneisen parameter*). Experimental measurements of the Grüneisen parameter of solids at ordinary pressures give similar results: the values of γ are typically on the order of 1 to 1.5, as compared with the high-pressure limit of 0.5. To be precise, some of the scaling laws which we have been using for the high-pressure limit are not strictly valid at low pressures, and so the thermodynamic behavior of solids at ordinary pressures frequently cannot be described solely in terms

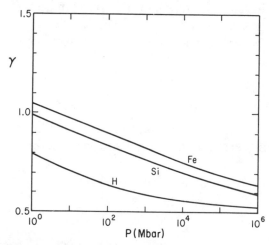

Fig. 3-3. Grüneisen parameter as a function of pressure for three elements.

of a single Grüneisen parameter and ω_m (which is known as the *Debye temperature* when expressed as a temperature).

All of the foregoing results are valid for a solid which is at a temperature well above its Debye temperature, so that quantization of the ion energy states can be neglected. The temperature must also be sufficiently low that the ions are still in the solid phase. As we have pointed out, the temperature in a planetary interior is generally high enough for quantum effects on ions to be negligible (with the possible exception of temperatures in certain small icy satellites in the outer solar system). If the temperature also exceeds the melting temperature of the solid, an alternative thermodynamic description of the material is needed. There are two problems here: one needs a theoretical expression for predicting the high-pressure phase boundary between a liquid and solid phase, and a theory is needed for the liquid phase.

Theory of Melting at High Pressure

Rigorously, the melting temperature as a function of pressure, $T_m(P)$, is given by the solution to the equation

$$\mu_L(T_m, P) = \mu_S(T_m, P), \qquad (3\text{-}27)$$

where $\mu_L(T, P)$ is the chemical potential of the ions in the liquid phase, as a function of temperature T and pressure P, and $\mu_S(T, P)$ is the same quantity for the solid phase. The problem with using this approach is that although $\mu_S(T, P)$ can be computed for the solid phase using Debye theory, the corresponding quantity in the liquid state must also be calculated. A very accurate physical description of both phases is needed in order for the results to be meaningful. In the following section, we will discuss some of the powerful theoretical techniques which are used to calculate the finite-temperature thermodynamics of high-pressure liquids (as well as solids). But first, let us consider a simple physical model which gives useful qualitative results for melting temperatures of various elements at high pressures.

We focus attention on a given ion in the high-pressure lattice, and assume that this ion executes excursions of amplitude about its equilibrium lattice site. A qualitative criterion for the melting of the lattice is then derived by assuming that the lattice melts when $\langle \xi^2 \rangle^{1/2}$ exceeds some critical fraction of the lattice spacing, where $\langle \xi^2 \rangle^{1/2}$ is the root-mean-square amplitude of the ion oscillations about equilibrium. This simple rule, which is due to Lindemann, asserts that a lattice ceases to exist when the amplitude of thermal oscillations become so large that neighboring ions begin to "overlap," and begin to have a significant probability of exchanging places. Lindemann's rule is expressed in a useful form by relating $\langle \xi^2 \rangle$ to the temperature. A simple statistical average over harmonic oscillations gives the result

$$\langle \xi^2 \rangle = T \langle \omega^{-2} \rangle (m_e/M), \tag{3-28}$$

in atomic units. Note that the atomic unit of $\langle \omega^2 \rangle$ is $e^2/a_0^3 m_e$. The Lindemann criterion states that melting occurs when

$$\langle \xi^2 \rangle / r_s^2 > \Delta^2, \tag{3-29}$$

where Δ is a universal constant with a value of about 0.1. Inserting the scaling relations given previously for the high-pressure (unscreened) solid, one finds that the Lindemann criterion then gives

$$T_m = KZ^2/r_s, \tag{3-30}$$

where K is a constant. This result turns out to be exact, i.e., more general than the Lindemann rule, in the limit of very high pressure. The constant K must be detemined by computer experiments, and is currently estimated to have a value of about 1/150.

Result 3-30 is extended to lower pressures by taking into account the effect of electron screening on the lattice vibrational modes. A rough estimate of this effect can be obtained by using Thomas-Fermi theory, and by assuming that all lattice frequencies scale with density like ω_m (not true in general). Using the Thomas-Fermi calculations of ω_m mentioned previously, we then obtain the results shown in Figure 3-4 for the high-pressure behavior of melting temperatures for various pure elements. All of these results have been scaled to match the correct limiting form of Eq. 3-30.

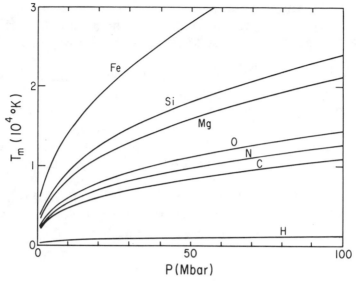

Fig. 3-4. Melting temperature as a function of pressure for several elements.

Theory of the Liquid State

Three main approaches are used for calculation of finite-temperature thermodynamics of liquids at high pressures. The first method, which is the simplest but also the least rigorous, assumed that the liquid phase is basically similar to the solid phase for purposes of computing most thermodynamic quantities. Certainly the major qualitative difference between a solid and a liquid is that the solid has long-range ordering of the atoms, while in a liquid the ordering is all short-range. This difference causes the shear modulus of a solid to be finite and the shear modulus of a liquid to be zero, but for the purposes of calculating the finite-temperature corrections to the internal energy and pressure, one may think of the liquid as having similar harmonic excitations of the nuclei about some momentary "equilibrium" positions, but with the long-wavelength component of such excitations absent. Thus we assume that the liquid has a Grüneisen parameter γ of about the same value as in the corresponding solid, and that the thermal corrections to the internal energy and pressure of the liquid are given by expressions similar to Eqs. 3-23 and 3-24, but multiplied by a correction factor $f\,(<1)$ which accounts for the absence of the long-range portion of the harmonic excitation spectrum. In this approach, one must rely upon the Lindemann law to derive the melting temperature of the solid. There is no real way to rigorously calculate f.

The second approach is known as thermodynamic perturbation theory,[9] and makes use of a rigorous solution for the thermodynamics of a liquid of particles having a specified form for the intermolecular interaction potential energy Φ_0. This specified liquid is known as the *reference state*. In such calculations, the reference state is frequently taken to be a liquid of hard spheres; i.e., particles which do not interact at all for separations greater than some critical distance equal to twice the hard sphere radius, and which repel each other infinitely at separations less than this distance. The Helmholtz free energy of the hard sphere liquid can be expressed in the form

$$F = F_{ideal} + F_{HS}(\eta), \tag{3-31}$$

where F_{ideal} is the Helmholtz free energy of an ideal gas and F_{HS} is the additional term produced by the interactions of the hard spheres, which is a function of the hard sphere packing fraction η.

The Gibbs-Bogolyubov inequality states that

$$F \leq F_0 + \langle \Phi - \Phi_0 \rangle_0, \tag{3-32}$$

where F is the exact Helmholtz free energy of an interacting system, F_0 is the Helmholtz free energy of some reference system, and $\langle \Phi - \Phi_0 \rangle_0$ is the difference between the configurational (interaction) energy of the system and the reference system, *averaged over configurations of the reference system*. Thus if we have a reference system for which F_0 is available in a convenient form, and for which we have available a convenient expression for the distribution of probabilities of various configurations, then a close approximation to the exact Helmholtz energy of the actual system can be generated by varying some parameter of the reference system so as to minimize the right-hand side of inequality in Eq. 3-32.

The hard sphere system is useful for such an application because one has available the hard sphere free energy

$$F_{HS} = F_0 = (4\eta - 3\eta^2)NkT/(1 - \eta)^2, \tag{3-33}$$

where $\eta = (\pi/6)(N/V)d^3$, and d is the hard sphere diameter. The probability of a given configuration of hard spheres, as a function of temprature, is also available in analytic form, which thus permits a convenient evaluation of $\langle \Phi \rangle$ as a function of η. The obvious choice for the variational parameter is η. Thus, using a prescribed interaction potential energy Φ for the actual liquid system, we adjust η to minimize the right-hand side of Eq. 3-32, and obtain an approximation to F for a specified value of T, V, and N. Despite the fact that the true interaction potentials are much "softer" than the hard sphere potential, the method appears to work well in most cases, and rapidly generates accurate thermodynamic parameters for many liquids.[10]

The most rigorous approach to calculation of the thermodynamic properties of a liquid is to directly calculate the Helmholtz free energy on a computer, using the true liquid structure consistent with a given interparticle potential. This approach, like the previous one, requires that the interaction energy of the system can be expressed as a sum over pairwise potentials. As we have discussed earlier, dielectric function theory applied to dense plasmas yields just such a potential.

A liquid is simulated on a computer by supposing that N atoms or molecules of the liquid are contained within a cubic box of side L. The density of particles is then N/L^3. Now the maximum number of particles which available computers can handle is on the order of 10^3, far fewer than the 10^{23} needed to correspond to a real macroscopic system. The trick which is used to evade this difficulty employs a mathematical stratagem to simulate an infinite system with a finite one. The basic cubic box is assumed to be repeated indefinitely in all directions by identical boxes, also containing N particles arrayed in just the same configuration as the basic box (Fig. 3-5). Any given particle then interacts not only with the other particles in the box, but also with their images in all of the other boxes (including the images of itself). The total energy of interaction per box can be shown to give a finite result, even for the most difficult case, the r^{-1} potential. Using this approach, experiments on computers have found that the average configuration energy per particle becomes insensitive to the total number of particles even for as few as 50 particles in the basic box.

Ensemble averages for liquids are computed by randomly choosing a sequence of configurations within the basic box, and by weighting each configuration by its appropriate Boltzmann factor, $\exp(-W/kT)$, where W is the configuration energy of the configura-

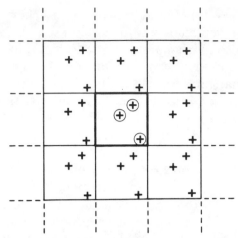

Fig. 3-5. Use of periodic boundary conditions to simulate an infinite system with a finite number of particles in a box.

tion. Special techniques are used to optimize the efficiency of choosing such configurations. The weighted averages are then used to compute a rigorous value of the Helmholtz free energy $F(T, L^3, N)$, which then yields all of the other thermodynamic variables of interest.[1]

Multicomponent Systems at High Pressure

In principle, the methods discussed above can be applied to the study of systems containing more than one type of atom. Such systems are of fundamental interest for planetary interiors because they are potentially capable of showing much more complicated behavior than simple one-component systems. New phases, corresponding to different mixing ratios of the components, may appear under certain circumstances. Even the most cursory glance at available geophysical and planetophysical evidence shows that such behavior of multicomponent substances has played a major role in shaping the elemental abundances which we now observe at the surfaces and in the atmospheres of the planets. Purely theoretical investigations of this type of behavior have made only limited progress in obtaining practically usable results, primarily because calculations of multicomponent phase behavior make extraordinary demands on the accuracy of the theoretical free energy. Only the simplest systems have so far been studied quantitatively, and even for them the results are not unambiguous. Therefore, the discussion which follows is primarily aimed at elucidating how in principle such calculations are carried out, and why, under conditions in planetary interiors, chemical separation of compounds is to be expected.

The calculation of phase equilibrium for a given theoretical Helmholtz free energy is straightforward in principle. Consider a two-component system, containing N_1 particles of species 1 and N_2 particles of species 2. Phase equilibrium occurs under conditions of constant temprature and pressure, and so it is convenient to introduce the Gibbs free energy $G = E - TS + PV = F + PV$, with the natural variables $G = G(T, P, N_1, N_2)$. A thermodynamically stable state is one in which the total Gibbs free energy of the system is at an absolute minimum. Let the temperature and pressure be fixed, and let the total number of particles $N = N_1 + N_2$ be fixed also. Define the number fraction $x = N_2/N$, and consider the function $G(x)$ as x varies between 0 and 1. Draw a straight line betwen $G(0)$ and $G(1)$. Let the excursions of $G(x)$ with respect to this line be $\delta G(x)$. Mathematically, we define the Gibbs free energy of mixing δG by

$$\delta G = G(P, T, x) - [xG(P, T, 1) + (1 - x)G(P, T, 0)]. \quad (3\text{-}34)$$

Clearly $\delta G(x)$ must be zero at $x = 0$ and $x = 1$. First suppose that $\delta G(x)$ is positive (except at its endpoints) and has a single maximum at some value of x. For a given overall value of x, the system may split into two subsystems with compositions x_a and x_b, say. The overall value of G will then be given by the weighted average of G_a and G_b, where $G_a = G(x_a)$, etc. For this example, the system can obviously reduce G the most by splitting into two subsystems with $x_a = 0$ and $x_b = 1$. The relative amounts of phase a and phase b are determined by the initial value of x. Similarly if $\delta G(x)$ is negative everywhere except at its endpoints, and has a single minimum at some value of x, then the value of G can only be increased by formation of separate phases, and therefore no phase separation can occur.

From the point of view of statistical mechanics, there are two main effects which compete to determine whether a two-component system is unstable to chemical separation. First, the ideal entropy of mixing contributes to δG the term

$$-TS_{\text{mixing}} = NkT[x \ln x + (1 - x) \ln (1 - x)], \quad (3\text{-}35)$$

which is obviously negative since $x < 0$. Thus this term acts to stabilize the system against phase separation. It can only come into play in a system where particles can freely interchange positions, such as a liquid or gas. Any opposing (i.e., positive) contribution to δG must come from the interparticle interactions, and must arise because the mixed system has a higher value of G than a linear combination of the endmembers. Usually, in a liquid, the ideal mixing term (Eq. 3-35) dominates δG near the endpoints ($x \sim 0$ and $x \sim 1$), while the positive interaction terms (which are also temperature-dependent in general) are more dominant in the middle. Thus a typical plot of δG versus x has a maximum near the middle and minima near the ends. Obviously the region of phase instability is then located in the middle range of x, which is bounded by constructing a double tangent to the vicinity of the two minima (Fig. 3-6). The mixing entropy term tends to dominate as the temperature increases, and thus the central maximum in δG gradually disappears with rising temperature. Correspondingly, the range of x subject to phase instability shrinks with rising temperature, and ultimately disappears at the critical temperature.

The limiting theory for high-pressure materials that we have discussed in this chapter is able to predict a fundamental characteristic of planetary interior matter — its tendency to separate into phases with different chemical compositions. Quantitative work requires that one accurately take into account the effects of electron screening and finite temperature. Thus far, theories have not achieved this degree of accuracy, and cannot make detailed predictions about phase boundaries of mixtures at high pressures. However, it is constructive to consider a fully pressure-ionized system at finite temperature. If the temperature is above the melting temperature but still low

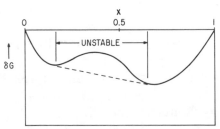

Fig. 3-6. Gibbs free energy of mixing as a function of composition, for a given temperature and pressure. In this example, the mixture is unstable to phase separation in the range shown.

enough so that the average ion kinetic energy is small compared with the average interaction energy, then calculations have shown that the average energy per electron is well represented by Eq. 3-10, plus small thermal corrections. This is true because the average potential energy $\langle P.E. \rangle$ is approximately the same in both the solid and liquid phases. Furthermore, Monte Carlo calculations for liquid mixtures of ionized material have shown that the average coulomb energy per electron of a mixture is closely approximated by a simple linear superposition of $\langle P.E. \rangle$ for pure phases:

$$\langle P.E. \rangle_{\text{mixture}} = -0.9 \, r_e^{-1}[(1 - x)Z_1^{5/3}$$
$$+ xZ_2^{5/3}]/[(1 - x)Z_1 + xZ_2]. \quad (3\text{-}36)$$

We can now form the Gibbs free energy G for the mixture by using Eq. 3-10, except that the temperature-dependent terms (including-TS) must be added to it. Actually, only one temperature-dependent term is really essential for obtaining a qualitative description of phase separation — the term for the ideal entropy of mixing.[12]

When Eqs. 3-35 (rewritten as a per-electron quantity) and 3-36, and the mixture generalizations of Eqs. 3-10 and 3-11 are combined to obtain δG for liquid mixtures, one finds that there exists a critical temperature T_c, above which the mixture "prefers" to be in a single liquid phase for any value of x, and below which the Gibbs free energy can be minimized by forming two liquid phases with differing values of x. Figure 3-7 shows some sample calculations for mixtures of liquid metallic hydrogen and the elements lithium through oxygen, at pressures of 10 megabars and 40 megabars. Note that for very low solubilities, the equilibrium concentration of the solute is proportional to $\exp(-A/T)$, where A is an activation energy. A typical temperature in the interior of Jupiter is indicated by a horizontal line. The results of this figure illustrate the point that chemical phase separation is always likely to be important under the conditions in planetary interiors, but they should not otherwise be taken very seriously, since the model ignores all electronic structure, which certainly persists in real materials at planetary interior pressures. Figure 3-8 shows the phase equilibrium curve for a mixture of hydrogen and helium at a pressure of 10 megabars. For each value of the temperature, the two corresponding points on the curve give the

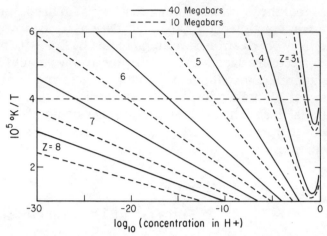

Fig. 3-7. Solubility of various elements in metallic hydrogen as a function of temperature, calculated using the fully pressure-ionized model (*not* an adequate approximation for planetary interiors). Horizontal dashed line corresponds to a typical Jovian interior temperature.

compositions of the two liquid phases which are in equilibrium.[13] Above the critical temperature of about 7500 °K, no phase separation occurs at all.

At somewhat lower temperatures, one must also consider the equilibrium between liquid and solid phases in multicomponent

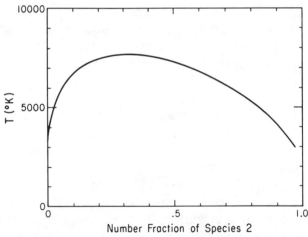

Fig. 3-8. Phase equilibrium of hydrogen (species 1) and helium (species 2) mixtures at 10 megabars, in the fully pressure-ionized model. Intersections of a horizontal line with the curve give the compositions of the two liquid phases which are in equilibrium.

systems. According to the Gibbs phase rule, the composition of a mixture *must* change across such a phase boundary. It is very difficult to predict this composition change theoretically, however. In a solid close to its melting temperature, the average interaction energy may differ substantially from the value given by the mixing law Eq. 3-36, depending upon the degree of disorder in the structure of the alloy. Similarly, the appropriate expression for the entropy of the solid may differ substantially from Eq. 3-35. It seems possible to address these points with appropriate computer simulations.

Serious difficulties are encountered when one attempts to include the effects of electron screening in chemical separation calculations. As should be evident by now, the tendency of a multicomponent system to form separate phases depends on rather delicate energy differences between combined and separated constitutents. Uncertainties in energies which are tolerable for the purpose of calculating bulk quantities such as the pressure and internal energy may render phase equilibrium calculations meaningless. So far, only the simplest systems have been theoretically studied, because only in them is it possible to apply perturbation theory to the electron wave functions under conditions of planetary interest. Thus, the hydrogen and helium system has been rather extensively investigated under the assumption that deviations of the electron wave functions from plane waves are small. Dielectric function theory, together with its generalization to higher order, indicates that liquid mixtures of hydrogen and helium behave much like the results shown in Fig. 3-8; the effect of electron screening is to slightly increase T_c above its value for the completely ionized system. On the other hand, calculations for the H-He system using Thomas-Fermi-Dirac theory indicate that T_c is much lower than the results of Fig. 3-8 would indicate.[13,14]

Suppose now that we wish to calculate the thermodynamic properties of a pure molecular component which consists of different atomic species, such as H_2O, for example. At low pressures (in atomic units) the material acts like a pure species, but at high pressure it becomes a multicomponent system. In general, systems such as these need to be studied experimentally at low pressures, and theoretical expressions are used as a guide to extrapolation of the experimental information to higher pressures. A common approximation to the pressure-density relation for a multicomponent system at high pressure makes use of the so-called "additive volume law." In this approximation, it is assumed that each different atom in the material is contained within an independent, isolated cell. The volume of the cell is constrained by the requirement that each cell be in pressure equilibrium with its neighbors. This assumption is evidently equivalent to the Wigner-Seitz approximation, but it can be used more generally to synthesize a multicomponent pressure-density relation from a one-component pressure-density relation generated by any means. Mathematically, we calculate the density from a prescribed pressure using the formula

$$\rho^{-1} = \sum_i X_i / \rho_i(P), \qquad (3\text{-}37)$$

where the sum is taken over all atomic constituents with mass fraction X_i, and $\rho_i(P)$ is the density-pressure relation for the ith atomic constitutent, which can be obtained from any theory. *Provided that all atomic constituents remain miscible in the high-pressure compound,* Eq. 3-37 is frequently an excellent approximation to the exact pressure-density relation evaluated for the multicomponent system. Figure 3-9 shows the results of calculations carried out for a mixture of equal numbers of hydrogen and helium atoms, with linear screening, in the metallic liquid phase. The ordinate is the difference

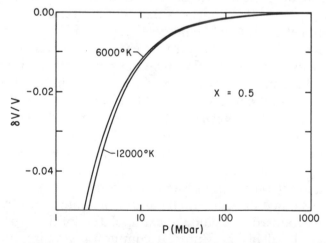

Fig. 3-9. Non-additivity of volume in an equal number mixture of hydrogen and helium, as a function of pressure. The mixture has a slightly smaller volume than the sum of the volumes of the two pure components.

between the volume of the mixture and the volume given by Eq. 3-37, expressed as a fraction of the volume. This quantity is plotted versus pressure for temperatures of 6000 °K (upper curve) and 12000 °K (lower curve). Such calculations can also be carried out for more elaborate models of mixtures; the results always indicate that deviations from the mixing law Eq. 3-37 are on order of 1 percent or smaller for pressures of 10 megabars and greater. This circumstance primarily follows from the dominance of the free electron pressure at high pressures.

SHOCK COMPRESSION EXPERIMENTS

Much basic information about the behavior of materials at pressures and temperatures characteristic of planetary interiors is derived from shock compression measurements. In such an experiment, a sample is placed in a chamber and a shock wave is transmitted through it, typically by impacting one wall of the chamber by a projectile. Propagation of the shock wave through the sample results in compression and heating of the sample. A major advantage of this technique is that pressures comparable to planetary interior pressures are readily obtainable, with the maximum achievable pressure depending upon details of the experimental arrangement and properties of the sample. Thus it is possible to test proposed theoretical equations of state of materials under conditions not unlike those in which the theoretical relations are to be applied. The principal disadvantage of the shock method is that it does not cleanly yield a specific point in (P, V, T) space for the sample. The problem is that shock compression is accompanied by a temperature increase. The pressure and volume of the postshock material can be directly determined from the shock parameters, but the temperature increase cannot. Unless the postshock temperature can be measured by some other technique such as optical pyrometry, it must be deduced by modeling the fundamental thermodynamic relations for the material.

In a shock compression experiment, the velocity of propagation of the shock v_s and the mass velocity of the postshock material v_p are the parameters which are actually measured. Let ρ_1 be the initial (uncompressed) mass density of the sample and let ρ be the final (shocked) mass density. Similarly, let E_1 and E be the internal energy per gram of material and P_1 and P the pressure, before and after shocking, respectively. The Hugoniot relations, which relate these quantities, are based upon the fact that the shock discontinuity itself does not create or destroy mass, momentum, or energy, although there is indeed a flux of all of these quantities through the shock. Conservation of the flux of each of these through the shock front then requires

$$\rho_1 v_s = \rho(v_s - v_p), \qquad (3\text{-}38)$$

$$v_s^2/2 + E_1 + P_1/\rho_1 = (v_s - v_p)^2/2 + E + P/\rho, \qquad (3\text{-}39)$$

and

$$P_1 + \rho_1 v_s^2 = P + \rho(v_s - v_p)^2. \qquad (3\text{-}40)$$

With a little algebra, v_s and v_p can be eliminated from the above relations to obtain

$$E - E_1 = (P + P_1)(\rho_1^{-1} - \rho^{-1})/2. \qquad (3\text{-}41)$$

We assume that E_1, P_1, and ρ_1 are all known, and thus the relations in Eqs. 3-38 through 3-40 can be used to find the parameters of the shocked state.

Application of differing impact energies to a sample prepared in state 1 produces excursions along a compression curve defined by Eq. 3-41, which is known as a *Hugoniot*. Although the pressure achieved along the Hugoniot can be made to increase by increasing the impact energy, a point of diminishing returns is reached, beyond which the sample tends to merely heat up without achieving further compression. This phenomenon can be understood in terms of the simple thermal equation of state introduced earlier (Eq. 3-24). Let us write for the internal energy per gram

$$E = E_0 + C_V T, \qquad (3\text{-}42)$$

where E_0 is the internal energy per gram at $T = 0$ and C_V is a (constant) heat capacity. The pressure is then given by

$$P = P_0 + \gamma \rho C_V T, \qquad (3\text{-}43)$$

equivalent to Eq. 3-24. Substituting these relations in Eq. 3-41, we then find

$$C_V T = (f \partial E_0 / \partial \ln \rho - 2E_0)/(2 - f\gamma), \qquad (3\text{-}44)$$

where $f = (\rho - \rho_1)/\rho_1$, and we have assumed strong shock compression such that $E \gg E_1$ and P_0 (the zero-T component of P) $\gg P_1$. For any reasonable zero-temperature internal energy, the numerator of Eq. 3-44 is positive, and so the temperature increases without limit as the compression parameter f approaches $2/\gamma$ from below. As we have seen, γ is typically close to unity for many materials at moderate pressures, and so it is impossible to attain more than a two- or threefold increase in density with a single shock compression.

Multiple shock experiments permit higher compressions to be attained than is possible with single shocks. Suppose that a sample has been shocked a single time from state ρ_1, P_1, and E_1 to state ρ, P, E. While it is still in this shocked state, another shock wave is passed through it, transforming it to state ρ', P', E'. It is usually convenient to achieve a second shock by reflecting the first shock wave back into the sample. Now the compression limit for the second shock state is again given by $f < 2/\gamma$, but this time with $f = (\rho' - \rho)/\rho$. If the first shock achieved, say, a twofold compression of the sample, then the second shock can provide another twofold compression. In principle, several shocks can be passed through the sample to produce very substantial compressions with only moderate heating. The temperature increase of the sample is ultimately bounded below by the temperature increase produced by adiabatic compression.

RESULTS FOR IMPORTANT PLANETARY MATERIALS

Hydrogen

As the most abundant and simplest of the elements, hydrogen assumes a special importance in the study of planetary interiors. Hydrogen is more amenable than any other element to a strictly theoretical treatment. Nevertheless, the high-pressure phase diagram of hydrogen is not yet adequately understood.

There are three principal sources of information about the high-pressure behavior of hydrogen: (1) theoretical studies for $T = 0$ and

$T \neq 0$, which are most accurate for $P > 10$ megabars; (2) experimental shock compression experiments attaining pressures of 0.8 megabars; (3) static compression experiments at room temperature, which reach pressures ~ 0.6 megabars. Thus there is a gap in our information about the behavior of hydrogen at pressures on the order of a megabar, which results from the fact that in this pressure range the H_2 molecule transforms in some manner from its low-pressure structure as a diatomic molecule into a pressure-ionized atomic form. Although various theoretical predictions about this transformation have been published, so far it has not been reliably observed under laboratory conditions.

The phase diagram for hydrogen is well understood at cryogenic temperatures and at moderate pressures accessible in the laboratory. Unfortunately, such conditions are not applicable to planetary interiors. The critical point which terminates the liquid-vapor transition is located at a pressure of 13 bar and a temperature of only 33 °K, and thus is not important in planetary physics. There is no planetary interior whose temperature profile passes close to this point. The melting temperature of solid H_2, which starts at 14 °K at 1-bar pressure, is probably likewise too low to be significant in any planetary interior (see Fig. 3-4).

Consider the following thought experiment. A sample of H_2 gas is at a temperature of ~ 100 °K and a pressure of one bar. Now compress the gas isothermally. Initially, the pressure increases in accordance with the ideal gas law, but with increasing density we begin to see deviations from this law due to the increasing importance of molecular interactions. Assuming that such interactions can be described by a pairwise additive potential expression in the form

$$\Phi = \sum_{i < i'} \phi(\mathbf{r}_i - \mathbf{r}_{i'}), \qquad (3\text{-}45)$$

where Φ is the total potential energy of the system (cf Eqs. 3-3 and 3-4) and $\phi(r)$ is the potential function for a pair of molecules. Since the H_2 molecule is not spherically symmetric, ϕ must be a function of the rotational orientations of the molecules as well as of the separation between their centers of mass. However, detailed studies of interactions between hydrogen molecules at high pressures and temperatures have shown that H_2 molecules generally rotate freely under conditions of planetary interest, so that ϕ can usually be replaced with its average overall relative orientations of the molecules. The spherically averaged function ϕ is shown in Fig. 3-10. As is typical of such curves, it displays a long-range attractive potential which results from mutual polarization of the molecules (Van der Waals force), followed by a weak minimum and then a steeply rising repulsive potential primarily produced by the overlap of the electron clouds. The minimum is responsible for the existence of condensed phases at low pressure and temperature, but is too shallow to be very significant at temperatures of 100 °K and higher.

Returning to the thought experiment, we find that an extra, nonideal component of the pressure begins to appear due to the decreasing volume of the gas, which causes the average distance between molecules to decrease. The corresponding point on the potential curve moves inward and up the rising repulsive potential. After sufficient compression has occurred, the ideal (thermal) contribution to the pressure, which was dominant at the beginning of the experiment, becomes quite small compared with the nonideal (potential) pressure. At some point, however, it is clear that the pairwise potential pressure which we calculate in this fashion must begin to grossly overestimate the actual nonideal pressure, for it neglects N-body interactions which tend to reduce the pairwise potential. That is, the

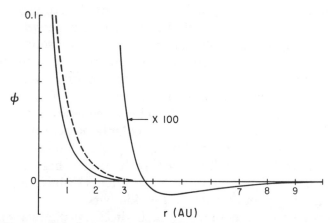

Fig. 3-10. Effective spherically-averaged two-body potential between H_2 molecules. The dashed curve shows the theoretical two-body repulsive potential, while the solid curve corresponds to the effective (empirical) potential as "softened" by many-body effects. The magnified tail of the curve shows the region of Van der Waals attraction.

presence of other molecules besides the pair which we are considering will change the shape of the overlapping electron clouds which produce the repulsion, making the clouds become more uniform. Eventually, the description of the material should go over to the fully pressure-ionized picture of highly compressed matter, which was discussed at the beginning of this chapter.

Figure 3-10 also shows how many-body effects act to weaken the interaction between pairs of molecules. The solid curve for $\phi(r)$ in this figure shows a semiempirical potential function which yields Hugoniots in good agreement with reflected shock measurements in H_2 to pressures of ~ 800 kilobars. The dashed curve shown in the region of the inner repulsive curve is a theoretically calculated inter-action potential for an isolated pair of hydrogen molecules. Note that it lies above the solid curve which fits the data. The effective potential is "softer" than the theoretical one because of three-body and N-body interactions. The participation of additional molecules serves to weaken the interaction between any given pair. The principal con-clusion that we draw from this comparison is that theoretical high-pressure equations of state for hydrogen (or any substance, for that matter) are sometimes inadequate, in certain pressure ranges, for quantitative description of the material.[15]

Diamond anvil static compression experiments have shed addi-tional light on the behavior of the H_2 molecule under pressure. An isolated H_2 molecule has a vibrational mode corresponding to stretching and compression of the bond length between the two protons. When the molecule is compressed in the solid phase, the frequency of this mode initially rises with pressure, as if the presence of adjacent molecules "stiffened" the vibration of a given molecule. But with further compression to around 300 kilobars, the initial frequency rise ceases, and is replaced by a frequency decline. This is taken to be an indication that the molecular bond betwen the protons is weakening. The actual disintegration of the molecule and its replacement with a metallic phase of hydrogen has not yet been observed in the diamond anvil experiments, which have so far achieved a maximum pressure of 600 kilobars.[16]

As we will discuss later in the book, hydrogen in planetary interiors occurs under pressure-temperature conditions ranging from one bar pressure and temperaure $\sim 50-150$ °K to ~ 50-megabar pressure and temperature $\sim 10,000-30,000$ °K. One finds in the literature quite a wide selection of proposed phase diagrams for hydrogen at high pressure,[17] and we will not adopt any particular one here. In any case, there is probably only one major phase boundary in hydrogen which is crossed for the pressure-temperature conditions given above. This is the phase boundary corresponding to the pressure-induced (and temperature-induced?) decomposition of the H_2 molecule, and its replacement with ionized hydrogen. The key issue for planetary interiors study is the location of this phase boundary, and whether it corresponds to an ordinary first-order phase transition.

In principle, the phase boundary between molecular and metallic hydrogen can be theoretically computed as follows. The empirical pair potential of Fig. 3-10 is used to compute thermodynamic functions in the molecular phase, while the dielectric function theory presented earlier is employed for the metallic phase. The phase boundary is then given by the equation

$$\mu_+(T, P) = \mu_M(T, P),\qquad(3\text{-}46)$$

where μ is the proton chemical potential, or Gibbs free energy per proton, and the subscripts $+$ and M refer to the metallic and molecular phases respectively. The free energies must be referred to the same energy zero point, such as the energy of an isolated hydrogen molecule, in which case μ_M must include half of the energy of dissociation of the molecule and the energy of ionization of the atom.

At $T = 0$, Eq. 3-46 determines a transition pressure which is independent of temperature. This pressure is highly sensitive to small uncertainties in the chemical potentials, and so cannot be considered well determined at present. Using the intermolecular potential of Fig. 3-10, one obtains a transition at 3 megabars, but with a probable uncertainty of a factor of two. Calculations at finite temperature indicate that this pressure does not change greatly as the temperature is increased to values of interest for planetary interiors ($\sim 10^4 \, °K$).

The hydrogen phase diagram must be considered uncertain because no phase boundary for pressures greater than a few hundred kilobars has been experimentally identified. The melting temperature for the solid phase of H_2 is theoretically estimated to be given by[17]

$$T_m \simeq 2800 \, \rho^2 \, °K,\qquad(3\text{-}47)$$

valid for $\rho \gtrsim 0.4$ g/cm^3. This temperature is probably below the prevailing temperatures in planetary interiors which contain highly-compressed hydrogen. Thus the H_2 must be treated as a supercritical fluid. At higher pressures than ~ 3 megabars, solid metallic hydrogen may be formed, but its melting temperature is very low, perhaps below 1000 °K (cf Fig. 3-4) because of the low ionic mass. Quantum effects may prevent solid metallic hydrogen from existing at all.

In the region of $P > 3$ megabars and $T \sim 10^3 - 10^4$ °K, we have liquid metallic hydrogen, which should resemble a molten alkali metal (since it is an alkali metal) in properties such as electrical and thermal conductivity. We do not know how the liquid metallic region interfaces with the supercritical molecular liquid. Return again to the thought experiment which begins with an H_2 gas at low density. Suppose that the gas is at first heated instead of compressed. At sufficiently high temperatures, say 10^5 °K, the molecules are dissociated and the atoms are ionized, forming a plasma. Now, keeping the temperature constant, compress the plasma to a pressure of, say, 100 megabars. The electrons will become degenerate, but no sudden phase transition has occurred, and the material is now in the metallic liquid phase, with the protons interacting strongly. Thus, if there is a first-order boundary between the molecular and metallic liquids, it must terminate at some critical temperature and pressure; the critical temperature is unlikely to exceed $\sim 10^5$ °K, and may be much lower. It seems quite possible that the liquid molecular and metallic phases of hydrogen grade continuously into each other, much as in the case of ordinary thermal ionization. As we shall see later, the question of the nature of this boundary is an important matter for the structure of Jupiter and Saturn interior models, since the presence of a first-order phase transition would require a discontinuity in the equilibrium concentration of other elements dissolved in the liquid hydrogen.

Helium

Because of its great volatility, helium is expected to exist in substantial amounts only in Jovian planet interiors, where it should be initially present mixed with hydrogen in approximately solar proportions (see Tables 2-1 and 2-2). Helium-enriched regions may be formed in those planets where liquid-liquid phase separation processes (discussed earlier) have occurred, but it is currently uncertain whether such processes of helium enrichment actually take place under planetary interior conditions. Experimental data on helium are therefore primarily needed for a dilute solution of helium in hydrogen (at a ratio of approximately one helium atom to 15 hydrogen atoms), at pressures on the order of a megabar and temperatures on the order of 10^4 °K. Unfortunately, because of the great technical difficulties of working with hydrogen-helium mixtures, no such data are available, nor is there much prospect for obtaining them in the near future. The situation is only slightly better for pure helium: measurements have only been made in the liquid and solid phases up to 20 kilobars. Thus our information on helium under conditions of interest is almost entirely theoretical at present.

The equation of state of helium at experimentally accessible pressures is computed by means of an empirical effective pair potential, which can be represented in the form of a so-called exponential-6 potential:[7]

$$\phi(r) = \epsilon\{0.845 \exp[13.1(1 - r/r^*)] - 1.845(r^*/r)^6\}, \quad (3\text{-}48)$$

where $\epsilon = 3.4 \times 10^{-5}$ a.u. of energy and $r^* = 5.61$ a.u. of distance. The total energy of a given configuration of helium atoms is then given by a sum of the form of Eq. 3-45. The phase boundary between liquid and solid helium can be computed by using the effective potential combined with liquid state theory to obtain the liquid chemical potential and pressure. The same variables are readily calculated in the solid state for an assumed lattice structure.

At pressures much greater than 20 kilobars, the energy of a configuration cannot be computed using a pairwise potential, because of many-body effects. The plane-wave limit computed earlier is reached in the limit of infinite pressure, but this limit is not useful for planetary applications. Figure 3-1 shows theoretical $T = 0$ high-pressure equations of state for helium, computed using (a) dielectric function theory carried to one higher order in perturbation theory; (b) linear-muffin-tin-orbital (LMTO) theory; (c) Thomas-Fermi-Dirac theory. The divergence between these three curves reflects the differing ways that the electron states are treated. Of these three pressure-density relations, the LMTO (b) curve probably has the most accurate treatment of the electron states and is thus preferable.[7]

Using LMTO theory, an effective pairwise potential has been

found for high-pressure helium. As in the case of hydrogen, many-body effects cause the high-pressure effective potential to be systematically weaker than the low-pressure potential. This effective potential has been used to extend liquid-state calculations into the high-pressure domain.

Helium is an insulating solid at pressures less than about 100 megabars because of the finite band gap between the filled 1s electron states and the available conducting states at higher energies. This band gap disappears at about 100 megabars, with the precise value depending on the stable lattice structure. These results indicate that metallic hydrogen, which exists in the Jovian planets at pressures greater than about 3 megabars, is not accompanied by metallic helium. However, one cannot automatically conclude that perturbation-theory calculations which start from plane-wave states are invalid under planetary interior conditions. The key issue, so far unresolved, is whether such expansions converge in the domain where helium is an insulator.

It turns out that the thermodynamic variables of helium at megabar pressures can be reasonably well-calculated by using an effective pair potential of the form

$$\phi(r) = 88912.1 \ \epsilon \ \exp(-11r/r^*), \qquad (3\text{-}49)$$

which is the "softened" form of Eq. 3-48, with ϵ and r^* having the same values as before.[7] The pressure-density relation for a mixture of hydrogen and helium can be obtained, to an accuracy of about 1% or better, from the "additive volume law," Eq. 3-37.

H$_2$O

Water ice is a significant component of planetary interiors in the outer solar system. Because of the high cosmic abundance of oxygen, accurate knowledge of the high-pressure behavior of H$_2$O ice is important for synthesizing the interior structure of objects with a relatively undepleted volatile component.

Detailed static investigations of solid H$_2$O high-pressure polymorphs have been caried out over the pressure range of $0-\sim100$ kilobars. This pressure range spans the maximum pressure range of interest for the interiors of the icy satellites of Jovian planets, and a corresponding temperature-pressure phase diagram[18] is shown in Fig. 3-11. Also shown is a band of typical estimated interior temperatures in icy satellites (see Chapter 8), which lies entirely in the stability fields of several solid ice phases. The common form of solid water, ice I, is supplanted at higher pressures by a complicated sequence of polymorphs, culminating in ice VII, which appears to persist to the highest temperatures and pressures investigated. Only the very largest icy satellites achieve high enough central pressures to produce ice VII. In all of these crystalline forms of ice, the H_2O molecule retains its identity, but is bonded in the solid in different ways. The density changes discontinuously across each of these phase boundaries, reaching a value of 1.66 g/cm³ for ice VII near its triple point with ice VI and VIII (compared with ice I's density of 0.92 g/cm³ at standard conditions). The high-pressure polymorphs ice VII and VIII have the structure of two interpenetrating cubic lattices of oxygen atoms, thus forming a body-centered cubic lattice of oxygens, each of which has its associated two hydrogen atoms. It is the arrangement of these hydrogen atoms which differs between ice VII and ice VIII. Theoretical considerations indicate that ice VII/VIII should be the terminal polymorph of solid H_2O at high pressure: it is not possible to pack the

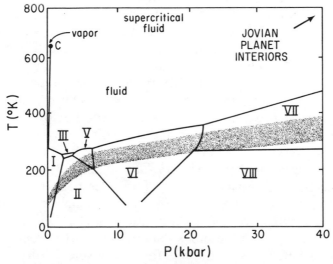

Fig. 3-11. Phase diagrams of water. *C* is critical point. Shaded zone shows approximate temperature distribution in icy satellites (see Chapter 8). Only the centers of the most massive satellites enter the ice VII phase. Water-rich regions in Jovian planets, if they exist, are in high-pressure fluid phases.

H_2O molecule more densely. However, with further compression the hydrogen bonds themselves will become nonexistent and a mixture of two parts hydrogen to one part oxygen, describable by a Thomas-Fermi-Dirac model for example, should become a more suitable description of the solid. We do not know whether ice VII transforms to such an atomic phase continuously or via another polymorphic phase transition.[18]

Interior models of icy satellites are calculated using the experimental information mentioned above. H_2O is thought to exist under more extreme conditions of temperature and pressure in the giant planets themselves, and a different data base is used for this purpose. As we shall see, H_2O in the giant planets may be at temperatures approaching $\sim 10^4$ °K, at pressures of $1-2$ megabars. The H_2O melting curve has not been experimentally determined for such extreme conditions, but is unlikely to exceed the oxygen melting temperatures shown in Fig. 3-4, which implies that H_2O is liquid in the giant planets.

The principal source of information about compressed H_2O at very high temperatures and pressures comes from shock wave measurements. These experiments are well suited to investigation of such conditions because they automatically produce temperatures on the same order as those expected in the deep interiors of the Jovian planets. Figure 3-12 shows results of two reflected shock experiments on water (error bars), reaching pressures of 1.3 and 2.3 megabars respectively.[19,20] Such pressures span the expected range for H_2O in Uranus and Neptune. Also shown are several theoretical zero-temperature isotherms for H_2O. The *dash-dot* curve is computed using an interpolation between static experimental data at low pressure and a version of TFD theory and the additive volume law Eq. 3-37 for a hydrogen-oxygen mixture at high pressure.[21] The theoretical curve with *short dashes* makes use of *ab initio* calculations of H_2O-H_2O intermolecular potentials, and therefore does not have the correct asymptotic high-pressure limit built in. Nevertheless, it is the only theoretical $T = 0$ pressure-density relation for H_2O which is approximately consistent with the shock data.[22] Note that a proposed $T = 0$ isotherm is ruled out if its pressure at a given density equals or exceeds the pressure of a shocked state (which is of course at a finite temperature).

Figure 3-12 compares the H_2O shock data further with two other $T = 0$ isotherms. The upper one (*solid*) is calculated directly from

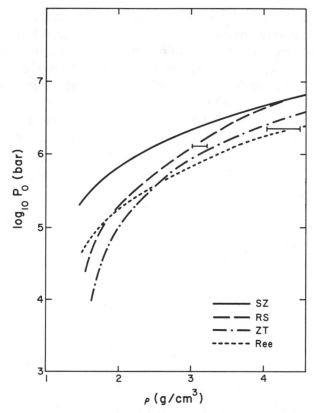

Fig. 3-12. Experimental shock-compression data points for water, compared with several theoretical zero-temperature pressure-density relations. [From Hubbard, W. B., and MacFarlane, J. J. Structure and evolution of Uranus and Neptune. *J. Geophys. Res.* **85:** 225–234 (1980); copyright American Geophysical Union.]

TFD theory, while the dashed curve makes use of interpolation between low-pressure experimental data and theoretical high-pressure relations.[23] The principal conclusion from these various comparisons between shock data and theoretical isotherms is that the actual compression curve of H_2O is far "softer" in the multimegabar pressure range than would be indicated by relatively simple theoretical relationships. This behavior may be caused by temperature-induced changes in the electronic structure of the molecules.

Further evidence for such changes comes from measurements of the electrical conductivity of water along the principal Hugoniot[24] (Fig. 3-13). Under room temperature conditions, pure water is slightly ionized, forming the ions H_3O^+ and OH^-. It appears that the high pressures and temperatures produced by shock compression may considerably enhance this ionization, increasing the electrical conductivity by several orders of magnitude. The saturation at a level of ~ 20 (ohm $\cdot$ cm)$^{-1}$ may be due to a combination of decreased ion mobility and lack of further ionization. Temperatures achieved in the shock compressions are not high enough to produce free electrons.

Liquid ammonia displays a similar increase in electrical conductivity along its principal Hugoniot (Fig. 3-13). The highest temperature reached along the plotted curve is estimated at ~ 2000 °K.

NH_3 and CH_4

Ammonia and methane are more volatile than H_2O, but may still be present in substantial amounts in certain Jovian planets and satellites. The high-pressure phase diagrams of these substances are not

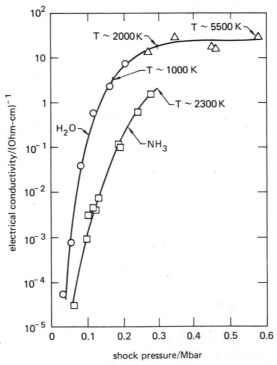

Fig. 3-13. Electrical conductivity of shock-compressed water and ammonia. [From Ross, M., Graboske, H. C., Jr., and Nellis, W. J. Equation of state experiments and theory relevant to planetary modeling. *Phil. Trans. R. Soc. Lond.* **A303**: 303–313 (1981). Copyright, the Royal Society and authors.]

well studied, but could be as complex as that of water. Figure 3-14 shows an approximate phase diagram for liquid and solid CH_4 based upon static compression studies at cryogenic temperatures and pressures less than 10 kilobars, and a series of diamond anvil measurements at room temperature spanning a pressure range from 0 to about 200 kilobars. The diamond anvil experiments show evidence for a possible further phase transition from methane V to methane VI in a pressure range from 118 to 187 kilobars (not shown in Fig. 3-14). This transition is not abrupt and may not be a first-order one.[25]

Recent data are available on the behavior of liquid ammonia and methane under shock compression. Results are summarized in Fig. 3-15. The solid curve on the right represents the principal Hugoniot of ammonia,[19] compressed from an initial liquid state at $\rho_1 = 0.693$ g/cm^3 and $T_1 = 230$ °K, while the dashed curve to the left of it shows a published theoretical zero-temperature isotherm, based in part upon a Thomas-Fermi-Dirac model of a nitrogen-hydrogen mixture.[21] Evidently the theoretical curve is not grossly inconsistent with the measurements, since it always gives a lower pressure for a given density. Calculations of the thermal component of the pressure give results that are in reasonable agreement with the difference between the two curves.

The left-hand solid curve shows the principal Hugoniot of meth-

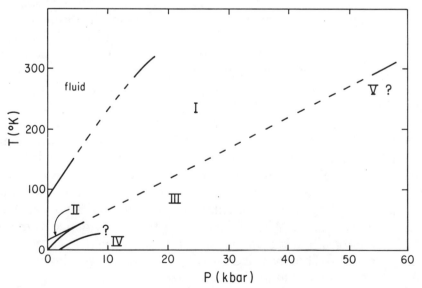

Fig. 3-14. Phase diagram of methane. [Data from Hazen, *et al.* (1980).]

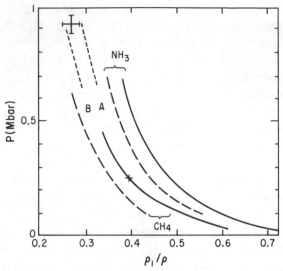

Fig. 3-15. Principal Hugoniots (solid curves) of ammonia and methane (data from Mitchell and Nellis, 1982, and Nellis, *et al.,* 1981). Dashed curves to their left are theoretical $T = 0$ pressure-density relations (from Zharkov and Trubitsyn, 1978). Error bars in upper left-hand corner show reflected shock point from initial state marked with cross. Dashed curves *B* and *A* are theoretical calculations of reflected shock Hugoniots from this point.

ane, starting from an initial state at $\rho_1 = 0.423$ and $T_1 = 111$ °K. Also shown (data point in upper left-hand corner) is a reflected shock state from the point marked on the principal Hugoniot.[10,26] This experimental point at 0.9 megabars has important implications for the high-pressure behavior of the CH_4 molecule, which we will now discuss. First of all, we note that a recent zero-temperature isotherm of CH_4 (*dashed* curve, left) is at least approximately consistent with the shock data. However, when detailed theoretical calculation of the principal Hugoniot is carried out, using empirically-determined intermolecular potentials, the high-temperature/high-pressure continuation of the Hugoniot follows curve A, which is inconsistent with the reflected shock data point. Curve A is calculated by assuming that the CH_4 molecule remains intact along the shock compression curve. However, if we assume that the CH_4 molecule decomposes into molecular hydrogen and elemental carbon at pressures greater than ~ 200 kilobars, curve B is obtained. Thus there is evidence from the shock data for a major breakdown of the CH_4 molecule at intermediate pressures. The phase assumed by the carbon is unknown, but may correspond to diamond powder.

Iron

Iron is an abundant constituent of solar composition material and is therefore expected to be abundant in the earth's interior, where seismic data suggest that it forms a massive core. As we will discuss in Chapter 7, inversion of seismic data for the earth leads eventually to an empirical pressure-density relation within the earth's interior. There are several discontinuities in the terrestrial pressure-density relation, but the most important of these occurs at a pressure of about 1.4 megabars, where the density jumps from about 5.5 g/cm^3 to about 10 g/cm^3 (see Fig. 3-17). The density then rises more or less smoothly with pressure to a maximum value of 12.5 – 13 g/cm^3 at 3.7 megabars at the center of the earth. This pressure-density relation, which is shown in Fig. 3-2, is generally believed to correspond to almost-pure iron, with an admixture of some lighter material which reduces the density slightly. Figure 3-2 also shows a theoretical $T = 0$ pressure-density relation for iron, computed using Thomas-Fermi-Dirac theory, and the experimental pressure-density relation obtained by reducing shock compression data to zero temperature by assuming a Grüneisen parameter and heat capacity. TFD theory works moderately well for this element at a pressure of a few megabars, only underestimating the density by about 20 percent in this pressure range, with smaller discrepancies at higher pressures. However, the experimental pressure-density relation in turn gives a density which is on the order of 10 percent *higher* than the terrestrial core value. Even after making small corrections for finite temperature in the core, there is fairly good evidence for the presence of some lower-density component alloyed with the iron in the earth's core. The theoretical and experimental evidence provide a strong circumstantial case that the core is predominantly composed of iron.

We will be concerned in Chapter 6 with the question of the possibility of dynamo action in the production of planetary magnetic fields. An essential ingredient for the operation of this mechanism is the existence of an electrically conducting fluid in a planetary interior. Since iron is the most abundant metallic element (other than metallic hydrogen), the question of its high-pressure phase diagram is of great importance for this problem. Although the $T = 0$ pressure-density relation for iron is now reasonably well-established over the pressure range of interest, much less is know about the melting curve.

Figure 3-16 shows a phase diagram for iron, which for pressures up to about 200 kilobars is obtained from static measurements using the diamond anvil technique. At temperatures of planetary interest, there are at least four different solid phases of iron, denoted as α, γ, δ, and ϵ. It is not known whether the γ and ϵ phases persist to the megabar pressure range. Direct extrapolation of the γ-liquid boundary to a pressure of 3 megabars is most difficult and uncertain, but has been estimated to pass through a temperature on the order of 4000 °K.[27,28]

Shock compression experiments on iron are capable of probing the multimegabar pressure range, but are of such brief duration that it is difficult to determine whether the compressed material is solid or liquid. One approach which can provide helpful, if somewhat ambiguous clues, is to look for abrupt changes in the velocities of rarefaction waves in the compressed sample. The velocity of such a longitudinal wave in an isotropic medium is given by

$$v_p{}^2 = (K + \frac{4}{3}\mu)/\rho, \qquad (3\text{-}50)$$

where K is the isothermal modulus of incompressibility, μ is the shear modulus, and ρ is the mass density. When the material enters the

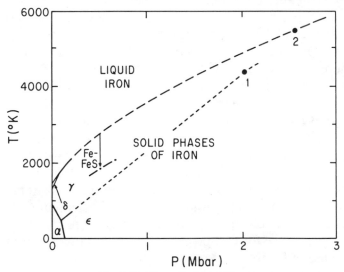

Fig. 3-16. Phase diagram of iron.

liquid phase, μ vanishes, and Eq. 3-50 reduces to the usual expression for the velocity of sound waves in a liquid or gas. Since μ and K are typically of a similar order of magnitude, the velocity of longitudinal waves should drop abruptly and substantially as the sample crosses a phase boundary.

Figure 3-16 shows the locations (1, 2) of *two* discontinuities observed in the velocities of rarefaction waves, which correspond to two points on the principal Hugoniot of iron. Tentatively, the first discontinuity can be interpreted as a crossing of the ϵ-γ boundary, while the second may be the actual melting of the γ-phase. The discontinuities could be caused by considerably more complicated phenomena, but the interpretation just given is roughly consistent with the Lindemann melting curve for iron shown in Fig. 3-4.

In addition to the behavior of pure iron, the behavior of a mixture of iron and sulfur is of considerable relevance to planetary interiors. Referring once more to Tables 2-1 and 2-2, we note that the cosmic proportions of iron and sulfur are about 74 percent Fe, 26 percent S. As we have also discussed in Chapter 2, under primordial nebular conditions and at temperatures below about 700 °K, H_2S gas reacts with iron to form FeS and H_2O gas reacts with iron to form FeO. The latter tends to be incorporated in magnesium silicates. Thus, planets which accreted from relatively low-temperature condensates could tend to have substantial amounts of sulfur in their iron cores. It turns out that an iron and sulfur alloy has *eutectic* behavior. That is, for a particular composition (the so-called eutectic composition), a sample melts completely into two liquid phases, and the temperature at which this occurs is lower than the melting temperature of either pure component. At 1-bar pressure, pure Fe melts at 1808 °K, pure FeS melts at 1469 °K, and a eutectic mixture melts at 1262 °K. The composition of the eutectic mixture is 27 percent sulfur. Thus, addition of sulfur in something like cosmic proportions can cause a substantial drop in the melting temperature of iron. This phenomenon persists to higher pressures. At the highest pressure investigated (100 kilobars), the melting temperature of the eutectic mixture relative to pure Fe is depressed by an amount approaching 1000 °K. Figure 3-16 indicates a plausible extrapolation of this trend to higher pressures. The eutectic composition seems to trend toward smaller amounts of sulfur with increasing pressure, and may be on the order of 10 percent S in the megabar pressure range.[29]

Rock

Consider a cloud of protoplanetary material having the overall chemical composition given in Tables 2-1 and 2-2. Under conditions of very low pressure ($\ll 1$ bar) and temperatures on the order of, say, 1500–500 °K, the material will have a gas component composed principally of H_2 and He, and a solid component composed of various compounds of abundant elements such as Fe, Mg, Si, and O (most of the O will still be bound up in gaseous H_2O, and most of the C and N will likewise be in gaseous molecules). The precise composition of the condensate will depend very critically on the precise thermodynamic conditions and details of the condensation process, but in any case we expect to find minerals such as free Fe, FeS, $(Mg,Fe)_2SiO_4$ (olivine), and $(Mg,Fe)SiO_3$ (pyroxene), where (Mg,Fe) means that either Mg or Fe ions can occupy lattice sites. The condensate thus basically consists of iron and silicates.

In dealing with planetary interiors, it is usual to treat the thermodynamic properties of the iron component separately from the thermodynamic properties of the silicate component. This procedure has a simple physical justification: when such a primordial condensate is subjected to enough heating to melt all of its components, the iron-sulfur liquid is immiscible in the silicate liquid, and being denser, is separated out gravitationally. It is thus natural to expect an iron core to be present in bodies which are large enough to have undergone this process (later we will give examples of what is meant by "large enough").

The point of all the preceding discussion is that we will define as "rock" that component of the previously-defined condensate which is left after formation of an iron core. Because we had the earth in mind when making this definition, we will further equate the average composition of "rock" to the average composition of the earth's lower mantle. Such material is used as a starting point for deriving models of silicate layers in the interiors of planets other than earth.

Figure 3-17 shows the average density profile in the earth's mantle, together with a pressure scale.[30] According to current interpretations of this profile, the upper mantle (depths < 1000 km) has been substantially modified in chemical composition by partial melting out of crustal material. The lower mantle, on the other hand, still has its

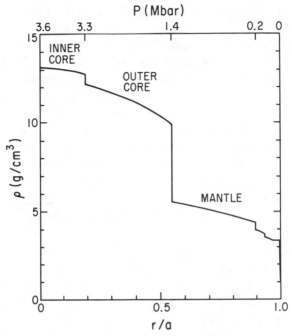

Fig. 3-17. Seismically-derived density profile of the earth's interior [from Dziewonski and Anderson (1981)]. Radius *r* is normalized to 6371 km. A corresponding pressure scale is shown at top.

primordial composition minus the iron in the core, but is undoubt-edly highly modified in its mineralogy by the high pressures. One of the principal objectives of geophysics is to establish, from such a profile and from suitable high-pressure experiments, what the chemi-cal composition of this presumably primordial layer might be. More-over, one may apply the empirical pressure-density relation of Fig. 3-17 to the interiors of other planets, but to do so intelligently, it is first necessary to obtain an adequate understanding of its dependen-ces on chemical composition and on other thermodynamic parame-ters. For example, in addition to the great discontinuity at the radius of the core at $r = 3480$ km ($r/a = 0.55$), which we understand to be caused by a major chemical discontinuity, we observe three more stepwise discontinuities at $r = 5701, 5971, 6151$ km (corresponding to $r/a = 0.895, 0.937, 0.965$), and a discontinuity in slope at 5771 km ($r/a = 0.906$). The final discontinuity at the base of the crust is understood to be a chemical discontinuity resulting from the formation of the crust by partial melting. What is the nature of the other discontinuities?

To first approximation, hypothetical uncompressed "rock" can be taken to be made up in large part of olivine, $(Mg,Fe)_2SiO_4$. Since Mg and Fe have similar ionic radii at ordinary pressures, olivine in general consists of a solid solution containing both ions in the available lattice sites. Although most of the earth's free iron is in the core, some iron remains bound up in the olivine as well as in other common minerals such as pyroxene, which includes compounds of the type $(Mg,Fe)SiO_3$. Analysis of the composition of magmas produced in the upper mantle implies that the ratio of magnesium to iron in the parent "rock" is initially about 8 by mass, although it is not known whether this ratio is constant throughout the mantle. Neglecting the iron content, then, "rock" can be considered to be pure Mg_2SiO_4 (forsterite) together with $MgSiO_3$ (enstatite).

At low pressures, silicates are made up of the basic ionic building block SiO_4, which is arranged with the oxygen ions forming a tetrahedron around the silicon ion. The total charge of this unit is -4, which is canceled by either combining it with pairs of cations of charge $+2$ each, as in olivine, or linking the tetrahedrons in a chain via shared oxygen ions and canceling the remaining -2 charge with $+2$ cations, as in pyroxene.

Olivine is not highly symmetrical. The basic cell is orthorhombic, having three mutually perpendicular axes of differing lengths. The relatively large oxygen ions are approximately in hexagonal closest packing, which is the arrangement of stacked billiard balls with identical alternate layers. The arrangement is not precisely close packed because of the differing lengths of the crystal axes, and thus there is room for a denser arrangement of ions. The denser arrangement which becomes favorable at sufficiently high pressure is called the spinel phase, and is produced when the oxygen ions assume a symmetric face-centered cubic close-packed form. The configuration differs from hexagonal closest packing in that the stacked billiard ball layers repeat every three layers rather than every two layers (the reader may find this discussion more understandable after experimenting with billiard balls or marbles!).

The olivine-spinel transition does not occur directly in pure forsterite: there is an intervening phase known as the β-phase, which is orthorhombic like olivine. This intervening phase disappears, depending on the temperature, for compositions close to the suspected mantle forsterite/fayalite ratio. At still higher pressures, there is evidence that the spinel phase decomposes into MgO and $MgSiO_3$ in perovskite structure. The latter, which is shown in Fig. 3-18, consists of a body-centered cubic structure with Si at the body centers, Mg at the corners, and O at the face centers. As is typical of high pressure phases, it has a high degree of symmetry which is ultimately due to the increasing uniformity of the electron distribution.

$MgSiO_3$ has several stable phases at various temperatures and pressures.[31] This mineral ends up as perovskite at high pressures, as seems natural. At intervening pressures one also has stable phases of β-phase and spinel-structure Mg_2SiO_4, combined with SiO_2 as stishovite. Stishovite is a high-pressure polymorph of quartz (low-pressure SiO_2), where the Si has six nearest-neighbor O's instead of the usual four. Ilmenite structure is yet another polymorph of $MgSiO_3$.

What then can we conclude about the pressure-density relation for the lower mantle (= "rock") diagrammed in Fig. 3-17, and its relation to a specified chemical composition? Unfortunately, there are still many more variables in the problem than there are constraints. The stability fields for the various polymorphs of "rock" are dependent

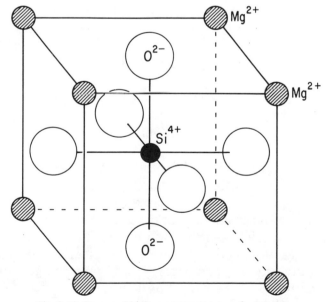

Fig. 3-18. Proposed high-pressure structure for $MgSiO_3$.

not only on temperature, which is poorly known in the earth's interior, but also on abundances of less-abundant constituents such as Ca, Al, Ti, etc. Thus there is not universal agreement about the nature of the upper-mantle discontinuities. The $r = 6151$ km discontinuity is approximately at the correct pressure and temperature to correspond to the forsterite $-$ β-phase transition, and the $r = 5971$ km discontinuity could correspond to the β-phase $-$ spinel transition, but it might also be associated with the transition of spinel $+$ stishovite to $MgSiO_3$ ilmenite. Major chemical changes may also be associated with all of these, and in particular with the $r = 5701$ km discontinuity.

Major information about rock pressure-density relations at pressures on the order of a megabar and higher comes from shock compression experiments.[32] Much of the research in this area is based upon a modular approach. One identifies the major constituents of rocks which are likely to retain their chemical identity at these very elevated pressures, and then synthesizes a pressure-density relation from the individual relations for the separate constituents, via the additive-volume law. The difficult part of this approach is to identify the appropriate chemical "modules." For example, it has been proposed that rock ultimately breaks down into a mixture of such oxides as SiO_2, MgO, and FeO. Thus one might hypothesize that forsterite breaks down according to

$$Mg_2SiO_4 \rightarrow 2\ MgO + SiO_2, \qquad (3\text{-}51)$$

and enstatite ultimately goes to

$$MgSiO_3 \rightarrow MgO + SiO_2. \qquad (3\text{-}52)$$

Shock data have been obtained for these substances up to pressures on the order of two megabars; the results do not confirm the hypothesis. The additive volume law was used to synthesize shock Hugoniots for forsterite and enstatite from the shock Hugoniots for periclase (MgO) and silica (SiO_2 — stishovite phase). The resulting compression curve was then compared with the experimental curve for artificially pure Mg_2SiO_4 and $MgSiO_3$, and found to give too low a density by about 5 – 10 percent. Thus it appears that these substances have adopted a different strategy for reaching closely packed configurations at megabar pressures. Rather than undergoing the reactions in Eqs. 3-51 and 3-52, it seems most plausible that enstatite simply remains in the perovskite phase up to at least multimegabar pressures, and that forsterite remains a mixture of MgO and $MgSiO_3$ (perovskite).

REFERENCES

1. Landau, L. D. and Lifshitz, E. M. *Statistical Physics.* London: Pergamon, 1969.
2. Salpeter, E. E. Energy and pressure of a zero-temperature plasma. *Astrophys. J.* **134:** 669–682 (1961).
3. Kirzhnits, D. A. *Field Theoretical Methods in Many-Body Systems.* Oxford: Pergamon, 1967.
4. Hubbard, W. B. and Slattery, W. L. Statistical mechanics of light elements at high pressures. I. Theory and results for metallic hydrogen with simple screening. *Astrophys. J.* **168:** 131–139 (1971).
5. Al'tshuler, L. V., Bakanova, A. A., and Trunin, R. F. Shock adiabats and zero isotherms of seven metals at high pressures. *Sov. Phys. JETP* **15:** 65–74 (1962).
6. McQueen, R. G., Marsh, S. P., Taylor, J. W., Fritz, J. N., and Carter, W. J. The equation of state of solids from shock wave studies. In *High Velocity Impact Phenomena,* pp. 293–568. New York: Academic Press, 1970.
7. Young, D. A., McMahan, A. K., and Ross, M. Equation of state and melting curve of helium to very high pressure. *Phys. Rev. B* **24:** 5119–5127 (1981).
8. Zharkov, V. N., and Trubitsyn, V. P. *Physics of Planetary Interiors.* Tucson: Pachart, 1978, p. 212.
9. Mansoori, G. A., and Canfield, F. B. Variational approach to the equilibrium thermodynamic properties of simple liquids. I. *J. Chem. Phys.* **51:** 4958–4967 (1969).
10. Ross, M., and Ree, F. H. Repulsive forces of simple molecules and mixtures at high density and pressure. *J. Chem. Phys.* **73:** 6146–6152 (1980).
11. Hubbard, W. B., and Slattery, W. L. Statistical mechanics of light elements at high pressure. I. Theory and results for metallic hydrogen with simple screening. *Astrophys. J.* **168:** 131–139 (1971).
12. Stevenson, D. J. Miscibility gaps in fully pressure-ionized binary alloys. *Phys. Lett.* **58A:** 282–285 (1976).
13. Stevenson, D. J. Thermodynamics and phase separation of dense fully-ionized hydrogen-helium fluid mixtures. *Phys. Rev* **12B:** 3999–4007 (1975).
14. MacFarlane, J. J., and Hubbard, W. B. Statistical mechanics of light elements at high pressure. V. Three-dimensional Thomas-Fermi-Dirac theory. *Astrophys. J.* **272:** 301–310 (1983).
15. Nellis, W. J., Ross., M., Mitchell, A. C., van Thiel, M., Young, D. A., Ree, F. H., and Trainor, R. J. Equation of state of molecular hydrogen and deuterium from shock-wave experiments to 760 kbar. *Phys. Rev. A* **27:** 608–611 (1983).
16. Sharma, S. K., Mao, H. K., and Bell, P. M. Raman measurements of hydrogen in the pressure range 0.2–630 kbar at room temperature. *Phys. Rev. Lett.* **44:** 886–888 (1980).
17. Stevenson, D. J., and Salpeter, E. E. The phase diagram and transport properties for hydrogen-helium fluid planets. *Astrophys. J. Suppl.* **35:** 221–237 (1977).

18. Franks, F. *Water: A Comprehensive Treatise.* New York: Plenum Press, 1972.
19. Mitchell, A. C., and Nellis, W. J. Equation of state and electrical conductivity of water and ammonia shocked to the 100 GPa (1 Mbar) pressure range. *J. Chem. Phys.* **76:** 6273–6281 (1982).
20. Hubbard, W. B., and MacFarlane, J. J. Structure and evolution of Uranus and Neptune. *J. Geophys. Res.* **85:** 225–234 (1980).
21. Zharkov, V. N., and Trubitsyn, V. P. *Physics of Planetary Interiors.* Tucson: Pachart, 1978, pp. 189–194.
22. Ree, F. Equation of state of water. *Rep. UCRL-52190,* Lawrence Livermore Lab., Livermore, Calif. (1976).
23. Reynolds, R. T., and Summers, A. L. Models of Uranus and Neptune. *J. Geophys. Res.* **70:** 199–208 (1965).
24. Ross, M., Graboske, H. C., Jr., and Nellis, W. J. Equation of state experiments and theory relevant to planetary modelling. *Phil. Trans. R. Soc. Lond. A* **303:** 303–313 (1981).
25. Hazen, R. M., Mao, H. K., Finger, L. W., and Bell, P. M. Crystal structures and compression of Ar, Ne, and CH_4 at 20 °C to 90 kbar. *Carnegie Inst. of Wash. — Yearbook,* **79:** 348–351 (1980).
26. Nellis, W. J., Ree, F. H., van Thiel, M., and Mitchell, A. C. Shock compression of liquid carbon monoxide and methane to 90 GPa (900 kbar). *J. Chem. Phys.* **75:** 3055–3063 (1981).
27. Liu L. G., and Bassett, W. A. The melting of iron up to 200 Kbar. *J. Geophys. Res.* **80:** 3777–3782 (1975).
28. Brown, J. M., and McQueen, R. G. Melting of iron under core conditions. *Geophys. Res. Lett.* **7:** 533–536 (1980).
29. Usselman, T. M. Experimental approach to the state of the core: Parts I and II. *Am. J. of Sci.* **275:** 278–303 (1975).
30. Dziewonski, A. M., and Anderson, D. L. Preliminary reference earth model. *Phys. Earth and Plan. Int.* **25:** 297–356 (1981).
31. Liu, L. G. Calculations of high-pressure phase transitions in the system MgO-SiO_2 and implications for mantle discontinuities. *Phys. Earth and Plan. Int.* **19:** 319–330 (1979).
32. Al'tshuler, L. V. and Sharipjanov, I. I. Additive equations of state of silicates at high pressures. *Bull. Acad. Sci. USSR Earth Phys.* **3:** 11–28 (1971).

4
Applications of Potential Theory to Interior Structure

GRAVITATIONAL POTENTIALS

The detailed external gravitational potential of a planet is a valuable source of information about the planet's interior structure. At a sufficiently great distance from the planet, the potential reduces to that of a point mass:

$$V = GM/r, \qquad (4\text{-}1)$$

where V is the gravitational potential, G is the gravitational constant, M is the mass, and r is the distance between the mass and the observer. Clearly this result contains no information about the planet other than its total mass.

Examining the planet at closer range and in more detail, we can expect to see deviations from Eq. 4-1, for no planet is precisely spherically symmetric. The deviations from Eq. 4-1 are produced by two general effects:

(1) Equilibrium response of the planet to perturbing potentials.
(2) Deviation of the planet from symmetry due to departures from hydrostatic equilibrium.

Study of the equilibrium response of a planet can give information about its interior structure, via the equation of hydrostatic equilibrium, which couples the hydrostatic pressure to the distribution of gravity within the planet. When the stress tensor is not isotropic, the planet is not in hydrostatic equilibrium, and effects of this will in general appear in the gravity field as well. Comparison of topography and close-range gravity gives information about the stress state of subsurface layers in the planet.

We start from the general expression for the potential of an arbitrary mass distribution $\rho(\mathbf{r})$:

$$V(\mathbf{r}) = \int_{\text{all space}} G\rho(\mathbf{r}') \, d^3r'/|\mathbf{r} - \mathbf{r}'|. \tag{4-2}$$

Equivalently, $V(\mathbf{r})$ satisfies Poisson's equation:

$$\nabla^2 V = -4\pi G\rho. \tag{4-3}$$

To calculate the potential within a planet, we will need techniques for solving Poisson's equation to successive approximations; these will be given later. For the moment, let us seek a general form for V in the region external to the planet, where $\rho = 0$:

$$\nabla^2 V_{\text{ext.}} = 0; \tag{4-4}$$

i.e., $V_{\text{ext.}}$ satisfies Laplace's equation.

The general solution to Laplace's equation in spherical coordinates can be written as

$$V(r,\theta,\phi) = \sum_{l=0}^{\infty} \sum_{m=-l}^{l} [\alpha_{lm} r^l + \beta_{lm} r^{-(l+1)}] Y_{lm}(\theta,\phi), \tag{4-5}$$

where the Y_{lm} are spherical harmonics, normalized such that the integral of the square of a spherical harmonic over all solid angles is unity, while the integral of the product of two different spherical harmonics over all solid angles is zero. Explicitly, the Y_{lm} are defined by

$$Y_{lm} = \sqrt{\frac{2l+1}{4\pi} \frac{(l-m)!}{(l+m)!}} \, P_l^m(\cos\theta) e^{im\phi} \tag{4-6}$$

where the P_l^m are the associated Legendre polynomials, defined by

$$P_l^m(x) = [(-1)^m (1-x^2)^{m/2}/2^l l!] \, d^{l+m}(x^2-1)^l/dx^{l+m}, \tag{4-7}$$

θ is the angle from the polar axis (colatitude), and ϕ is the azimuthal angle (longitude). Unfortunately, these rather tedious definitions need to be given because of the various ways in which external gravity fields are expressed. For example, note that the P_l^m are an orthogonal but not normalized set. Sometimes the external gravity field is expressed via coefficients multiplying the product $P_l^m e^{im\phi}$; such an expansion is said to be *unnormalized.* Other normalizations are not unknown, and so great care must be taken in comparing the coefficients in the harmonic expansions of different planets, or different representations of a single planet. The index l, which describes the rapidity of potential variations in latitude, is called the degree of the harmonic, while m, representing the rate of variation in longitude, is called the order.

In Eq. 4-5, we have to set $\alpha_{lm} = 0$ so that the potential vanishes at infinity. Now we need to identify the coefficients β_{lm} with the mass distribution within the planet. Since $r > r'$ (because the potential is measured outside the planet), one can use the expansion

$$1/|\mathbf{r} - \mathbf{r'}| = \sum_{l=0}^{\infty} (r'^l/r^{l+1})P_l(\cos \gamma) \qquad (4\text{-}8)$$

where γ is the angle between $\mathbf{r}$ and $\mathbf{r'}$. This can be further decomposed:

$$P_l(\cos \gamma) = (4\pi)/(2l+1) \sum_{m=-l}^{l} Y_{lm}^*(\theta',\phi')Y_{lm}(\theta,\phi), \qquad (4\text{-}9)$$

and so Eq. 4-2 yields

$$V(\mathbf{r}) = \int_0^{\infty} Gr'^2 \, dr' \int_0^{\pi} \sin \theta' \, d\theta' \int_0^{2\pi} d\phi' \rho(r').$$

$$\sum_{l=0}^{\infty} \sum_{m=-l}^{l} r^{-(l+1)} [4\pi/(2l+1)] \, r'^l \, Y_{lm}^*(\theta',\phi')Y_{lm}(\theta, \phi). \qquad (4\text{-}10)$$

Comparing Eqs. 4-5 and 4-10, we have

$$\beta_{lm} = [4\pi G/(2l+1)] \int_0^{\infty} r'^{l+2} \, dr' \int_0^{\pi} \sin \theta' \, d\theta'$$

$$\int_0^{2\pi} d\phi' \rho(r',\theta',\phi')Y_{lm}^*(\theta',\phi'). \qquad (4\text{-}11)$$

For example, $\beta_{00} = GM\sqrt{4\pi}$, and so to lowest order $V = GM/r$, since $Y_{00} = (4\pi)^{-1/2}$.

A standard expansion which is used in planetary physics (the so-called "unnormalized" expansion) is

$$V(\mathbf{r}) = (GM/r) \left\{ 1 + \sum_{l=2}^{n} \sum_{m=0}^{l} (a/r)^l P_l^m(\cos \theta) \right.$$

$$\left. [C_{lm} \cos m\phi + S_{lm} \sin m\phi] \right\}, \tag{4-12}$$

where a is a reference radius (usually the average equatorial surface radius, or else the equatorial radius at a specified atmospheric pressure), n is the limiting degree of the expansion, the C_{l0} are called the zonal harmonics, and the C_{lm}, S_{lm} are called the tesseral harmonics. These coefficients are actually dimensionless multipole moments of the mass distribution in the planet. The zonal harmonics are frequently denoted with the letter J:

$$J_l = -C_{l0}. \tag{4-13}$$

Note that placing the origin at the center of mass requires $J_1 = C_{11} = S_{11} = 0$. Assume from now on that this has been done.

In the following discussion, we will adopt the convention of choosing the planet's polar axis (the axis from which the angle θ is measured) so that it coincides with the planet's spin axis. In a relaxed state, planets spin about their axis of maximum moment of inertia, as can be proved from the second law of thermodynamics. The proof proceeds as follows. If A, B, and C are the principal moments of inertia of the planet, ordered such that $C > B > A$, then the rotational kinetic energy of the planet is given by

$$K_{\text{rot.}} = (M_A^2/A + M_B^2/B + M_C^2/C)/2, \tag{4-14}$$

where $M_A = A\omega_A$ is the angular momentum about the x-axis, etc. Clearly for any M_A, M_B, M_C,

$$(M_A^2 + M_B^2 + M_C^2)/C \le (M_A^2/A) + (M_B^2/B) + (M_C^2/C). \tag{4-15}$$

Because $M_A^2 + M_B^2 + M_C^2 = M^2 = $ constant, the minimum value of the rotational kinetic energy occurs when $M_A = M_B = 0$, and $M_C = M$. According to the second law of thermodynamics, the planet will tend to minimize $K_{rot.}$, because if the planet acts as a closed system, the total energy, which is the sum of the internal (microscopic) energy $E_{int.}$ and the macroscopic rotational kinetic energy, must be constant. We have

$$E_{int.} = E_{total} - K_{rot.}, \tag{4-16}$$

and since the entropy S is given by $S = S(E_{int.})$, where $S(E_{int.})$ is an increasing function of $E_{int.}$, the maximum value of S occurs for the minimum value of $K_{rot.}$.[1]

The argument needs to be modified somewhat for the case where A, B, and C are dependent on the rotation rate of the planet, but for a slow rotator such as Venus, no modification is necessary. Note that we have only shown that the thermodynamic end state of a planet's rotation is rotation about the axis of maximum moment of inertia. We have said nothing about how rapidly a planet may approach this end state; this depends upon the dissipative properties of the planet's interior during the imposition of periodic strain states.

The second-degree gravitational multipoles of a planet are related to the planet's moments of inertia. Define three diagonal moments of inertia A, B, and C, corresponding respectively to the three cartesian axes x, y, and z, with the origin at the planet's center of mass:

$$A = \int \rho \, d^3r \, (z^2 + y^2), \tag{4-17}$$

$$B = \int \rho \, d^3r \, (z^2 + x^2), \tag{4-18}$$

$$C = \int \rho \, d^3r \, (x^2 + y^2). \tag{4-19}$$

It is straightforward to use Eqs. 5, 11, and 12 to show that

$$Ma^2 J_2 = -\int r'^2 \, d^3r' P_2(\cos \theta') \rho(r')$$
$$= C - (A + B)/2. \tag{4-20}$$

Similarly, one may show that

$$Ma^2 C_{22} = (B - A) \cos (2\Delta\phi)/4, \tag{4-21}$$

and

$$Ma^2 S_{22} = (B - A) \sin (2\Delta\phi)/4, \tag{4-22}$$

where $\Delta\phi$ is the difference in longitude between the x-axis and the axis of the minimum moment of inertia. When the x, y, and z axes are chosen so as to diagonalize the inertia tensor with the diagonal components ordered such that $C > B > A$, the components C_{21}, S_{21}, and S_{22} are all zero (the expansion of Eq. 4-12 has already been written with C_{21} and S_{21} absent). It should be borne in mind that the coordinate system on a planet is normally not chosen in this way, however. The z-axis is always taken to be the rotation axis, and this normally coincides closely with the axis of greatest moment of inertia. However, the origin of planetary longitude is frequently chosen with respect to some physical feature and usually does not coincide with the x-axis. Thus S_{22} is always finite, although C_{21} and S_{21} will vanish if the z-axis coincides with the axis of maximum moment of inertia. In solutions for the coefficients in Eq. 4-12, the coefficients C_{21} and S_{21} are *assumed* to be zero, and the other coefficients found accordingly. For planets which rotate sufficiently rapidly ("sufficiently rapidly" will be defined later), one finds that $|B - A| \ll |C - A|$ and $|B - A| \ll |C - B|$, so that one may write $Ma^2 J_2 \cong C - A \cong C - B$. Note that the external gravity field of a planet does not give the principal moments of inertia of the planet, but only differences of these quantities. Nevertheless, it is possible to derive the moments of inertia themselves with an additional piece of information. This information is the rate of forced precession of the planet's rotation axis under the action of an external torque. Such a precession has been observed for the earth and moon, and is proportional to the so-called precession constant

$$H = (C - A)/C. \tag{4-23}$$

With a knowledge of H and J_2, it is then possible to derive C/Ma^2 and A/Ma^2 for the planet. Such a direct measurement of these quantities has not been carried out for any planetary body other than the earth and the moon.

STUDY OF THE INTERIOR STRUCTURE OF PLANETS THROUGH GRAVITY FIELD MEASUREMENTS

The gravity field of a planet contains two principal components which are related to the planet's interior structure. First, if the planet is entirely in a state of hydrostatic equilibrium, the external gravity field reflects only the planet's adjustment to its own gravitational and rotational potentials, and its adjustment to the potentials of any external perturbing masses such as the sun and satellites. This component is normally quite "smooth" since the perturbations are "smooth," and the terms of high degree and order in the expansion of Eq. 4-12 are usually quite small or absent. For example, as we shall see below, only the even zonal harmonic terms are present in the potential of a rotating planet in hydrostatic equilibrium, and their coefficients (J_{2l} or C_{2l0}) diminish rapidly with increasing l. The response of the planet to a perturbation such as rotation can in principal be used to obtain information about the planet's internal mass distribution.

The second component of the gravity field reflects departures from hydrostatic equilibrium. In general, internal stresses will cause a planet to deviate from a state of hydrostatic equilibrium at some level, and internal dynamics can cause departures from the equilibrium state even in a liquid planet. The external gravity field of a planet is actually a superposition of components produced by hydrostatic adjustments and by internal stresses. There is no foolproof way to disentangle these two components. However, it sometimes turns out that one component is dominant, as in the case of Venus, where the planet rotates so slowly that the response to rotation is buried in the nonhydrostatic terms, or in the case of Saturn, where the extremely rapid rotation and liquid structure of the envelope insure that the nonhydrostatic terms are essentially negligible.

In the following, we will first consider the hydrostatic component of a gravity field. Then, we examine the information provided by the nonhydrostatic component, assuming that it can be separately measured.

Some Classical Results

We define hydrostatic equilibrium to be the state in which a fluid obeys the following equation:

$$\nabla P = -\mathbf{g}\rho \tag{4-24}$$

where P is the (isotropic) pressure, $\mathbf{g}$ is the gravitational plus rotational acceleration in a frame where the fluid is locally at rest, and ρ is the fluid's mass density. If $\mathbf{g}$ can be derived from a potential U, then

Eq. 4-24 can also be written

$$\nabla P = \rho \nabla U. \tag{4-25}$$

Since ρ is a scalar, Eq. 4-25 requires that surfaces of constant P and constant U coincide. It then follows that ρ must also be constant on such a surface, which is called a *level surface*. Let V be the gravitational potential acting on the fluid, and let W be any other potential, such as the rotational potential, so that $U = V + W$. Now since V exterior to the planet's surface is known in the form of the expansion of Eq. 4-12, and W is also assumed to be known, it is possible to solve the equation $V + W = $ constant for a given external level surface expressed in the form $r = r_P(\theta, \phi)$, where r is the distance of the level surface from the center of mass; the solution is usually normalized such that r coincides with some standard planetary radius at a fixed value of θ and ϕ. The equation $r = r_P(\theta, \phi)$ and the expansion for V (Eq. 4-12) contain equivalent information about the external gravity field.

Before discussing generalized solutions for the structure of planets in hydrostatic equilibrium, let us consider a simplified model (the *Maclaurin spheroid*). Consider an incompressible fluid with density ρ_0, rotating as a solid body, and in a state of hydrostatic equilibrium. Let the angular rotation rate be ω. The structure of such a planet is defined entirely by the shape of its external surface. Let the shape of this surface be given by

$$r_P = \sum_{l,m} \delta r_{lm} Y_{lm}(\theta, \phi), \tag{4-26}$$

where the average radius of the planet $a = (4\pi)^{1/2} \delta r_{00}$. We then interchange the limits of integration in Eq. 4-11 so that integration over r' is performed first. Because of the constancy of ρ for $r < r_P$, the upper limit on the r' integral can be replaced with r_P and ρ can be replaced with ρ_0 and taken outside all integrals. To lowest order in δr_{lm}, and for l and m not simultaneously equal to zero, one easily finds

$$V_{lm} = \frac{3}{2l+1} \frac{GM}{a} \frac{\delta r_{lm}}{a}, \tag{4-27}$$

where the V_{lm} are harmonic coefficients of the gravitational potential expressed in the form

$$V = \sum_{l,m} V_{lm}(a/r)^{l+1} Y_{lm}(\theta,\phi) \qquad (4\text{-}28)$$

(cf Eqs. 4-5 and 4-12). Equation 4-27 is particularly useful because it enables us to express the external gravitational potential of the Maclaurin spheroid in terms of its surface relief, or *vice versa*.

Now let W be the rotational potential which acts on the planet:

$$W = W_{\text{rot.}} = (\tfrac{1}{2})r^2\omega^2 \sin^2(\theta), \qquad (4\text{-}29)$$

which can be written in the form

$$\begin{aligned}
W_{\text{rot.}} &= (\tfrac{1}{3})r^2\omega^2[1 - P_2(\cos\theta)] \\
&= (\tfrac{1}{3})r^2\omega^2[1 - (4\pi/5)^{1/2}Y_{20}(\theta)].
\end{aligned} \qquad (4\text{-}30)$$

Letting b be the polar radius of an external level surface and a its equatorial radius, the potential at the pole ($\theta = 0$) is

$$U = V = (GM/b)[1 - J_2(a/b)^2 - J_4(a/b)^4 - \ldots] \qquad (4\text{-}31)$$

The equation of a level surface is then given by the solution to

$$\begin{aligned}
(GM/r)&[1 - J_2(a/r)^2(4\pi/5)^{1/2}Y_{20}(\theta) - \ldots) \\
&+ (\tfrac{1}{3})r^2\omega^2[1 - (4\pi/5)^{1/2}Y_{20}(\theta)] \\
&= (GM/b)[1 - J_2(a/b)^2 - \ldots],
\end{aligned} \qquad (4\text{-}32)$$

which implicitly defines the function $r = r_p(\theta)$. To lowest order in J_2 and ω^2, one finds

$$\delta r_{20} = -a(4\pi/5)^{1/2}(J_2 + q/3) \qquad (4\text{-}33)$$

for the leading correction to the zeroth-order spherical surface of the planet, where

$$q = \omega^2 a^3/GM \qquad (4\text{-}34)$$

is a dimensionless measure of the centrifugal potential. Equation 4-33 is general, and does not require ρ to be constant. Now since

$$V_{20} = (4\pi/5)^{1/2}J_2(GM/a),\qquad(4\text{-}35)$$

Eqs. 4-35 and 4-33 can be substituted into Eq. 4-27 to solve self-consistently for J_2 in terms of q, with the result

$$J_2 = q/2,\qquad(4\text{-}36)$$

valid to lowest order in q. Equation 4-36 is valid only for constant ρ. As a general result, however, in hydrostatic equilibrium the leading correction to the spherically symmetric part of the gravitational potential is proportional to q. We will consider the extent to which real planets conform to this law.

In order to deal with real planets, two generalizations of Eq. 4-36 are necessary. First, the coefficient of $\frac{1}{2}$ on the right-hand side of Eq. 4-36 is valid only for a uniform density fluid (the so-called Maclaurin spheroid). For more general planetary structures with an internal density variation, the coefficient may be either greater than or less than 0.5, although a value greater than 0.5 requires an unphysical increase of density toward the planet's surface. We will presently consider how this coefficient may be calculated for an arbitrary planetary model.

The second generalization, which is sometimes required for rapidly-rotating giant planets, is the continuation of the expansion for the J_{2n} and the δr_{lm} in powers of q. In general, one finds that

$$J_{2n} = \sum_{l=0}^{\infty} \Lambda_{2n,l}\, q^{n+l},\qquad(4\text{-}37)$$

where the $\Lambda_{2n,l}$ are dimensionless response coefficients (for example, $\Lambda_{2,0} = 0.5$, $\Lambda_{4,0} = -0.536$ for the Maclaurin spheroid).

Expansions 4-26 and 4-37 are rapidly-converging series in powers of q, since the coefficients of the various powers of q are all of order unity or smaller. Thus, for a planet in strict hydrostatic equilibrium, the second-degree zonal harmonic J_2 and the oblateness

$$f = (a - b)/a \simeq (3/2)J_2 + (\tfrac{1}{2})q\qquad(4\text{-}38)$$

are of order q. The higher zonal harmonics (J_4, J_6, etc.) are of higher order in q and therefore are small compared with J_2. Because of the planet's axisymmetry, the moments of inertia A and B are equal, and are smaller than C by an amount of the order of q (cf Eq. 4-20).

Table 4-1 presents some results for the J_2's of a number of solar system bodies. Also presented are the values of q for each body. The latter values are quite small for Mercury, Venus, and the moon because these bodies have been tidally despun over the age of the solar system, resulting in very slow rotation at the present time. Thus the observed zonal harmonics are orders of magnitude larger than the values expected in hydrostatic equilibrium, and therefore have nothing to do with adjustments to hydrostatic equilibrium, at least at the current epoch.

Table 4-1. Values of J_2 and q for Planets and Satellites*

OBJECT	J_2	q
Mercury	$(8 \pm 6) \times 10^{-5}$	1.0×10^{-6}
Venus	$(6 \pm 3) \times 10^{-6}$	6.1×10^{-8}
Earth	1.0826×10^{-3}	3.5×10^{-3}
Moon	2.024×10^{-4}	7.6×10^{-6}
Mars	1.959×10^{-3}	4.6×10^{-3}
Jupiter	1.4733×10^{-2}	0.089
Saturn	1.646×10^{-2}	0.153
Uranus	3.352×10^{-3}	0.035
Neptune	4×10^{-3}	0.04?

* Values for q are calculated using the parameters given in Table 1-1, except that, for various reasons, different values of a(=normalizing radius for J_2) have been adopted for some of the Jovian planets. These values are: $a = 71,398$ km for Jupiter; $a = 60,000$ km for Saturn; $a = 26,200$ for Uranus. These values were also used to calculate q for these bodies.

References for values of J_2 are as follows:

Mercury: Esposito, P. B., Anderson, J. D., and Ng, A. T. Y. Experimental determination of Mercury's mass and oblateness. *Space Res.* **17**: 639–644 (1977).

Venus: Ananda, M. P., Sjogren, W. L., Phillips, R. J., Wimberly, R. N., and Bills, B. G. A low-order gravity field of Venus and dynamical implications. *J. Geophys. Res.* **85**: 8303–8318 (1980).

Earth: Kaula, W. M. *Theory of satellite geodesy.* Blaisdell: Waltham, Mass., 1966.

Moon: Bills, B. G. and Ferrari, A. J. A harmonic analysis of lunar gravity. *J. Geophys. Res.* **85**: 1013–1025 (1980).

Mars: Christensen, E. J. and Balmino, G. Development and analysis of a twelfth degree and order gravity model for Mars. *J. Geophys. Res.* **84**: 7943–7953 (1979).

Jupiter: Null, G. W. Gravity field of Jupiter and its satellites from Pioneer 10 and Pioneer 11 tracking data. *Astron. J.* **81**: 1153–1161 (1976).

Saturn: Hubbard, W. B., MacFarlane, J. J., Anderson, J. D., Null, G. W., and Biller, E. D. Interior structure of Saturn from Pioneer 11 gravity data. *J. Geophys. Res.* **85**: 5909–5916 (1980).

Uranus: Elliot, J. L. Rings of Uranus: occultation results, in *Uranus and the Outer Planets* (G. Hunt, ed.). Cambridge University Press: Cambridge, 1982.

Neptune: Peale, S. J. The gravitational fields of the major planets. *Spa. Sci. Rev.* **14**: 412–423 (1973).

Rapidly-rotating Maclaurin Spheroids

As we shall see below, a number of planets and satellites appear to have been significantly despun over the age of the solar system. We will consider bodies in hydrostatic equilibrium in the limit of rapid rotation, as this limit may provide useful boundary conditions on poorly understood processes of the origin of planetary rotation rates. For the purposes of this discussion we introduce a dimensionless measure of the rotation rate of a Maclaurin spheroid. Let the dimensionless rotation rate be

$$\Omega^2 = \omega^2/(2\pi G\rho). \qquad (4\text{-}39)$$

(Note that $\Omega^2 = 2q/3$ for small q; cf Eq. 4-34.) Now as q increases from very small values, the oblateness of the spheroid (defined by Eq. 4-38) also increases. But the function $\Omega^2(f)$ reaches a maximum at $f = 0.64$ and $\Omega^2 = 0.22$, and Ω^2 then declines with a further increase in f. The reason for this behavior is as follows. For small q, the planet's shape and mass distribution are only slightly changed from the nonrotating values. Ω^2 cannot exceed the critical value of 0.22 — angular momentum continues to increase with increasing f beyond this point, but the moment of inertia increases rapidly enough due to the flattening of the planet into a disc-shaped object so that omega decreases with further increase of the angular momentum. As more and more angular momentum is added to the spheroid, it approaches a disc-like shape. However, this limit is never attained by a real planet. At $\Omega^2 = 0.19$ and $f = 0.42$, a Maclaurin spheroid reaches the so-called "point of bifurcation," where a new sequence of hydrostatic equilibrium configurations become possible. These configurations are called the Jacobi ellipsoids, and they exist for all angular momenta greater than the angular momentum of the point of bifurcation. Jacobi ellipsoids are cigar-shaped triaxial ellipsoids, and rotate about the axis of greatest moment of inertia. At the point of bifurcation, the two semiaxes in the plane of rotation become equal and the body becomes axially symmetric, reducing to a Maclaurin spheroid. Since a Jacobi ellipsoid has a smaller mechanical (kinetic plus potential) energy than a Maclaurin spheroid with the same angular momentum, any dissipation will cause a Maclaurin spheroid to evolve into a Jacobi ellipsoid if the angular momentum of the body is higher than the critical value at the point of bifurcation. Jacobi ellipsoids rotate more and more slowly as angular momentum is added to them, and they "stretch" into more and more slender configurations. At a somewhat higher angular momentum than the value at the point of bifurcation, the Jacobi ellipsoids become dynamically unstable, and it is speculated that fission into a binary object may occur at this point.[2]

All of the above considerations indicate that there should be an upper envelope to the distribution of angular momenta in solar system bodies. Assume that the envelope is defined by $\Omega^2 = 0.2$, and let the angular momentum of a planet be given by $L = \alpha M a^2 \omega$, where α is a numerical coefficient of order unity; we will take $\alpha = 0.3$. Then the envelope is defined by the following relation between the angular momentum per unit mass L/M (in this chapter only, we will use the customary symbol L to denote angular momentum; elsewhere, it will be reserved for luminosity, in units of ergs/sec) and the mass M of the body:

$$\log (L/M) = -4 + (\tfrac{2}{3}) \log M, \qquad (4\text{-}40)$$

where cgs units are used, and we have assumed that all planetary bodies have the same density of 1 g/cm³ (the logarithm of the density would only appear in Eq. 4-40 with a coefficient of $-\tfrac{1}{6}$). Figure 4-1 shows a plot of specific angular momenta vs. mass for a number of planets. All of the objects lie well below the envelope defined by Eq. 4-40, but tend to follow it, as might be expected for objects which dispose of that amount of angular momentum necessary to permit their formation as individual bodies. Details of the accumulation process are also undoubtedly important in establishing the empirical $\log (L/M)$ vs. $\log M$ relation, which appears to have a slope on the order of 0.9 rather than $\tfrac{2}{3}$. Several bodies fall well below the general trend, and appear to have lost an excess amount of angular momentum. We will return to this point later in this chapter.

Deviations from Hydrostatic Equilibrium

A Maclaurin-like model can also be useful for understanding how the structure of a planet's gravity field can be influenced by deviations from hydrostatic equilibrium within the planet's interior. Continue

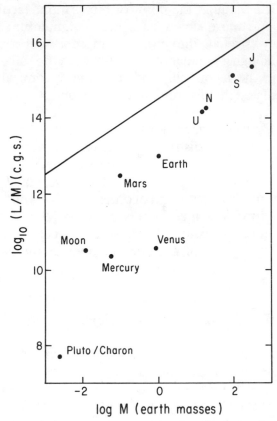

Fig. 4-1. Specific angular momentum versus mass for solar system bodies. Straight line corresponds to Eq. 4-40.

to suppose that the planet has a uniform density ρ_0, but now assume that the planet's surface is not coincident with a level surface. Let the shape be expanded in spherical harmonics as in Eq. 4-26, but now the δr_{lm} are not self-consistently determined by the gravity field; they are instead defined by the arbitrary surface topography. The harmonic coefficients of the gravitational potential are still given by Eqs. 4-27 and 4-28. In hydrostatic equilibrium, the shape coefficients δr_{lm} decrease rapidly in absolute value as l increases, and they are identically zero for $m \neq 0$. Thus the shape of a planet in hydrostatic equilibrium is a smooth figure of revolution, differing only slightly from an ellipsoid. This behavior is reflected in the shape of the gravity potential as expressed by the coefficients J_2. But real planets have topographic relief, sometimes as much as several kilometers, with respect to the level surface defined by the gravitational potential. Thus the δr_{lm} do not decrease with increasing l, and the δr_{lm} with $m \neq 0$ are also finite. Then, according to Eq. 4-27, the harmonic coefficients of the external gravity field may not decrease rapidly with l or m. One might expect that terms in the gravitational potential with l sufficiently small and $m = 0$ would correspond approximately to hydrostatic equilibrium. However, for sufficiently large l and for any $m \neq 0$, the gravitational potential is dominated by departures from hydrostatic equilibrium.

If the parameter q is extremely small, as in the case of Mercury, Venus, and the moon, there exists no range of l for which hydrostatic equilibrium dominates the shape of the gravity field. In this case even the $l = 2$ terms are primarily determined by departures from hydrostatic equilibrium, and the moment of inertia differences $B - A$, $C - A$, and $C - B$ tend to be of the same order of magnitude.

Table 4-2 lists moment of inertia differences for the terrestrial planets, to the extent that they are known. The "despun" objects, Mercury, Venus, and the moon, clearly differ from Mars and the earth. The gravity fields of the latter are dominated by response to the rotational potential, but Mercury, Venus, and the moon are not described by Eq. 4-37.

First consider Venus, where the gravity field is now reasonably well-known, and where the rotational response is unimportant. As an opposite limit, one might try to apply Eq. 4-27 using for the δr_{lm} the measured topographic relief on the planet. If the surface layers which are assumed to produce the gravitational irregularities differ in density from the planet's mean density, then the right-hand side of Eq. 4-27 must be multiplied by a correction factor (ρ_s/ρ_0), where ρ_s is the assumed density of the planet's surface layers and ρ_0 is the mean

Table 4-2. Dimensionless Moment of Inertia Differences for Terrestrial Planets*

OBJECT	$(B - A)/Ma^2$	$(C - A)/Ma^2$
Venus	7.08×10^{-6}	9.53×10^{-6}
Earth	7.21×10^{-6}	1.08×10^{-3}
Moon	8.91×10^{-5}	2.47×10^{-4}
Mars	2.53×10^{-4}	2.09×10^{-3}

* See references to Table 4-1.

density. This procedure produces a model gravity field for Venus which has qualitatively correct spatial variations, but the variations themselves are much too exaggerated (see Chapter 8). Local gravity "highs" and "lows" computed from the topographic model are in the right places, but are typically a factor of 3 to 10 too large in absolute value.

Although a planet may have substantial topography, the topography may not be expressed in the gravity field via Eq. 4-27 if the relief is isostatically compensated. Equation 4-27 assumes that there is no lateral density variation in the planet. But now consider an infinite slab of material with density ρ_0; floating on this slab is another finite slab of density ρ_1. If we are far from the edge of either slab, then the gravitational acceleration at a given altitude h above the slab is given by

$$g = 2\pi G \int_0^h \rho(z) \, dz, \qquad (4\text{-}41)$$

where the integral is taken along a vertical column through the slab or slabs. Let point A be located above the infinite slab with density ρ_0 and far from the finite slab with density ρ_1. Let point B be located at the same altitude but above the middle of the finite slab (the finite slab is assumed to have a horizontal extent great compared with the altitude of point B). Note that the gravity will be the same at point A and point B, i.e. there will be no gravity anomaly, if the column-averaged densities below points A and B are the same. But this is just the condition for the slab of density ρ_1 to be buoyantly supported, i.e. isostatically compensated. In real planets with more complicated geometries than infinite slabs, the situation is more complicated, but there is nevertheless a strong tendency for gravity anomalies to be reduced by isostatic compensation. Thus there is a considerable degree of compensation in Venus, but the low-degree gravity map is nonetheless correlated with the topography. The same is true for Mars and the moon.

There are two representations that are conventionally used to graph the external gravitational field of a planet. In the first method, we take the radial gradient of the full gravitational potential Eq. 4-12 at some standard radius above the planet's surface (typically 100 km, which lies above all undulations of the surface). We then take this radial component of the gravitational acceleration and subtract the radial gradient of the Newtonian potential GM/r; the remainder is called the *gravity disturbance*. The value of the gravity disturbance is usually expressed in *milligals* (one $gal = 1$ cm/s^2). The distortions of the planet's gravitational potential caused by the various multipole components of its mass distribution can then be represented by a contour map of this scalar quantity.

Alternatively, one can plot the shape of the equipotential surfaces themselves. For this purpose, we adopt some standard value of the potential V for which we wish to make a contour map. Typically, this will be simply GM/a, where a is the mean radius of the planet. Substituting this value on the left-hand side of Eq. 4-12, we then have an implicit equation for $r(\theta, \phi)$, whose contours can then be plotted as a function of θ and ϕ. For the earth, a map of this type is called a *geoid*. It shows the elevation or depression of the surface equipotential with respect to the planet's mean surface. In this book we will present maps of this type for other planets as well. The geoid for Mars is sometimes known as an *areoid,* but corresponding terms for other planets, such as *lunoid* or *cytheroid,* are not commonly used (because they sound preposterous, in all probability!).

In a planet with a significant amount of distortion due to rotation, the second-degree harmonic $C_{2,0}$ usually dominates the map of the gravity disturbance or potential. In order to clearly display the effect of nonhydrostatic terms, such maps are usually referenced to the combined effects of GM/r and the $C_{2,0}$ term.

Neither the gravity disturbance map nor the geoid provides a full description of the planet's external potential, but such map provides a convenient method for quickly assessing whether there is a correlation between gravity and topography, as would be the case if there is little isostatic compensation. The geoid contour tends to be elevated in regions where the inward radial gravity disturbance is enhanced (i.e. where there is extra uncompensated mass). Higher-degree contributions to the gravity disturbance map are weighted by $(l + 1)$ relative to the corresponding contributions to the geoid, as is evident from Eq. 4-12.

Figure 4-2 shows a relatively low-resolution map of the geoid (corresponding maps for other planets are presented in Chapter 8). This plot is shown at approximately the same resolution — sixth degree and order — as the best available plots for the other terrestrial

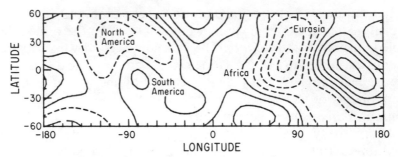

Fig. 4-2. The geoid, plotted to sixth degree and order, at a contour interval of 20 meters. Dashed lines correspond to negative heights. The first solid contour is the zero contour.

planets, although the terrestrial potential has now been mapped to the thirtieth degree and order.[3] Note that continents and most major mountain ranges are for the most part invisible on this representation of the earth's gravitational potential. This is a fundamental difference between the earth and the other terrestrial planets. Apparently the degree of isostatic compensation in the earth is far greater than in the outer layers of even Venus, a planet similar in size and density. This circumstance may be related to the presence of an asthenosphere and tectonic activity in the earth's outer layers, and may be evidence for the lack of an analogous situation in any other planet.

The Hydrostatic Limit

We now return to the case of a planet which responds hydrostatically to the rotational perturbation. As discussed above, departures from hydrostatic equilibrium will cause expansion Eq. 4-37 to ultimately break down for sufficiently large values of l. But to the extent that Eq. 4-37 is valid, the coefficients $\Lambda_{2n,l}$ are functions of the planet's interior structure. Since the J_2 and q can be experimentally determined, we can constrain the range of possible models by comparing theoretical values of the J_{2l} with measured ones. In the following, we will discuss the method of calculating the coefficients $\Lambda_{2n,l}$, and the information which these numbers contain about the planet's interior structure. The fundamental equation governing the structure of a planet in hydrostatic equilibrium is set up by first calculating the total potential by integrating the equation of hydrostatic equilibrium, Eq. 4-25:

$$U = U_s + \int_{\rho_s}^{\rho} (dP/d\rho')(d\rho'/\rho'), \qquad (4\text{-}42)$$

where we defined the reference level to be the level where $\rho = \rho_s$, and U_s is the value of the total potential on this reference level. This can be chosen arbitrarily. To create a closed equation, we now use the definition of the potential given in Eq. 4-2, and add to it the perturbing potential W to obtain the total potential U. This expression is then equated to the right-hand side of Eq. 4-42:

$$W + G \int d^3r' \, \rho(\mathbf{r}')/|\mathbf{r} - \mathbf{r}'| = U_s + \int_{\rho_s}^{\rho} (dP/d\rho')(d\rho/\rho'). \qquad (4\text{-}43)$$

This equation can also be expressed in differential rather than integral form.

When the perturbation W is absent, or when W is a function of r only, Eq. 4-43 reduces to a one-dimensional equation, and is readily solved. In the general case, however, W is a function of all three spatial variables and therefore so is $\rho(\mathbf{r})$. There are two cases of principal interest. If the perturbing potential is produced by rotation alone, then we substitute Eq. 4-30 for W. Since W is a function of r and θ only, the problem reduces to a two-dimensional one — hence the absence of $m \neq 0$ terms in the expansion Eq. 4-12. The other case of interest is tidal distortion, which we discuss below. By a suitable choice of coordinates, W can be expressed for this case in essentially the same form as Eq. 4-30:

$$W_{\text{tidal}} = (GM_p r^2/R^3) \, P_2(\cos \theta'), \qquad (4\text{-}44)$$

where M_p is the mass of the tidal perturber, which is at a distance R from the center of the planet, and θ' is the angle between $\mathbf{r}$ and the planet-perturber axis. Note that the spatial dependence of W_{tidal} is very similar to that of W_{rot}. Again, the problem reduces to a two-dimensional one. By means of a simple transformation, response coefficients calculated for rotational perturbation can also be used for tidal response. When tidal and rotational perturbations act simultaneously, we have a three-dimensional problem in principle. However, the tidal perturbation is usually so small that it can be treated in the linear approximation only; i.e. the tidal response can be simply superimposed on the rotational one. General procedures for obtaining solutions to Eq. 4-43 are too technical to discuss in detail here.[4] One convenient procedure expands the level surfaces within the planet in a form similar to Eq. 4-26, so that the δr_{lm} become functions of depth in the planet. A hierarchy of integro-differential equations for the δr_{lm} can then be set up. An alternative procedure expands the density itself in multipole components, again generating a series of integral equations to various orders in q. In either case, the multipole moments J_{2n} can be readily calculated either from Eq. 4-11 or by comparing the surface δr_{lm} with the solution of Eq. 4-32. The equations become quite tedious in either method to high order in q. However, the first-order solution, proportional to q, has some interesting properties which can be discussed here. The lowest-order solution to Eq. 4-43 (proportional to ρ_0 or q^0) can be expressed in the form

$$r^{-1} \int_0^r x^2 \, dx \, \rho_0 + \int_r^a x \, dx \, \rho_0 = Z(\rho_0)/(4\pi G), \qquad (4\text{-}45)$$

where ρ_0 is the density distribution in the absence of perturbation and $Z(\rho)$ is defined by

$$Z(\rho) = \int_{\rho_s}^{\rho} (dP/d\rho')(d\rho'/\rho'). \qquad (4\text{-}46)$$

The solution to Eq. 4-45 gives the structure of an unperturbed spherical planet. The next order in q gives the first-order correction due to the perturbation. After expanding to first order in the oblateness, the result can be expressed in terms of Clairaut's differential equation:

$$d^2e/dr^2 = (6e/r^2) - (2\rho_0/rS_0)(e/r + de/dr), \qquad (4\text{-}47)$$

where e is the oblateness of a level surface within the planet, reducing to $e = f$ at the surface. The quantity S_0 is available from the solution to Eq. 4-45, and is defined by

$$r^3 S_0 = \int_0^r r'^2 \, dr' \, \rho_0(r'). \qquad (4\text{-}48)$$

The boundary conditions imposed on regular solutions of Eq. 4-47 are

$$r(de/dr) + 2e = 5q/2 \qquad (4\text{-}49)$$

at $r = a$ (where $e = f$), and

$$de/dr = 0 \qquad (4\text{-}50)$$

at $r = 0$.[4]

It is not possible to solve Eq. 4-47 analytically for an arbitrary density distribution. This is not a serious difficulty in the computer age, but there does exist a classical approximation which is commonly used and which gives one a good feeling for the type of constraint imposed by Eq. 4-47 on the planetary structure. Let $\eta = d \ln e/d \ln r$. Then Eq. 4-47 can be rewritten in the form

$$d\{r^5 S_0 (1 + \eta)^{1/2}\}/dr = 5 S_0 r^4 F(\eta), \qquad (4\text{-}51)$$

where

$$F(\eta) = [1 + \eta/2 - \eta^2/10]/(1 + \eta)^{1/2}. \qquad (4\text{-}52)$$

Since the unknown oblateness function still appears on both sides of Eq. 4-51, apparently little progress has been made. However, it turns out that F is very close to unity for realistic values of η. The Radau-Darwin approximation sets $F = 1$ in Eq. 4-51. After a few more manipulations, the approximate formula can be written

$$C/Ma^2 = (\tfrac{2}{3})[1 - (\tfrac{4}{5})(1 + \eta_s)^{1/2}], \qquad (4\text{-}53)$$

where η_s, the surface value of η, is given by Eq. 4-49 and C is the polar moment of inertia given by Eq. 4-19. Now because of the boundary condition (4-49) on Clairaut's equation, the value of η_s is just $2.5q/f - 2$. One can then express f in terms of J_2 and q using Eq. 4-33, with the final result

$$C/Ma^2 = (\tfrac{2}{3})\{1 - (\tfrac{2}{5})[5/(3\Lambda_{2,0} + 1) - 1]^{1/2}\}, \qquad (4\text{-}54)$$

the famous Radau-Darwin aproximation. This equation tells us that the response coefficient $\Lambda_{2,0}$ is essentially a function of the planetary moment of inertia only. Thus if the planet is in hydrostatic equilibrium, knowledge of J_2 gives us not only the difference in the polar and equatorial moments of inertia but the moments of inertia themselves. If the planet is not in hydrostatic equilibrium to the extent that $\Lambda_{2,0}$ can be determined, then the precession constant must be determined in order to separately measure the moments of inertia.

It must be emphasized that the Radau-Darwin formula is an approximation. In general, there is no unique relationship between $\Lambda_{2,0}$ and C/Ma^2. However, it turns out that the Radau-Darwin relation becomes exact in the limit of a uniform-density body, i.e. a Maclaurin spheroid. In this case, $C/Ma^2 = 0.4$, which is also implied by Eq. 4-54. However, in the limit $C/Ma^2 \to 0$, the Radau-Darwin approximation becomes seriously deficient. There is, of course, a further approximation in Eq. 4-54; namely, it is valid only to order q.

As a matter of practice, the Radau-Darwin relation is reasonably satisfactory for the earth and Mars, because neither planet is strongly centrally-condensed (C/Ma^2 approx. $= 0.38$ for Mars and 0.33 for the earth) and q is small for both. The Radau-Darwin relation is no more than a crude first approximation for the outer planets, however. So far, precession constants are only available for the earth and the moon, and so the only method which can so far be applied to obtain C/Ma^2 for the other planets is the Radau-Darwin approximation. This method has been applied to Mars, but the principal problem concerns the extent to which Mars is in hydrostatic equilibrium. Since q for Mars is about 5×10^{-3}, one expects the Martian J_2 to be of order q and J_4 to be of order q^2, or about 2×10^{-5}. But the low-order nonhydrostatic terms in the Martian gravity field are also of order q^2, and so hydrostatic theory is inapplicable for Mars to order q^2. Essentially the same result is obtained for the earth's gravity field, where J_3 and J_4 are of the same order of magnitude, and similar to q^2. In the case of Mars, Reasenberg[5] has argued that the value of C/Ma^2 derived from the Radau-Darwin formula may be "contaminated" by nonhydrostatic contributions from Mars' figure, and has attempted to correct for these, basically by removing the Tharsis region.

TIDES

There are two important aspects to the response of a planet to a tidal perturbation. First, the equilibrium response of the planet to the tidal perturbation (Eq. 4-44) can be described in much the same way as the planet's response to the rotational perturbation; much of the same mathematical formalism can be used. However, there is a basic difference between the tidal and rotational perturbations: the tidal perturbation varies over a much shorter time period than does the rotational perturbation, unless the planet's tidal and rotational periods are synchronous. In effect, the rotational bulge remains a static feature of the planet (with the exception of some small effects which we will discuss later), while the tidal bulge moves around the planet, following the sub-perturber point. The tidal bulge is a dynamic feature of the planet, while the rotational bulge is a static feature. Dissipation and time-dependent relaxation to equilibrium are therefore much more important for tides than for rotation.

Tidal Torques

We begin by deriving Eq. 4-44. Consider a planet of mass M which is orbited by a perturbing object of mass M_P at a vector position $\mathbf{R}$ from the planet's center of mass. Place the origin of coordinates at the center of mass of the planet, and define a cartesian coordinate system with the z-axis normal to the planet-perturber orbital plane, the x-axis pointing toward the perturber, and the y-axis normal to the other two. We now wish to write down the perturbing potential which acts on the planetary material at position $\mathbf{r}$ in this reference frame. There are two components to the perturbing potential. The first component is the gravitational potential produced by the perturbing mass, $GM_p/|\mathbf{R} - \mathbf{r}|$. In addition there is another component which results from the fact that the chosen frame of reference is not an inertial one. There is an inertial "force" acting in this frame due to the acceleration of the planet's center of mass by the perturber. We can represent this "force" (which is analogous to the centrifugal "force" in a rotating coordinate system) by an additional term in the perturbing potential which yields a uniform force along the x-axis equal to the force exerted by the perturber. This additional term must obviously equal $-(GM_p/R^2)x$. Thus the total perturbing potential which appears in the noninertial frame is

$$W = GM_p/|\mathbf{R} - \mathbf{r}| - (GM_p/R^2)x, \qquad (4\text{-}55)$$

where the terms are respectively the potential due to the perturbing mass, and a potential due to acceleration of the frame along the line of centers. Letting θ' be the angle between the vectors $\mathbf{r}$ and $\mathbf{R}$, we now use the expansion of Eq. 4-8 for $1/|\mathbf{R} - \mathbf{r}|$, expanding up to the P_2 term. The P_1 term exactly cancels the second term in Eq. 4-55. As a result we find

$$W = (GM_p r^2/R^3)P_2(\cos\,\theta') = W_{\text{tidal}}, \qquad (4\text{-}56)$$

as given by Eq. 4-44. We will now consider the response of the planet to W_{tidal}, using the same machinery which was set up to derive the response to $W_{\text{rot.}}$. Recall that $W_{\text{rot.}}$ has two terms: one proportional to r^2 and the other proportional to $r^2 P_2(\cos\,\theta)$. The first term is spherically symmetric and only leads to a radial expansion of the planet, while the second term produces rotational flattening. Functionally, the second term is identical to W_{tidal}, and so all of the preceding theory for the accommodation of a planet to a rotational perturbation can be directly applied to a tidal perturbation by making allowance for the different coefficient of $r^2 P_2$. This is most conveniently carried out by making the substitution

$$q \rightarrow q_{\text{tidal}} = -3M_p a^3/(MR^3) \qquad (4\text{-}57)$$

everywhere in the previous discussion of the response to a rotational potential. Also bear in mind that the angle θ' is measured from the line of centers, while θ is measured from the rotation axis. With these straightforward substitutions, all of the expressions valid for hydrostatic equilibrium in the presence of rotation can be taken over to the case of hydrostatic equilibrium tides. The gravitational potential of a tidally-perturbed planet can be expressed by the expansion of Eq. 4-5, and to first order in q (rotational or tidal) we can break the full expansion into tidal pieces and rotational pieces and simply add them together, because of the linearity of Eq. 4-3. Thus, the tidal response of the external potential can be expressed by an expansion of the form of Eq. 4-12, but with the substitution $\theta \rightarrow \theta'$. All terms with m $\neq$ 0 vanish because of symmetry. Let $J_{2,t}$ represent the first zonal harmonic produced by the planet's tidal response. It is given by Eq. 4-36 (for a constant-density planet) or by Eq. 4-37 (in general), and is negative since q_{tidal} is negative. One can then apply Eq. 4-33 or 4-38 to find the planet's shape. We note that it is prolate, with the long axis pointed along the line of centers. It is sometimes convenient to define a dimensionless number which gives the constant of proportionality between the perturbing potential W_{tidal} and the tidal displacement of the planet's surface. Let $g = GM/a^2$ be the unperturbed surface gravity, and let δr be the (small) displacement of the planet's surface due to tides. Then we define

$$\delta r = hW_{tidal}/g, \qquad (4\text{-}58)$$

where h is a so-called tidal Love number. Using the foregoing theory for the Maclaurin spheroid, it is straightforward to show that $h = \frac{5}{2}$. In general, h will be smaller than this limiting value for two reasons. First, the planet will generally have some degree of central concentration, and second, the planet may not be able to relax to hydrostatic equilibrium over a tidal cycle. If the planet cannot relax to hydrostatic equilibrium but on the contrary acts as an elastic body over a tidal cycle, then the Love number is given by

$$h = (\tfrac{5}{2})/[1 + 19\mu/(2ga\rho_0)], \qquad (4\text{-}59)$$

where this formula is valid for a planet with uniform density ρ_0 and elastic shear modulus μ. The deformation of the planet's surface and other aspects of its response to tides are thus smaller than they would be in hydrostatic equilibrium, because elastic forces tend to maintain the planet's spherical shape.

The tidal response of a planet also affects its own external gravitational field. In analogy to the expansion Eq. 4-12, we write

$$V_{tidal, induced} = -J_{2,t}(GM)(a^2/r^3)P_2(\cos \theta'). \qquad (4\text{-}60)$$

Next, consider the force in the θ' direction, per unit mass, on a test particle at position $\mathbf{r}$. This is given by

$$F_{\theta'} = -(1/r)(\partial V_{tidal, induced}/\partial\theta'), \qquad (4\text{-}61)$$

so the tidal torque per unit mass is given by

$$T_t = -r\,(1/r)(\partial V_{tidal, induced}/\partial\theta')$$
$$= 3J_{2,t}GM(a^2/r^3)\cos \theta' \sin \theta'. \qquad (4\text{-}62)$$

For small θ',

$$T_t \simeq 3J_{2,t}GM(a^2/r^3)\theta'. \qquad (4\text{-}62a)$$

Next, define the response coefficient $\Lambda_{2,0}$ by

$$J_{2,t} = \Lambda_{2,0}q_{tidal}, \qquad (4\text{-}63)$$

where for a liquid or perfectly elastic planet, $\Lambda_{2,0}$ would be the same response coefficient as the one defined by Eq. 4-37. Thus $\Lambda_{2,0}$ equals ½ for the liquid Maclaurin spheroid, and

$$\Lambda_{2,0} = (½)/[1 + 19\mu/(2ga\rho_0)] \qquad (4\text{-}64)$$

for a uniform density elastic sphere.

However, real planets are sometimes neither liquid nor perfectly elastic over a broad range of frequencies. Such planets may be capable of fluid deformations over very long time periods, such as those over which the rotational period changes substantially, and thus their response to rotation may approximate that of a liquid body. At the same time, the much higher-frequency variations in the tidal perturbation may elicit an elastic response rather than a purely hydrostatic one. Thus the tidal and rotational $\Lambda_{2,0}$'s may in some cases be different.

Another dimensionless Love number k is frequently used in place of the tidal $\Lambda_{2,0}$. Define

$$k = V_{tidal,\,induced}/W_{tidal}, \qquad (4\text{-}65)$$

evaluated at the surface of the planet. It then follows that

$$k = 3\Lambda_{2,0}, \qquad (4\text{-}66)$$

to lowest order in V_{tidal}. Sometimes k or the equivalent tidal $\Lambda_{2,0}$ are estimated from the quantity J_2/q. However, this may not always give an accurate result because (1) for rapidly rotating planets, J_2 has a significant component of order q^2, and (2) for solid planets, the tidal $\Lambda_{2,0}$ may be substantially smaller than the rotational $\Lambda_{2,0}$ for reasons just given.

Substituting the definition of q_{tidal} (Eq. 4-57) in Eq. 4-62a, we then obtain

$$T_t \simeq 9\Lambda_{2,0}GM_p(a^5/R^6)\theta', \qquad (4\text{-}67)$$

the tidal moment which acts on a unit mass at a distance R and a small angular offset θ' from the line of centers.

Since by definition $\theta' = 0$ along the line of centers, the torque between the tidal bulge and the perturbing body vanishes. However, this result is only true if there is no dissipation in the system. The other aspect of the dynamic nature of tides now comes into play. If the planet is not rotating synchronously with the tidal period, its interior is being subjected to a periodic tidal deformation over each tidal cycle. For the solid terrestrial planets, such deformation is small and takes the form of a periodic variation in the elastic strain state of the body of the planet. Such variation is always accompanied by dissipation, and by analogy with the theory of a forced, damped classical oscillator, we expect the tidal response of the planetary interior to lag in phase behind the periodic tidal perturbation.

The equation of motion of a one-dimensional forced oscillator with damping takes the form

$$d^2x/dt^2 + 2\lambda dx/dt + \omega_0^2 x = (f/m)\cos(\gamma t), \qquad (4\text{-}68)$$

where x is the space coordinate, λ is the damping coefficient, ω_0 is the natural frequency of the oscillator, m is its mass, and f and γ are the amplitude and frequency of the external driving force. The general solution to Eq. 4-68 takes the form

$$x = ae^{-\lambda t}\cos(\omega' t + \text{const.}) + b\cos(\gamma t + \delta). \qquad (4\text{-}69)$$

Here a is an arbitrary constant and $\omega' = (\omega_0^2 - \lambda^2)^{1/2}$; the first term in Eq. 4-69 is transient and is not of interest here. The second term represents the steady-state response of the oscillator to the driving force. It has an amplitude given by

$$b = \frac{f}{m[(\omega_0^2 - \gamma^2)^2 + 4\lambda^2\gamma^2]^{1/2}}, \qquad (4\text{-}70)$$

and a phase shift given by

$$\tan \delta = 2\lambda\gamma/(\gamma^2 - \omega_0^2). \tag{4-71}$$

In all cases of interest in the present solar system, the tidal frequency γ is substantially less than any natural planetary frequency, and so the appropriate version of Eq. 4-70 is $b = f/(m\omega_0^2)$. Likewise, since the damping is presumed weak, one has $\tan \delta = -2\lambda\gamma/\omega_0^2$. The planetary tidal bulge is therefore offset by some small negative amount from the line of centers. This angular offset depends linearly on the damping coefficient and the frequency of the perturbation. In place of the phase shift δ, the reciprocal quantity Q is frequently used, where

$$Q = 2\pi E_0 / \oint (-dE/dt)\, dt, \tag{4-72}$$

E_0 is the peak elastic energy stored in the oscillator, and the integral evaluates the energy dissipated over one complete cycle of the oscillator. Q and δ are related by $\delta = Q^{-1}$ for small δ. If the elastic properties of the interior of a solid planet are analogous to the properties of a damped harmonic oscillator, one then has for the total tidal moment acting on a perturber of mass M_p:

$$M_p T_t = 9\Lambda_{2,0} G M_p^2 (a^5/R^6)\delta, \tag{4-73}$$

where the direction of this tidal moment is indicated in Fig. 4-3. Obviously there is an equal and opposite moment on the planet. If the planet rotates more rapidly than the tidal bulge, i.e. if the perturber's orbit is outside the synchronous orbit radius, then the effect of dissipation is to transfer angular momentum from the planet's rotation to the perturber's orbital motion. Part of the kinetic energy of planetary rotation is transferred to the perturber's orbital motion, and the remainder is dissipated in the planet as heat. This process continues as the planet's rotational rate approaches the perturber's orbital rate from above, and ceases as these rates become equal.

If, on the other hand, the perturbing body is within the synchronous orbit radius, the planetary rotation is slower than motion of the tidal bulge around the planet, and there is a tidal torque which acts to remove angular momentum and energy from the perturber's orbit and add them to the rotation of the planet. The excess kinetic energy

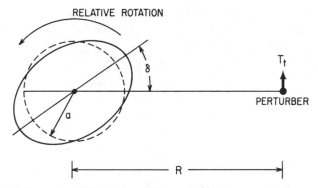

Fig. 4-3. Geometry of tidal torques. The angle δ is shown greatly exaggerated.

is again dissipated as heat in the planet. Such evolution cannot proceed indefinitely. It must end when the perturber evolves sufficiently close to the primary to impact upon it or be disrupted by tidal forces.

Although the tidal dissipation described by Eq. 4-73 *usually* makes only a small instantaneous contribution to the planet's energetics, it can have substantial cumulative effects after many eons. These effects can be used to constrain the physical state of the planetary interior, as parameterized via Q or δ.

We will now consider a number of examples of tidal evolution. First let us examine the planets which seem to be rotating slowly according to Fig. 4-1. A characteristic time for tidal spin-down is defined by dividing the planet's primordial rotational angular momentum by the tidal torque given by Eq. 4-73. For the inner planets, such as Mercury and Venus, the most effective source of tidal torque is the sun. Let the primordial angular momentum be

$$L_{\text{prim}} = I\omega_p \simeq Ma^2\omega_p, \qquad (4\text{-}74)$$

where I is the moment of inertia and ω_p is the primordial rotation rate. Figure 4-1 suggests that the latter is probably not a strong function of planetary mass, and probably corresponds to a rotation period on the order of 10 hours. Thus the characteristic time from Eqs. 4-73 and 4-74 is, to order of magnitude

$$\tau_{\text{tidal}} \sim Ma^2\omega_p/[GM_s^2(a^5/R^6)\delta], \qquad (4\text{-}75)$$

where M_s is the mass of the sun, R is the mean distance between the planet and the sun, and we have taken $9\Lambda_{2,0} = 3k \simeq 1$. The latter assumption is reasonably valid for earth-sized bodies, but is not correct for small, solid objects. For such objects, the tidal time scale should be multiplied by a correction factor on the order of $19\mu/(2ga\rho_0)$—see Eq. 4-64. This factor can be quite large for very small objects such as Mars' satellite Phobos. Adopting $\mu \sim 10^{12}$ dyne/cm^2 (the most uncertain parameter in the correction), we find that the correction factor due to Phobos' rigidity is on the order of 10^7. As we shall see, even this large correction does not change the conclusion that most small, solid satellites should have been despun by tidal interactions with their primary over the age of the solar system.

Equation 4-75 is clearly quite sensitive to R. But the principal unknown parameter in Eq. 4-75 is δ. The conventional procedure for adopting δ is to assume that laboratory measurements on the dissipative properties of rock deformation can be applied directly to tidal dissipation. Since laboratory dissipation measurements are made at substantially higher frequencies than tidal frequencies, attention must be given to the frequency dependence of Q or δ. As Eq. 4-71 shows, δ is expected to have a frequency dependence, and in the limit $\gamma \rightarrow 0$, δ should depend linearly on γ. Moreover, δ should pass through zero at the same time as γ passes through zero, since the tidal torque must reverse sign if the planetary rotation changes sign. It is therefore almost certainly incorrect to assume that δ has a constant and more or less universal value, although much of the published tidal lore hinges on this assumption. Based on laboratory data on the Q of rocks, typical values of δ are on the order of 10^{-2} to 10^{-3}, and we will therefore compile tidal spin-down times based upon the assumption that $\delta = 10^{-2}$, subject to the concerns stated above. Table 4-3 gives results for these times for a number of solar system bodies. In the case of the tidal spin-down of the earth by the moon, the actual deceleration can be measured, and thus δ can be directly determined. It is found to be about 2°, substantially outside the range of δ indicated above. This result is thought to be caused by extra dissipation in oceanic tidal currents.

Although the hypothesis of constant Q for all planetary interiors is a rather extreme one, the results presented in Table 4-3 are in

Table 4-3. Characteristic Times for Tidal Despinning of Planets and Satellites by a Perturber

(It is assumed in all cases that $\delta \simeq 0.01$ and that the object has a primordial rotation period of 10 hours. The present distance between the planet and perturber has been used. In cases where this distance was substantially smaller in the past, the tidal despinning time was also substantially shorter. A correction for rigidity, discussed in the text, has not been applied. For very small, solid objects, this correction could be as large as a factor of 10^7.)

PLANET	PERTURBER	τ_{tidal} (YEARS)
Mercury	Sun	1.8×10^9
Venus	Sun	7.4×10^{10}
Earth	Sun	5.5×10^{11}
Earth	Moon	1.1×10^{11}
Moon	Earth	1.0×10^7
Mars	Sun	4.8×10^{12}
Deimos	Mars	35
Phobos	Mars	0.1
Io	Jupiter	1.9×10^2
Callisto	Jupiter	7.4×10^5
Titan	Saturn	5.2×10^5
Oberon	Uranus	5.9×10^5
Triton	Neptune	7.3×10^4
Pluto	Charon	3.1×10^5

considerable agreement with observation.[6] Note, first of all, that essentially every close natural satellite in the solar system is predicted to have been significantly spun down over the age of the solar system ($\simeq 4.6 \times 10^9$ years). In most cases this has led to a rotation state which is synchronous with its orbital period about the primary. In the case of the Martian satellites Phobos and Deimos, the despinning time is extremely short if rigidity of the satellite is neglected. Because of the strong R-dependence in the tidal torque formula, only planets quite close to the sun are predicted to have undergone significant tidal despinning over the age of the solar system. The tidal despinning time found for Mercury is consistent with the implication of Fig. 4-1 that Mercury has lost a significant amount of rotational angular momentum over the age of the solar system.

The situation is less clear in regard to Venus. According to Table 4-3, under the stated assumptions the tidal spin-down time for Venus is considerably greater than the age of the solar system, and yet Venus is not only rotating very slowly, it even has a retrograde rotation. The theory which we have outlined cannot really account for this result without considerable embellishment. In any case, it seems that Venus has or had considerably more dissipation than would correspond to the nominal figure of $Q \sim 10^2$.

On the other hand, the tidal theory works fairly well for the earth-moon system, which is one of the reasons why it may be justifiably applied to the interiors of other planets. The theory predicts that the earth's rotational rate is changing only very slightly due to tidal friction at present, although this effect was much more pronounced in the past, when the earth and moon were closer together (if the earth had more rotational angular momentum in the past, the moon's orbit had correspondingly less, and hence was closer to the earth). It also implies that the moon was tidally despun by the earth quite early in its history. However, if the earth had the same value of Q in the distant past as it has today ($Q \simeq 15$), a detailed calculation shows that the moon was within the Roche limit only about 2×10^9 years ago. We will discuss a similar problem with tidal dissipation in Jupiter's interior in Chapter 8.

The present rotation rates of Mercury and Venus merit further discussion. Although both planets rotate slowly in the sense defined above, neither has reached a state where the period of rotation has become equal to the orbital period, the so-called 1:1 commensurability familiar in the case of the moon. For a circular orbit, there is then no tidal torque and thus no change in the rotation rate; a planet or satellite can remain in this state indefinitely, once it is achieved. The 1:1 state is also stable for an eccentric orbit, provided that the object's gravity field has a permanent tesseral harmonic component.

Other stable spin-orbit commensurabilities are possible for an eccentric orbit, and in fact Mercury exhibits one such resonance (so far the only one detected in the solar system other than the 1:1 resonance). Mercury rotates exactly three times in an inertial frame during two complete orbits about the sun in the same frame, and is thus in a 3:2 spin-orbit resonance. For Mercury to have achieved this rotational state, its interior must have two principal properties: the material must be dissipative to tidal distortions, i.e. have a finite Q, and there must be a permanent tesseral harmonic component in Mercury's gravity field.

Because of the substantial eccentricity of Mercury's orbit about the sun ($e = 0.206$), it is possible for the *average* tidal torque on Mercury to be zero although the planet does not rotate synchronously. Figure

4-4 shows how this can occur for an eccentric orbit. The figure plots the angular rate of motion of Mercury in its orbit with respect to the sun, i.e., df/dt, where f is the true anomaly of Mercury's orbit, and the angular rotation rate of Mercury itself, both plotted as a function of time. Mercury's rotation rate is constant and equal to $1.5n$, where n is the average value of df/dt. The shaded areas represent the interval of time during perihelion when the angular velocity of Mercury with respect to the sun exceeds the planet's own rotation rate. Because of the eccentricity of its orbit, there is therefore an interval during Mercury's day when the sun ceases its normal motion to the west, stops, moves for a time to the east, and then continues its normal westward course. But this means that the instantaneous value of γ which appears in Eq. 4-71 is therefore reversed in sign, and so δ is reversed in sign. The instantaneous peak tidal bulges on a dissipative Mercury are normally slightly to the east of the subsolar point and $180°$ away, but near perihelion, in the shaded region on Fig. 4-4, the

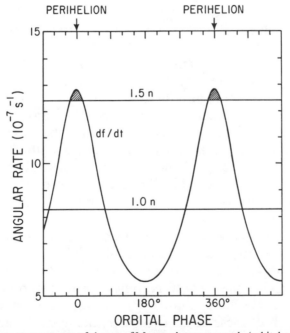

Fig. 4-4. The instantaneous rate of change of Mercury's true anomaly (orbital angular position) df/dt as a function of orbital phase f. Horizontal lines correspond to rotation rates of $1.5\ n$ (n = mean value of df/dt)—the actual rotation rate—and $1.0\ n$, which corresponds to a lunar-type resonance.

bulges shift to the other side of the subsolar point, and the torque given by Eq. 4-73 reverses. Mercury is despun everywhere in its orbit except the shaded region. In the shaded region, solar tides act to transfer angular momentum back to the planet's spin from its orbit. The "spin-up" period of Mercury's orbit is of course much briefer than the "spin-down" period, but because of the R^{-6} factor in the torque, the integrated torques can balance over the entire orbit, leading to no net change in the rotational state. But to lock in such a state over the age of the solar system, we also require that Mercury have an intrinsic "preferred" orientation with respect to the sun. Suppose that Mercury's external gravity field has a tesseral term, C_{22}, corresponding to the expansion of Eq. 4-12, and we set $S_{22} = 0$ by a suitable choice of axes. Using the same procedures as in the case of Eq. 4-62, we find that the sun exerts a torque on Mercury equal to

$$M_s T = -(\tfrac{3}{2})(GM_s/R^3)(B - A)\sin 2\phi, \qquad (4-76)$$

where ϕ is the longitude of the sun with respect to the axis of minimum moment of inertia. This torque, produced by the permanent figure of Mercury, acts in addition to the tidal torque. It is this torque which gives rise to a preferred orientation of Mercury. Let the gravitational potential energy associated with the interaction of Mercury's permanent figure and the sun be E_g. A state of minimum or maximum E_g occurs when the torque is equal to zero. The minimum (stable equilibrium) occurs when $\phi = 0$, when the axis of minimum moment of inertia points toward the sun, and this minimum is deepest when $\phi = 0$ at perihelion. In general, E_g will vary with time as the planet rotates and orbits the sun in an eccentric orbit. Figure 4-5 shows the variation of E_g for an eccentric orbit and various ratios of the planet's rotation rate to mean orbital motion. If these quantities differ substantially from a ratio of small whole numbers, as in the case of the uppermost curve, then the value of E_g averaged over many orbits is zero. The middle curve shows the case for Mercury, where the planet rotates precisely three times during two orbital periods. The bottom curve shows a lunar-like $1:1$ resonance in an eccentric orbit. In both of these latter two cases, the average value of E_g is negative. The presence of dissipation causes a system to decay into a state with a minimal total energy, and thus a resonance tends to be the end state for a system with tidal dissipation. Note that a finite

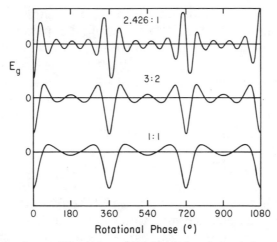

Fig. 4-5. Gravitational potential energy associated with the rotational phase of a planet with a permanent mass asymmetry, moving in an elliptical orbit, as a function of rotational phase, for three different ratios of the rotational rate to mean orbital rate.

value of the eccentricity is not essential for the existence of the 1 : 1 resonance, but is required for all higher-order resonances. In the 3 : 2 resonance, one end or the other of the axis of minimum moment of inertia points toward the perturber at closest approach.

If the difference of the moments of inertia $B - A$ becomes too small, then the planet ceases to have a permanent gravitational "handle" for the perturber, and there will be no preferred orientation. In order for a resonance to be stable, it is sufficient that the maximum average torque due to the permanent figure be greater than the torque due to tidal dissipation. A value of $|B - A| > 10^{-8} Ma^2$ comfortably satisfies the requirement for Mercury to remain captured into a 3 : 2 resonance, for reasonable values of Q. Unfortunately, we do not have enough data from the Mariner 10 Mercury flybys to determine the value of $B - A$ for Mercury. However, the minimum value of $|B - A|/Ma^2$ required for stability is at least a thousand times smaller than the value of $|B - A|/Ma^2$ for the moon, and so there can be little doubt that the requirement for a stable resonance is satisfied. There is also now no doubt that Mercury is in the resonance: its rotational period of 58.6461 ± 0.005 days is ⅔ of the orbital period of 87.96935 days.[7]

If Mercury began its existence rotating considerably faster, the chances are not necessarily 100% that it would be captured into the

3:2 resonance. Depending on the precise rotational phase as the resonance is approached and the values of other parameters, in some cases the planet could continue to spin down toward the 1:1 resonance. Calculation of the probability of capture into a given resonance state is quite model-dependent, so that it is difficult to make a precise statement about whether Mercury's present rotational state is extremely unlikely.

Venus does not display a rotational resonance with respect to the sun. When the first accurate rotational period for Venus was determined from radar reflections off surface features, it was suspected that Venus might be in a rotational resonance involving the earth. For a sidereal rotation period of 243.16 days retrograde, Venus rotates backwards precisely five times between successive conjunctions with the earth. Thus the same feature on Venus would be pointed toward the earth at each closest approach to the earth. From a theoretical point of view, it is not easy to see how the relatively weak average torque produced by interaction between the earth and Venus' C_{22} gravity term could lock the planet in such a high-order resonance. In order for the resonance to be stable, one also requires that the solar tidal torque on Venus not exceed this torque, which requires that $|B - A| > 10^{-2} Ma^2/Q$. Since we now know from Pioneer Venus gravity measurements that $|B - A|/Ma^2 = 7 \times 10^{-6}$, this means that $Q > 1400$ is required for Venus. With such a large value for Q, it would be difficult to argue that solar tides have played much of a role in reducing Venus' rotation rate to its present low value. This problem has now apparently been removed by more accurate determination of Venus' rotation rate. With a retrograde rotation period of 243.0084 days (value adopted by Pioneer Venus project, cf Table 1-1), the subearth point on Venus at successive conjunctions with the earth gradually drifts to the east. The closeness of the period to a resonant value could be a coincidence.

Tidal Heating

Whenever tides are moving in a planet's rest frame, the periodic distortion of the planet's figure is accompanied by dissipation of energy, as described by the parameter Q (Eq. 4-72). The irreversible transformation of mechanical energy into heating of the planet's interior must be taken into account in the planetary energy balance, along with other sources of heating such as radioactivity, gravitational contraction, and insolation. Although it is straightforward to write down an equation governing this process, as with all other tidal dissipation problems, the essential physics is bound up in the parameter Q. We begin with Eq. 4-72, slightly rewritten:

$$dE/dt = -(2\pi/\tau)E_0/Q, \qquad (4\text{-}77)$$

where $-dE/dt$ is the rate of dissipation of mechanical energy in the body, averaged over one tidal cycle of duration τ. Thus, $-dE/dt$ represents a (positive) source of energy in the planet. The magnitude of this term can be estimated if we can determine E_0, the peak elastic energy stored in the tidal deformation, and Q. We will assume that the planet is solid and elastic, so that the peak strain energy per unit volume is given, to order of magnitude, by

$$E_0/\text{vol.} \sim \mu u^2, \qquad (4\text{-}78)$$

where μ is the shear modulus and u is a typical shear strain in the body, which is assumed to be of constant density and therefore incompressible. To order of magnitude,

$$u \sim \delta r/a, \qquad (4\text{-}79)$$

where δr is given by Eq. 4-58. Thus, to order of magnitude,

$$E_0 \sim \mu a^3(h^2 M_p^2 a^6/M^2 R^6), \qquad (4\text{-}80)$$

and so

$$-dE/dt \sim (2\pi/\tau)\mu a^3(h^2 M_p^2 a^6/M^2 R^6 Q). \qquad (4\text{-}81)$$

Tidal heating turns out to be quite unimportant for the interior energetics of any terrestrial planet. The rate of transformation of mechanical energy is too slow at present (see Table 4-3) for this process to compete with any other source of interior heating.

Even if a planet or satellite is spun down to a state of zero net torque in a $1:1$ resonance, tidal dissipation may still continue, because in an eccentric orbit the value of R depends on time, and thus the tidal strain (Eq. 4-79) is also time-dependent. The fluctuating part of the tidal strain is now given to order of magnitude by

$$u \sim e \, \delta r/a, \tag{4-82}$$

where e is the orbital eccentricity, and so formula 4-81 becomes

$$-dE/dt \sim ne^2\mu a^3(h^2 M_p^2 a^6/M^2 R^6 Q). \tag{4-83}$$

when $n = (2\pi/\tau)$ is the mean orbital motion. Assuming $\mu \gg g\rho a$ (valid for a small, rigid satellite), the Love number becomes

$$h \sim g\rho a/\mu, \tag{4-84}$$

and with the further substitution from Kepler's law

$$n^2 \sim GM_p/R^3, \tag{4-85}$$

we find

$$-dE/dt \sim n^5 e^2 a^7 \rho^2/\mu Q. \tag{4-86}$$

This dimensional relationship states that tidal dissipation could be a potentially important source of energy in a satellite or planet with a small orbital period, relatively large dimensions, sufficiently large orbital eccentricity, and sufficiently small Q. It turns out that these conditions are not satisfied for any object in the inner solar system, so that tidal heating can be neglected in comparison with mechanisms which we discuss in the following chapter. However, the situation is quite different for Jovian planet satellites (see Chapter 8).[8]

REFERENCES

1. Landau, L. D. and Lifshitz, E. M. *Statistical Physics.* London: Pergamon, 1969.
2. Tassoul, J.-L. *Theory of Rotating Stars.* Princeton: Princeton University Press, 1978.
3. Gaposchkin, E. M. Global gravity field to degree and order 30 from Geos 3 satellite altimetry and other data. *J. Geophys. Res.* **85:** 7221–7234 (1980).

4. Zharkov, V. N., and Trubitsyn, V. P. *Physics of Planetary Interiors.* Tucson: Pachart, 1978.
5. Reasenberg, R. D. The moment of inertia and isostasy of Mars. *J. Geophys. Res.* **82:** 369–375 (1977).
6. Goldreich, P., and Soter, S. Q in the solar system. *Icarus* **5:** 375–389 (1966).
7. Goldreich, P., and Peale, S. Spin-orbit coupling in the solar system. *Astron. J.* **71:** 425–438 (1966).
8. Peale, S. J., Cassen, P., and Reynolds, R. T. Melting of Io by tidal dissipation. *Science* **203:** 892–894 (1979).

5
Heat Flow

THE PLANETARY ENERGY BUDGET

The energy balance of planets is greatly affected by transfer of solar energy, which may also play an important but indirect role in internal processes. As we observe a planet from space, we may define its total luminosity L_p to be the total energy emitted by the planet at all wavelengths, per unit time. There are in general three components to L_p:

$$L_p = L_r + L_t + L_i,$$ (5-1)

where L_r represents solar power incident on the planet and reflected directly into space by the planet's atmosphere and surface layers without changing its spectral characteristics, L_t represents the remainder of the incident solar power which is absorbed by the planet's atmosphere and surface layers, converted into heat, and re-emitted into space as infrared radiation, and L_i represents the planet's "intrinsic" luminosity, which originates from energy sources in the interior. Even L_i may ultimately be produced by sources outside the planet, however, as in the case of tidal or electromagnetic heating.

It is in principle straightforward to separately determine the three components of L_t. Given the radius of the planet, a, and its distance from the sun, R, the solar power incident on the planet is given by

$$L_r + L_t = L_0 \, (\pi a^2 / 4\pi R^2),$$ (5-2)

where L_0 is the solar luminosity. Conveniently, the spectral wavelengths of the components L_r and L_t are well-separated. The reflected component L_r consists of photons corresponding to the spectral emission curve of the solar photosphere, at a temperature of approximately 6000 °K, and thus is strongly peaked at visual wavelengths between 0.4 μm and 0.8 μm. The other two components of the planetary emission, L_t and L_i, have a spectral distribution corresponding roughly to a characteristic equilibrium temperature of the planet's surface or optically-thick atmospheric layer, ranging from about 50 °K in the outer solar system to about 300 °K in the inner solar system. The corresponding wavelengths range from $\sim 10^2$ μm to about 5μm, and are thus very different from the wavelengths of the reflected component. A suitable experiment can measure the value of L_r by integrating over the reflected energy from the planet at visual wavelengths and over all possible scattering angles. Note that it is normally necessary to perform this important measurement from a spacecraft, because when we observe a distant outer planet from the earth, we see only the component of sunlight which is scattered back in the earth's and sun's direction. From Jupiter and the other outer planets, the earth and sun are never separated by more than about 10°. The component of backscattered light which is observed from earth may not be typical of the power scattered into other angles.

Assume then that L_r has been measured and L_t calculated using Eq. 5-2. It is frequently convenient to define the *Bond albedo:*

$$A = L_r/(L_r + L_t). \qquad (5\text{-}3)$$

Evidently A is the fraction of the total solar energy incident on the planet which is immediately scattered into space, while $1 - A$ is the fraction which is thermalized in the planet's atmosphere and surface layers. A related quantity, T_s, can also be defined:

$$4\pi a^2\sigma T_s^4 = L_t, \qquad (5\text{-}4)$$

where σ is the Stefan-Boltzmann constant and T_s is the equilibrium temperature that the planet would have if it radiated as a perfect black body and produced a total thermal photon power equal to L_t. It is also assumed that the absorbed solar radiation is distributed uniformly over the planet's surface. Although planets do not radiate as black bodies, the approximation is valid enough that T_s turns out to be quite similar to a typical surface temperature or optically thick atmospheric temperature for all planets. Values of T_s for all planets are given in Table 5-1. Note that for Uranus and Neptune, planets which have not yet been visited by spacecraft, the values of A are obtained from theoretical models.

TABLE 5-1. Heat Flow Parameters for Some Solar System Objects*

OBJECT	T_e (°K)	T_s (°K)	H_i (erg/cm²/s)	L_i/M (erg/g/s)
Sun	5770		6.2×10^{10}	1.9
Carbonaceous chondrite				4×10^{-8}
Mercury		441	?	?
Venus		238	?	?
Earth		246	62	5.3×10^{-8}
Moon		274	17	8.7×10^{-8}
Mars		216	?	?
Jupiter	124.4 ± 0.3	109.5 ± 1.6	5400 ± 400	1.7×10^{-6}
Io	106 ± 1	100	1500 ± 300	6.7×10^{-6}
Saturn	95.0 ± 0.4	82.5 ± 1.3	2000 ± 140	1.5×10^{-6}
Uranus	58 ± 2	57	<180	$<2 \times 10^{-7}$
Neptune	55.5 ± 2	46	285	2×10^{-7}
Pluto		33 to 43	?	?

* Sources of data are as follows:
   Sun: Allen, C. W. *Astrophysical Quantities*. London: Athlone Press, 1973.
   Carbonaceous chondrites: Stacey, F. D. *Physics of the Earth*. New York: Wiley & Sons, 1969.
   Earth: Stacey, F. D., Zharkov, V. N., and Trubitsyn, V. P. *Physics of Planetary Interiors*. Tucson: Pachart, 1978.
   Moon: Keihm, S. J. and Langseth, M. G. Lunar thermal regime to 300 km. *Proc. 8th Lunar Sci. Conf.* 1: 371–398 (1977).
   Jupiter: Hanel, R. A., Conrath, B. J., Herath, L. W., Kunde, V. G., and Pirraglia, J. A. Albedo, internal heat, and energy balance of Jupiter: preliminary results from the Voyager infrared investigation. *J. Geophys. Res.* 86: 8705–8712 (1981).
   Io: Pearl, J. C. and Sinton, W. M. Hot spots of Io. In *Satellites of Jupiter* (D. Morrison, ed.). Tucson: University of Arizona Press, 1982, pp. 724–755.
   Saturn: Hanel, R. A., Conrath, B. J., Kunde, V. G., Pearl, J. C., and Pirraglia, J. A. Albedo, internal heat flux, and energy balance of Saturn. *Icarus* 53: 262–285 (1983).
   Uranus: Fazio, C. G., Traub, W. O., Wright, E. L., Low, F. J., and Trafton, L. The effective temperature of Uranus. *Astrophys. J.* 209: 633–637 (1976).
   Neptune: Lowenstein, R. F., Harper, D. A., and Moseley, H. The effective temperature of Neptune. *Astrophys. J.* 218: L145–L146 (1977).

It is possible to measure $L_t + L_i$ using a similar technique to that used to measure L_r. In this case one integrates the emission spectrum of the planet over all angles and all *infrared* wavelengths. The result of this measurement is sometimes summarized by an effective temperature T_e, analogous to T_s, defined by

$$4\pi a^2 \sigma T_e^4 = L_t + L_i, \tag{5-5}$$

so that the internal power is given by

$$L_i = 4\pi a^2 \sigma (T_e^4 - T_s^4). \tag{5-6}$$

The above method is only feasible if L_t and L_i are comparable in magnitude. If L_i is much smaller than the thermalized sunlight component, L_t, then Eq. 5-6 requires us to calculate L_i by differencing two large and nearly equal quantities. This problem is encountered for all objects except the Jovian planets Jupiter, Saturn, and Neptune. The method outlined above is applied to the latter planets via remote sensing from a spacecraft flyby or orbiter. But if the intrinsic luminosity of a planet is too small to measure in this way, a different and more indirect approach is required. This approach works as follows.

If heat is transported by conduction in the outer layers of a planet, the heat flux vector $\mathbf{H}$ (in ergs/cm²/s) is given by the equation of thermal conduction

$$\mathbf{H} = -K \nabla T, \tag{5-7}$$

where K is the thermal conductivity of the material. If a global average value $\langle H \rangle$ can be obtained, and if H can be measured at a deep enough layer so that it represents only the intrinsic flux from the deep planetary interior (H_i), one then finds L_i from the relation $L_i = 4\pi a^2 \langle H_i \rangle$. The actual measurement of H is made by determining the vertical temperature gradient in subsurface rocks and then computing H_i using formula 5-7 and a known value of the thermal conductivity.

Since we are applying Eq. 5-7 under conditions where $L_t \gg L_i$, it is desirable to penetrate into the planet below the layer where sunlight is thermalized in order to obtain an accurate measurement of L_i. In a planet with a transparent atmosphere (or no atmosphere at all), the critical depth in the solid surface is defined by the thermal skin depth, given by

$$l \sim (\chi/\omega)^{1/2}, \tag{5-8}$$

where χ is the thermal diffusivity of the surface material, given by $K/\rho C_p$, where K is the ordinary thermal conductivity and C_p is the heat capacity per gram, and ω is the angular frequency of the variation of solar power input. For depths closer to the surface than l, there is a varying component of heat flow due to daytime input of solar energy and nighttime loss of this energy. For typical values of the thermal conductivity and heat capacity, the skin depth turns out to be about 1 m for 24-hour variations, increasing to about 20 m for yearly cycles, and about 2 km for 10^4-year cycles.

Let us now consider a specific example: the earth. Using the technique outlined above, geophysicists have found the global average intrinsic heat flux to be 62 erg/cm^2/s, corresponding to $L_i = 3.17 \times 10^{20}$ erg/s.[1] This figure is to be contrasted with the heat received directly from the sun, $L_t + L_r = 1.72 \times 10^{24}$ erg/s, a figure approximately 5000 times greater. The problem is further complicated by the fact that a few random measurements of heat flow at the earth's surface would not serve to characterize the global value to better than a factor of two, as values of the local heat flux typically vary by at least that amount. The values are also correlated with geological structures.

With these comments in mind, it now becomes clear why we have as yet very little information about the global energy balance of most planets in the solar system. The terrestrial planets, which we may expect to have values of H_i somewhat comparable to the earth's, will therefore not have intrinsic luminosities which are measurable from space. But even a surface lander will not suffice: ideally, one should drill a borehole to below the thermal skin depth, and then measure the temperature gradient and thermal conductivity within this borehole. So far this difficult and important experiment has only been attempted on the moon, at the Apollo 15 and Apollo 17 landing sites. The results gave $H_i = 21$ erg/cm^2/s at Apollo 15 and 14 erg/cm^2/s at Apollo 17.[2] Since lunar heat flow measurements have only been carried out at these two sites, the value of L_i for the moon cannot be considered well-determined. If the values are typical of the moon as a whole, we have $L_i = 7 \times 10^{18}$ erg/s, or about $\frac{1}{50}$ of the earth's luminosity.

Results for the intrinsic luminosities of some solar system objects are given in Table 5-1.

PLANETARY ENERGY SOURCES

Table 5-1 contains examples of planets with very different types of energy sources; some may be simultaneously important in the same object. Other mechanisms, now extinct, may have been important in some of these objects during the evolution of the early solar system. The following is a list of potential planetary energy sources:

(a) Release of gravitational energy.
(b) Loss of heat.
(c) Release of nuclear binding energy through nuclear fusion.
(d) Release of nuclear binding energy through radioactive decay.
(e) Tidal and rotational dissipation of kinetic energy.
(f) Ohmic dissipation of electrical currents.

We will now give a semi-quantitative description of each of these processes.

First, let us write down an expression for the heat released per second by a gram of planetary matter which is undergoing an arbitrary transformation:

$$\epsilon = \epsilon_{ex} + \epsilon_N - dE/dt - Pd(\rho^{-1})/dt, \tag{5-9}$$

where ϵ is the total rate of heat release in ergs/g/s, ϵ_{ex} is the heat dissipated in the body due to external effects (such as mass accretion, tidal flexing, ohmic dissipation, etc.), ϵ_N is the energy released by changes in nuclear binding energy, dE/dt is the rate of change of internal energy per gram, and $Pd\rho^{-1}/dt$ is the work done per gram per unit time on the sample. The intrinsic luminosity of the planet is then obtained by integrating Eq. 5-9 over all mass elements dm of the planet:

$$L_i = \int_0^M (\epsilon_{ex} + \epsilon_N)dm - (d/dt) \int_0^M E \, dm + \int_0^M (P/\rho^2)(d\rho/dt) \, dm. \tag{5-10}$$

All of the mechanisms listed above are included in this equation; we are now merely concerned with identifying them for bookkeeping purposes. The last term in L_i is exactly equal to the rate of release of gravitational energy by the object's contraction (release of gravitational energy by infalling matter is assumed to be included in ϵ_{ex}), provided that it evolves through a sequence of hydrostatic equilibrium configurations. If the planet is not in hydrostatic equilibrium, we must include another term which corresponds to the release of elastic energy, but this term normally makes a very small contribution to the global energy balance.

The gravitational binding energy is given by

$$E_G = -(\tfrac{1}{2}) \int G\rho V \, d^3r, \qquad (5\text{-}11)$$

where V is the gravitational potential defined by Eq. 4-1 and the integral is taken over the volume of the planet. For simplicity, consider a spherical, nonrotating planet which is in hydrostatic equilibrium, so that we can write Eq. 4-24 in the form

$$dP/dr = -Gm\rho/r^2, \qquad (5\text{-}12)$$

where $m(r)$ is the mass enclosed within a spherical shell of radius r. In considering the general evolution of a planet, with changes in the radius and density, it is less confusing to adopt m as the independent variable rather than r, because a mass element dm by definition contains a constant amount of planetary matter during the evolution of the body. We have

$$dm = 4\pi r^2 \rho dr, \qquad (5\text{-}13)$$

and so Eq. 5-12 is rewritten

$$dP/dm = -Gm/4\pi r^4. \qquad (5\text{-}14)$$

Equation 5-11 can be integrated by parts and then dV/dr can be replaced with $\rho^{-1}dP/dr$, followed by the definition in Eq. 5-13, to obtain

$$E_G = -\int_0^M G(m/r)dm = \int_0^M 4\pi r^3 (dP/dm)dm, \qquad (5\text{-}15)$$

using Eq. 5-14. Integrating this equation by parts again, we find an alternate expression,

$$E_G = -3 \int_0^M (P/\rho)dm, \qquad (5\text{-}16)$$

valid for an object in hydrostatic equilibrium.

We now consider an evolutionary change in which the planet remains in hydrostatic equilibrium but the pressure and density change by infinitesimal amounts. From Eq. 5-15, we have, for the change in gravitational binding energy,

$$\delta E_G = - \int_0^M Gm\delta(1/r)dm = \int_0^M (Gm/r^2)\delta r \, dm$$

$$= - \int_0^M 4\pi r^2(dP/dm)\delta r \, dm = - \int_0^M \delta(4\pi r^3/3)(dP/dm)dm.$$

$$(5\text{-}17)$$

Integrating the latter expression by parts, we have

$$\delta E_G = \int_0^M P\delta[d(4\pi r^3/3)/dm]dm = \int_0^M P\delta(1/\rho)dm. \quad (5\text{-}18)$$

Thus, subject to the assumption that the planet evolves from one state of hydrostatic equilibrium to another, the change in gravitational energy is just equal to the work done on the sample.

The gravitational energy release and internal energy release in an evolving planet in hydrostatic equilibrium are closely related. As the interior temperature changes, the pressure changes, and therefore the density changes as hydrostatic equilibrium is maintained. Consequently both internal energy and gravitational energy change as the temperature changes. Let us look more closely at this process.

Consider two extreme limits for the pressure and energy equations of state. First, suppose that the planet is at a sufficiently low temperature (in practice, a temperature which is low enough so that kT is small compared with a characteristic electronic excitation energy of the material) for the equations of state to be written in the form

$$E = E_0 + AT, \qquad (5\text{-}19)$$

$$P = P_0 + BT, \qquad (5\text{-}20)$$

where E_0 is the internal energy of the material at zero temperature and P_0 is the pressure at zero temperature, while AT and BT are assumed to be small linear (in T) corrections to these quantities. These expressions are normally reasonably good approximations for most planetary interior conditions (see Chapter 3). The rate of change of internal energy is then given by

$$dE/dt = dE_0/dt + d(AT)/dt, \qquad (5\text{-}21)$$

and the rate of doing work on the material is

$$P\,d(\rho^{-1})/dt = P_0\,d(\rho^{-1})/dt + BT\,d(\rho^{-1})/dt. \qquad (5\text{-}22)$$

We have supposed that the planet is close to zero temperature in the sense that Eqs. 5-19 and 5-20 are valid. Therefore, if we consider two equilibrium configurations, one calculated at precisely zero temperature and the other calculated at the current value of T, we may to good approximation assume that the density at a given point in the planet at pressure P changes by some small linear amount proportional to T as the temperature drops from T to zero. Thus, we write

$$d\rho/dt = -c\,dT/dt, \qquad (5\text{-}23)$$

where c is a constant of proportionality.

Substituting Eq. 5-23 in Eq. 5-22 and adding Eqs. 5-21 and 5-22, we obtain a final expression for the heat evolved by a gram of planetary matter:

$$\begin{aligned}
-dE/dt - Pd(\rho^{-1})/dt &= -dE_0/dt - d(AT)/dt \\
&+ (P_0/\rho^2)d\rho/dt - (BT/\rho^2)c\,dT/dt \\
&= -d(AT)/dt - (BT/\rho^2)c\,dT/dt, \quad (5\text{-}24)
\end{aligned}$$

since the terms in E_0 and P_0 exactly cancel each other. Now we have supposed that the temperature is very low, and thus terms of second order in the temperature are negligible. Expanding A in power of T and denoting by A_0 the low-temperature limit of A, the lowest-order result is

$$-dE/dt - Pd(\rho^{-1})/dt = -A_0\,dT/dt. \qquad (5\text{-}25)$$

This calculation shows that when the temperature is low in the sense defined above, the increase in density produced by cooling results in an increase of gravitational binding energy, but that most of this energy is diverted into driving up the value of E_0 and is not available for radiation by the planet. The energy which is released by the planet is primarily derived from cooling of the thermal reservoir.

Now consider the opposite limit, not applicable to planets, where internal temperatures are so high that the material behaves as an ideal gas. The pressure equation of state is then written as

$$P = R_g \rho T / \mu, \tag{5-26}$$

where R_g is the gas constant and μ is the mean molecular weight. Thus Eq. 5-16 for the gravitational binding energy becomes

$$E_G = -3R_g \int_0^M (T/\mu) \, dm. \tag{5-27}$$

The internal energy per gram of the gas can be written

$$E = 1.5 \, R_g T / \mu + E_m, \tag{5-28}$$

where the first term represents the average translational internal energy and the second term is the contribution to the internal energy from molecular degrees of freedom. Combining Eqs. 5-27 and 5-28 in Eq. 5-10, we find that the object's luminosity is given by

$$L_i = \int_0^M (\epsilon_{ex} + \epsilon_N) dm + 1.5 R_g \int_0^M dm \, d(T/\mu)/dt$$

$$- \int_0^M dm \, dE_m/dt. \tag{5-29}$$

If there is no contribution from ϵ_{ex} or ϵ_N, and if E_m remains constant, then this expression says that a positive luminosity is accompanied by an *increase* in interior temperatures with time. That is, the object heats up and contracts as it loses heat to space. This behavior is opposite to that in the low-temperature limit, where the object cools off as it loses heat to space. Now it may happen that the molecular internal energy per gram (E_m) changes during the process of contraction. If E_m increases during contraction, then Eq. 5-29 says that the luminosity is diminished by the rate that energy is stored in E_m. If the term in E_m becomes too significant, the object may not be able to evolve in a quasistatic manner between successive hydrostatic equilibrium configurations, and collapse may occur. This process is not presently relevant to any object in the solar system, but it may have played a role in the formation of some solar system bodies.

Next, we consider another limiting case for gravitational energy release: accretion heating of a solid protoplanet. Consider a protoplanet which is being assembled from widely dispersed planetesimals. Assume that the planetary material is incompressible, with a density ρ and a heat capacity per gram at constant pressure C_p. First assume that the accreting material is initially at zero temperature; we will then derive a correction which allows for a finite starting temperature.

After the protoplanet has built up to a radius r and a mass $m(r)$, the surface layers will have heated up to a temperature $T(r)$. An expression for $T(r)$ is found by considering the budget for the energy which is added per unit surface area of the protoplanet following addition of a mass layer with thickness dr. The increase in internal energy per unit area is given by $\rho C_p T(r)dr$. The gravitational energy which is added to unit area is $\rho[GM(r)/r]dr$, but the amount which is available for raising the internal energy is equal to this term minus the energy per unit area which is radiated away over the period of time dt during which the layer of thickness dr is accreted. Assuming that the material radiates as a black body at temperature $T(r)$, this comprises $\sigma T(r)^4$. Thus the accretional temperature profile is given by

$$\rho[GM(r)/r]dr = \sigma T(r)^4 dt + \rho C_P T(r)dr. \qquad (5\text{-}30)$$

The accreted material is not actually at zero temperature, because it will have been heated by the primordial sun and by radioactive decay. This initial temperature is assumed to be constant and equal to T_b (the so-called "base" temperature). Then the increase in internal energy per unit area will be given by $\rho C_P[T(r) - T_b]dr$. The gravitational energy which is added to unit area is $\rho[GM(r)/r]dr$ as before, and the portion of this which is radiated into space is the total energy radiated less the amount which would be radiated anyway by the planetesimals at their base temperature, $\sigma[T(r)^4 - T_b^4]dt$. Thus the complete equation for the accretional temperature profile is[3]

$$\rho[GM(r)/r]dr = \sigma[T(r)^4 - T_b^4]dt + \rho C_P[T(r) - T_b]dr. \quad (5\text{-}31)$$

This equation assumes that there is no compressional heating (this will be small for planets in the terrestrial mass range), and that accretion proceeds so rapidly that conductive heat transport is insignificant (also a good assumption for planetary sized bodies). There are two unknown parameters in Eq. 5-31: the rate of growth of the planet dr/dt, and the base temperature. Of these two parameters, dr/dt is by far the more critical, for if the accretion time is short compared with the characteristic time for radiation of the internal energy, most of the potential energy of accretion is available for heating the planet, which can then attain an initial interior temperature on the order of several thousand degrees. A common expression for representing dr/dt takes the form

$$dr/dt = ct^2 \sin\gamma t, \quad (5\text{-}32)$$

where c and γ are constants, one of which is constrained by the requirement that r be equal to the present planetary radius at $t = $ present. Equation 5-32 has no fundamental theoretical basis, but represents a rapid initial growth rate (proportional to t^3) which reaches a maximum and then declines.

The nuclear contribution to heat production, ϵ_N, fundamentally arises from the release of nuclear binding energy, either by stepwise conversion of very light elements to intermediate elements (in the vicinity of iron) via fusion, or by stepwise conversion of very heavy elements to intermediate elements via nuclear decay. The two processes have very different dependences on physical conditions, however. Fusion is the primary source of energy release in the sun, and occurs at a significant rate when the temperature and density are high enough for two protons to penetrate their mutual coulomb barrier and react to form a heavier nucleus. The height of the barrier is defined by the mutual coulomb energy at the point where the separation between the two nuclei is equal to r^*, where r^* is the distance at which the coulomb repulsion disappears. Empirically, the barrier height is found to correspond to a temperature on the order of 10^9 °K for protons, and higher values for other elements. Reactions can proceed at considerably lower temperatures, on the order of 10^7 °K in the case of the center of the sun, but there is no possibility of fusion at temperatures on the order of 10^4 °K which are typical of planetary cores (we will return to this point in discussing the early history of the Jovian planets).

Radioactive heating produced by nuclear decay is, on the other hand, a most important process in planetary interiors. The spontaneous decay of a nucleus is essentially unaffected by the physical environment, and so this source of heat depends only on the abundances of unstable nuclei. We may write

$$\epsilon_N = \sum_i \psi_i x_i, \tag{5-33}$$

for the heat generated per gram of matter by radioactive decay. Here the sum is taken over all radioactive species, x_i is the mass fraction of the ith radioactive species in the material, and the ψ_i's are constants characteristic of the species, which are independent of physical conditions such as temperature, pressure, or density. The x_i's are functions of time, however, because a decay event which releases a quantity of heat into the sample also changes the number of unstable nuclei. The equation which governs the time dependence of the x_i's is written

$$dx_i/dt = -\lambda_i x_i + s(t), \tag{5-34}$$

where λ_i is the decay constant for the ith species, and the term $s(t)$ denotes the contribution to the number of species i from all other species which can decay into species i. For the reactions of interest to us, species i is not significantly replenished by such decays, and so we may set $s(t) = 0$. Then the concentration x_i simply decreases exponentially with a characteristic half-life (defined as the time it takes the concentration to decrease by a factor of two) given by

$$T_{1/2} = \ln 2/\lambda_i, \tag{5-35}$$

and therefore the heat generation rate decreases with the same half-life. Although a number of unstable nuclei are found in planetary matter, only a few are sufficiently abundant to be of significance in a planet's global energy balance. These are the two uranium isotopes U^{235} and U^{238}, the thorium isotope Th^{232}, and the potassium isotope K^{40}. In addition, studies of the Allende meteorite have indicated that the extinct isotope Al^{26} may have been present in the early solar system. Because of the relatively large cosmic abundance of aluminum (Tables 2-2, 2-1), decay of Al^{26} may have been a very significant component in the heat budgets of early solar system bodies. However, because Al^{26} has a half-life of only 8×10^5 years, it now plays no role as a planetary energy source.

Figure 5-1 shows ϵ_N in carbonaceous chondrite material as a function of time since the origin of the solar system.[4] The present value of ϵ_N for such material is about 4×10^{-8} erg/g/s, although it may have been nearly an order of magnitude higher at the time of formation of the solar system, primarily because of a much greater abundance of K^{40}. However, potassium is a relatively volatile element and is therefore substantially depleted in bodies which have been substantially heated, such as the moon. For such bodies, the heat generation rate produced by radioactive decay would have been

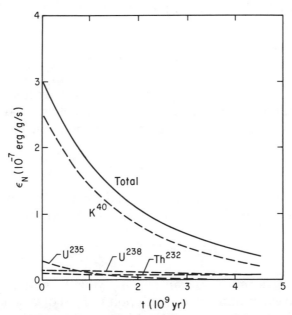

Fig. 5-1. Heating rate due to radioactivity in carbonaceous chondrite material, as a function of time since the origin of the solar system.

more nearly constant over the age of the solar system. Figure 5-1 does not show a possible heating "spike" due to rapid decay of Al^{26} near the beginning of the solar system.

We have now discussed all of the "traditional" mechanisms for heating planets. Recent spacecraft observations, as well as meteorite research, have led researchers to believe that certain other mechanisms may be (or have been) important as well.

Planetary heating due to tidal dissipation has been discussed in Chapter 4. This mechanism appears to be significant for the Jovian satellite Io and possibly other outer-planet satellites (see Chapter 8). In principle, energy could also be released by the change in rotational state associated with tidal spin-down, via a similar mechanism, but such changes would normally be so slow that this version of the mechanism is unlikely to be significant for any object.

The final mechanism which we will discuss—also a "nontraditional" one—is ohmic heating. The concept behind this mechanism is that interplanetary space is penetrated by magnetic fields, and that the motion of a conducting planet through these fields leads to the appearance of an electric field in the planet's frame of reference. The magnitude of the field and its direction are readily calculated from the equation

$$\mathbf{E} = \mathbf{v} \times \mathbf{B}/c, \qquad (5\text{-}36)$$

where $\mathbf{E}$ is the electric field vector, $\mathbf{v}$ is the velocity vector, and c is the speed of light (Gaussian units). Now if the planet "sees" the field $\mathbf{E}$, a current density j appears in the body, as given by the equation

$$\mathbf{j} = \sigma_c \mathbf{E}, \qquad (5\text{-}37)$$

where σ_c is the electrical conductivity of the body. The power per unit volume produced in the body by ohmic dissipation is then given by

$$\rho\epsilon_{ex} = \mathbf{j} \cdot \mathbf{E} = \sigma_c E^2 = \sigma_c |\mathbf{v} \times \mathbf{B}|^2/c^2. \qquad (5\text{-}38)$$

Although these formulas are simple, applying them to real planetary bodies involves a number of difficult points. First of all, it is necessary that the body not have a magnetosphere, so that the interplanetary field lines can permeate its interior. Next, there must be a closed path available so that the induced electrical currents can flow through the planet's interior, out into the interplanetary medium, and somehow return to the other side of the planet to complete the circuit.

Inserting numbers in Eq. 5-38 and using a typical rock conductivity of 10^8 s^{-1}, one finds that the product vB must be on the order of 10^2 gauss · km/s in order for ϵ_{ex} to be in an interesting range, e.g. large enough so that an asteroid-sized object (diameter on the order of 100 km) could be melted over the age of the solar system. Now v is basically given by the velocity at which the solar wind convects field lines past the planet. This number is unlikely to have changed greatly from its present value of 400 km/s in the vicinity of the earth. Since the present value of the interplanetary field magnitude is only a few $\times$ 10^{-5} gauss, this mechanism clearly could not be important at present. Was it ever important? If during the early stages of the sun's existence there was a strong interplanetary magnetic field on the order of a gauss (possibly during the T-Tauri phase), then small rocky bodies might have been partially melted by this mechanism.[5]

MECHANISMS OF HEAT TRANSPORT

As we have seen above, the diffusion of heat by thermal conduction is governed by Eq. 5-7. This equation is valid whenever the energy is transported by particles which have a velocity distribution close to the thermodynamic equilibrium value as defined by a local temperature, and where the temperature itself varies slowly through the material. By "slowly" we mean that the temperature changes by only a small fraction of itself over a particle mean free path.

The particles which carry the energy may be free electrons or heavier particles, or they may be wave excitations such as photons or phonons. Frequently several conduction mechanisms operate simultaneously, in which case the effective thermal conductivity is just the sum of the individual conductivities. The most efficient conductors of heat are the particles with the highest velocities, which in thermal equilibrium are the less massive ones. Thus materials with large numbers of free electrons (metals) are normally good heat conductors. Likewise, photons are efficient transporters of heat, provided that the opacity of the material is low enough for them to be able to propagate.

In the main, conduction is not a very effective process for changing global temperature distributions in planets over geological time intervals. This can be understood by considering Eq. 5-8 written in the form

$$l \sim (\chi t)^{1/2}, \tag{5-39}$$

where l is the distance over which the temperature has changed significantly due to conduction and t is the time during which this change occurs. A planet-sized object takes a much longer time to change its temperature distribution by conduction than does an asteroid-sized object. Again using a value $\chi \sim 10^{-2}$ cm^2/s, typical of rocky materials, and setting t equal to the age of the solar system, we find that l becomes equal to a few hundred kilometers. This result indicates that a body of asteroidal dimensions can change its interior temperature distribution significantly by heat conduction, but that a much larger body cannot. However, we must be careful in making this statement because χ depends on the composition of the material. For metallic materials appropriate to planetary cores, the value of χ is on the order of 10^{-1} cm^2/s, and so the critical dimensions for significant thermal alteration by conduction become larger. Under such circumstances, one can normally ignore photon transport, because a photon of frequency ω cannot propagate in a plasma with a plasma frequency $\omega_p > \omega$. For typical metallic electron densities, this condition is satisfied for all photons except hard x-rays. It is for basically this reason that metals are opaque and reflect photons.

In general, mechanisms of heat conduction in planetary interiors are not very well understood, particularly in nonmetallic high-pressure phases. In nonmetallic substances, e.g. silicates, there are essentially no free electrons, and energy is transported by less efficient carriers. Phonon transport is dominant at low temperatures. Energy is carried by elastic lattice waves which propagate until they are scattered by imperfections in the lattice or by anharmonic effects. The anharmonic contribution to the scattering increases with temperature, causing the lattice thermal conductivity to be a decreasing function of temperature. At more elevated temperatures, high-energy thermal photons are created, which begin to take over a substantial fraction of the energy transport. The results for uncompressed silicates can be represented by the expressions

$$K = K_L + K_R, \tag{5-40a}$$

where K_L is the lattice thermal conductivity, K_R is the radiative thermal conductivity,

$$K_L = 4.184 \times 10^7/(30.6 + 0.21\ T), \qquad (5\text{-}40b)$$

$$K_R = 0\ (\text{for } T < 500),\ 230\ (T - 500)\ (\text{for } T > 500),\ (5\text{-}40c)$$

T is expressed in °K, and the units of thermal conductivity are ergs/cm/s/°K.[6,7] Since the silicate mantles in terrestrial planets are not strongly compressed, these results may be adequate as a first approximation for this application. The thermal diffusivity can be computed by using the value 1.2×10^7 ergs/g/°K for the heat capacity per gram C_P and ~ 3.3 g/cm³ for the density ρ.

Let us summarize our conclusions thus far. There is frequently an order of magnitude or more of uncertainty in the value of the thermal conductivity, but this uncertainty does not always play an overwhelming role in thermal history calculations, because the diffusion of heat in bodies of planetary dimensions is normally a slow process. In a few cases, however, the problem can be a significant one: a case in point is the question of the cooling of Mercury, a relatively small planet, and the possibility that the planet still has a liquid core (see below). Another example of a crucial question which has to be settled by thermal evolution calculations is whether liquid water ever existed in significant amounts within the icy outer planet satellites (also see below).

There is a process which, if active, can render moot most of the uncertainty in the thermal conductivity. This process is convection. When a body is subjected to a thermal gradient, under certain circumstances the matter does not remain stationary, but begins to form macroscopic currents (as opposed to the random microscopic motions of individual particles). Such convection currents always ensue when the temperature gradient exceeds a certain critical value. A very considerable enhancement in the transport of heat out of the planet can be expected when convection is present, because of the organized character of the motion. The heat transport is *not* governed by an equation similar to Eq. 5-7. To describe the transport, one actually needs two equations: one to determine the critical value of the temperature gradient at which convection begins, and one to determine the heat flux as a function of the temperature gradient and other physical parameters, beyond this critical value.

The simplest case occurs for a material of vanishingly low viscosity and thermal conductivity, the so-called *ideal fluid*. Such a model may apply to deep atmospheric layers which are opaque to thermal radiation, or to the liquid interiors of the Jovian planets, but would not in general be appropriate to other regions of planetary interiors. Suppose an element of fluid is displaced upward by a small distance r. Because the conductivity is assumed to be very small, the element will not initially exchange heat with its surroundings, although it quickly adjusts to pressure equilibrium, and so the density of the element is given by

$$\rho = \rho_0 + [(\partial \rho / \partial P)_S - (d\rho / dP)_0](dP/dr)\delta r, \qquad (5\text{-}41)$$

where the partial derivative with subscript S is taken at constant entropy, while the subscripts 0 refer to ambient conditions. If the quantity in brackets is less than zero, the fluid is unstable to such perturbations, for buoyant forces will continue to carry the element away from its original position (the argument obviously applies equally to negative, i.e. downward, values of δr). If the fluid is inviscid, there is nothing to damp these unstable motions, and so convection will occur precisely at the point where the bracketed quantity first becomes negative. This criterion can also be expressed in terms of a critical temperature gradient:

$$(dT/dP)_0 > (\partial T / \partial P)_S \qquad (5\text{-}42)$$

for the onset of convection.

We have now obtained the first of the equations needed to describe convective heat transport for an ideal fluid. But what is the rate of transport as a function of the temperature gradient? In order to derive this relation, one must make some additional assumptions about the behavior of the fluid element as a function of time. Now although the conductivity of the fluid is assumed to be small, it must ultimately play a role in the dissipation of individual convecting fluid elements, and so one assumes that there exists a certain "mixing length" l over which fluid elements maintain their identity before they dissipate. The energy transported by a fluid element can then be related to the mean convective velocity attained by the element and to its temperature contrast with the background. The theory which is derived on this basis is somewhat analogous to the theory of heat transport by particles with a certain mean velocity and mean free path on the order of l, except that the velocity is itself a function of l. The resulting formula is[8]

$$H = C_P[T\Psi/P]^{3/2}|(\partial\rho/\partial T)_P|^{1/2} (g\rho l)^2, \qquad (5\text{-}43)$$

where H is the convective heat flux, C_P is the heat capacity per gram at constant pressure, g is the local gravity, and

$$\Psi = (P/T)[(dT/dP)_0 - (\partial T/\partial P)_S]. \qquad (5\text{-}44)$$

This formula (the so-called mixing length equation) has the first equation built in, for Ψ must be positive in order for H to have a real value. It expresses the heat flux in terms of the temperature gradient, but in a rather complicated manner. The value of l is not really known from first principles, but it is customary in the application of this equation to set l equal to a local pressure or density scale height. Actually, this assumption is not critical because it normally turns out that formula 5-43 implies an extremely small value of Ψ for any physically meaningful value of H. Thus, for example, in the earth's atmosphere we have $H \sim 10^5$ erg/cm²s, and taking parameters appropriate to the lower troposphere, one finds $H \sim 10^{10}\ \Psi^{3/2}$ erg/cm²s for $l \sim 10$ km. Another application of formula 5-43, this time to a planetary interior, will be discussed in Chapter 8.

Our conclusions about the nature of heat transport in inviscid fluids are summarized as follows. Let energy flow be analogous to electrical current in a circuit, where temperature plays the role of voltage. We consider two points in space, one at temperature T_1 and the other at temperature T_2, with $T_1 > T_2$. Energy (current) flows from T_1 to T_2 along paths with characteristic resistances, which represent various mechanisms of thermal conductivity. Convection acts as a special type of circuit: we may think of this branch as having a "voltmeter" which determines whether $T_1 - T_2$ is greater than some critical temperature difference in accordance with the relation in Eq. 5-42. If this temperature difference is not exceeded, then the convective "circuit" remains open (no convection occurs, and heat must be transported by ordinary conductive processes). But once the critical temperature difference is exceeded, the convective "circuit" closes, and heat ("current") flows through this branch with negligible resistance. It follows that the temperature difference will stabilize at a value very slightly larger than the critical value, whatever the value of H. Thus, in an inviscid fluid with a sufficiently large heat flow to maintain convective instability, the temperature gradient will be very close to the adiabatic temperature gradient.

In the interior of a solid planet, the viscosity and thermal conductivity are of course significant, and a better theory is needed to describe convective heat transport. The critical parameter for the onset of convection has been calculated for the following problem. Suppose we have a plane distribution of uniform, essentially incompressible medium at rest in a gravity field g. The upper surface of the medium is maintained at temperature T_2 and the lower surface is maintained at temperature $T_1 > T_2$. Suppose the fluid density varies because of thermal expansion according to the formula

$$\rho = \rho_0(1 - \alpha T), \qquad (5\text{-}45)$$

where ρ_0 is the average density and α is the thermal expansion coefficient. Suppose also that the fluid has a viscosity η and a thermal diffusivity χ. Convection will be stable provided that the rate at which buoyant forces liberate kinetic energy is not exceeded by the rate at which viscous forces dissipate it. This criterion is expressed in terms of the Rayleigh number,

$$Ra = \alpha(T_1 - T_2)g\rho_0 d^3/(\chi\eta), \qquad (5\text{-}46)$$

where d is the thickness of the layer. If $Ra > Ra_{cr} \sim 10^3$, the layer becomes unstable to convection, while if Ra is less than this critical value, energy transport is by conduction. How does this criterion relate to the criterion of Eq. 5-42 for an inviscid, nonconducting material? In Eq. 5-45, it is assumed that the layer is sufficiently thin that self-compression of the material can be neglected, while in Eq. 5-42 the self-compression is an essential aspect of the criterion. Thus a more elaborate model is needed in order to obtain the Schwarzschild criterion as a limiting case of the Rayleigh criterion. Equation 5-45 must be applied with some caution to a thick planetary layer where the density change due to gravitational compression is substantial.

Having presented the first equation, which determines the onset of convection in a viscid, conducting material, we now turn to the second equation, which expresses the heat flux in terms of the temperature gradient. For a material with a finite thermal conductivity, a certain fraction of the heat flux through the layer will be transported by conduction. If the Rayleigh number is less than the critical value, this fraction is unity, while it steadily decreases from unity as Ra increases beyond the critical value. Define the Nusselt number Nu to be the ratio of the heat flux carried by convection plus conduction to the heat flux which would be carried by conduction alone if the system had the same values of T_2 and T_1. Then one may use the following expression,[9] which has been established by laboratory experiments and by calculations, to obtain the convective heat flux:

$$Nu = (Ra/Ra_{cr})^{\beta}, \qquad (5\text{-}47)$$

where β is a constant on the order of 0.3. This law appears to be valid for vigorous convection, i.e. for $Ra \gg Ra_{cr}$, but becomes less reliable as Ra/Ra_{cr} approaches unity. In summary, although finite viscosity and thermal conductivity prevent a convecting planetary interior from acting as if it had infinite conductivity, as in the case of the inviscid fluid, nevertheless a considerable enhancement in the heat flux occurs as Ra/Ra_{cr} becomes large. It is sometimes difficult to calculate Ra accurately when the effective value of the viscosity has a temperature dependence of the form $\eta \propto \exp(-A/T)$, as may occur for temperatures below the solidus temperature.

It is an experimental fact that given enough time, any material will flow like a liquid. Expressed quantitatively, one may say that if a nonhydrostatic stress (e.g., a shear stress) is applied to a solid body, that body will initially deform elastically in accordance with Hooke's law. The deformation of the body (the strain) is proportional to the applied stress. However, over a period of time τ, the body will flow or deform inelastically in such a way as to relieve the applied stress. Over periods of time which are long compared with τ, it is a better description to say that the stress is proportional not to the strain but to the strain *rate*, i.e. to the rate at which the body is being deformed. The proportionality constant is in this case just the viscosity η. Thus, in this limit, one may define a viscosity for a solid.

Experimental and theoretical research[9] has found that rocks are able to plastically deform under an applied strain rate, and that under certain conditions, convection can proceed even in solid planetary mantles. The condition for this to occur is that the characteristic time τ be shorter than geological time periods. If τ is greater than the age of the solar system, the material is unable to transport energy by convection: the viscosity is essentially infinite. It is important to note that a solid planetary mantle may be transporting energy by convection and still appear solid to experimental investigations which have time constants much shorter than τ. For example, the mantle would be able to transmit shear seismic waves, and would act as a solid even for long-period oscillations having periods on the order of an hour.

As mentioned above, the effective viscosity of a solid tends to have a strong temperature dependence. Thus there is a critical temperature T_c, above which the solid is able to flow over geological time periods, and below which it acts as a rigid body. This critical temperature is not well known, and undoubtedly depends on composition and pressure, but is thought to lie in the range $\sim 1100 - 1300$ °K for rock at low ($P \lesssim 100$ kilobar) pressures. Thus we see that there is a natural division in the rocky envelope of a terrestrial-type planet. The outermost layers of the envelope are always at a temperature $T < T_c$, and therefore must always transport energy by conduction. Below the $T = T_c$ isotherm, solid-state convection can also carry energy if the Rayleigh number is large enough. Accumulated stresses should be lower in this region. The region of the planet which lies above the $T = T_c$ isotherm is called the *lithosphere.* This layer acts like a rigid solid, and relieves stresses by faulting (fracturing), rather than by flowing. It is in this region of the planet that the principal deviations from hydrostatic equilibrium should occur.

The concept of a lithosphere is important because it allows us to draw conclusions about the interior temperature distribution of a planet from observations of its surface geology and gravity anomalies.

THE SURFACE BOUNDARY CONDITON

Although convective energy transport can remove heat from a planetary interior much more rapidly than simple thermal diffusion, the energy must ultimately escape into space in the form of photons, a process which obviously involves a mechanism other than convection. Depending on the efficiency of transport mechanisms within the planet, the outermost layer where this process occurs may turn out to be the "bottleneck" for escape of internal heat. The problem is also complicated by the fact that it is this outer layer which is subject to the influence of incoming solar radiation.

First let us consider a terrestrial planet with a relatively thick, opaque atmosphere, such as the earth. The equilibrium temperature of the "photosphere" will be approximately equal to T_e as given by Eq. 5-5; since $L_t \gg L_i$, this temperature is essentially defined by the solar input. Although the local surface temperature varies with time of day, season, etc., the time scale for the global thermal evolution of the planet is very much longer than any of these periods, so that the surface temperature can be represented by a secular average value. The equation for the evolution of the planet's interior temperature distribution, assuming conductive heat transport, is then

$$-\nabla \cdot K \nabla T = \rho \epsilon, \qquad (5\text{-}48)$$

where ϵ is given by Eq. 5-9 and depends upon time derivatives of thermodynamic quantities. The boundary conditions for the solution of Eq. 5-48 are then the initial temperature distribution, the regularity of the solution at the center of the planet, and the surface temperature as determined by the solar influx and the intrinsic planetary heat flow (the latter is a negligible factor in the surface temperature at present but may have been important during earlier epochs).

A different approach is needed for the Jovian planets. These planets have no well-defined solid surface which radiates heat into space. Instead, one can use the concept, borrowed from stellar astrophysics, of an effective photosphere, which is that layer in the atmosphere which is just deep enough so that thermal infrared photons which are radiated from it have a probability of $1/e$ of propagated out of the atmosphere without being absorbed. Below this layer, the atmosphere rapidly becomes very opaque to infrared radiation, the diffusion approximation for radiation holds, and the energy flux is given by an equation of the form of Eq. 5-7, with K defined by the photon opacity. At these same layers, however, not much deeper than the photosphere, the instability criterion (Eq. 5-42) is fulfilled and the atmosphere becomes convectively unstable. Thus, in rapid succession going deeper in a Jovian planet atmosphere, we have the photosphere, a radiative zone, and the troposphere (convective zone). Now we have just argued that a convective zone in a liquid or gas can carry a large heat flux without deviating significantly from the adiabatic gradient. Therefore, the temperature profile in the troposphere should be essentially isentropic, following a relation of the form

$$T = T_0 \, (P/P_0)^{(\gamma - 1)/\gamma}, \qquad (5\text{-}49)$$

where γ is the ratio of specific heat at constant pressure to specific heat at constant volume, assumed constant in this example, and T_0 is the starting temperature for the adiabat at the starting pressure P_0. Equation 5-49 defines the temperature distribution in the deeper atmospheric layers, except that we do not know the appropriate value of T_0. It is obvious that the latter must be obtained from a solution for the temperature profile in the radiative layers of the planet, i.e. by solving the conduction Eq. 5-7 (or a more complicated set of transport equations in the optically thin part of the atmosphere) subject to the condition that the total energy flowing into space be equal to a specified value. Evidently, the upper radiative zone of the planet's atmosphere serves as a "bottleneck" for the planet's loss of internal heat. The temperature profile in the troposphere, which may extend to very deep layers, is almost entirely dependent upon the temperature gradient in the outer few scale heights of the atmosphere.

The temperature gradient in the photospheric layers is defined not just by the heat flow from the interior but also by the heat flow generated by converted sunlight. After photons from the sun strike the planet's atmosphere, a fraction of them are reflected into space as defined by Eq. 5-3, while the remainder are converted into heat in the atmosphere. Where does this conversion take place? Undoubtedly it occurs at those atmospheric layers which are beginning to be opaque to photons of visual wavelength, layers which are somewhat deeper than the photosphere (where the atmosphere is beginning to be opaque to photons of *infrared* wavelength). It is just this process which is responsible for the atmospheric *greenhouse* phenomenon — solar radiation enters a planetary atmospheric, is converted into heat at a relatively deep layer, and then escapes slowly to the photosphere because the infrared opacity of the atmosphere is higher than the opacity at visual wavelengths. This "backwarming" effect is also significant for the internal energetics of the Jovian planets because it affects the surface boundary condition for convective transport of energy in the interior. This is because the photospheric temperature profile, which determines the starting temperature for the troposphere, is in turn defined by the *total* heat flux through the photosphere, the sum of the heat flux from the deep interior and the heat flux from converted sunlight. Consider the following limiting cases for the effect of the surface boundary condition on the temperature profile in the deep interior of a planet. For case (a), generally applicable to terrestrial bodies, the energy transport in the interior is entirely by conduction. The effect of the surface boundary condition on interior temperatures is minimal. In case (b), probably applicable to the Jovian planets, the interior temperature profile is governed by the temperature at the top of the troposphere, which in turn is determined by the total heat flux through the radiative layers above this point, since solar energy is converted to heat within the troposphere. In case (c), solar energy is converted to heat well above the troposphere or photosphere, and thus plays no role in the planet's interior heat transport. Cases intermediate between (b) and (c) can of course occur. Applications of these principles to specific planetary interiors will be considered in Chapter 8.

REFERENCES

1. Zharkov, V. N. and Trubitsyn, V. P. *Physics of Planetary Interiors.* Tucson: Pachart, 1978.
2. Keihm, S. J. and Langseth, M. G. Lunar thermal regime to 300 km. *Proc. 8th Lunar Sci. Conf.* **1**: 499–514 (1977).
3. Johnston, D. H., McGetchin, T. R., and Toksoz, M. N. The thermal state and internal structure of Mars. *J. Geophys. Res.* **79**: 3959–3971 (1974).
4. Stacey, F. D. *Physics of the Earth.* New York: Wiley, 1969, p. 244.
5. Sonett, C. P., Colburn, D. S., Schwartz, K., and Keil, K. The melting of asteroidal-sized bodies by unipolar dynamo induction. *Astrophys. and Space Sci.* **7**: 446–488 (1970).
6. Schatz, J. F. and Simmons, G. Thermal conductivity of earth materials at high temperatures. *J. Geophys. Res.* **77**: 6966–6983 (1972).
7. Toksoz, M. N., Hsui, A. T., and Johnston, D. H. Thermal evolutions of the terrestrial planets. *The Moon and Planets* **18**: 281–320 (1978).
8. Schwarzschild, M. *Structure and evolution of the stars.* Princeton: Princeton University Press, 1958.
9. Schubert, G., Cassen, P., and Young, R. E. Subsolidus convective cooling histories of terrestrial planets. *Icarus* **38**: 192–211 (1979).

6
Planetary Magnetism

Some planets in the solar system are observed to have strong intrinsic magnetic fields (on the order of a gauss or more at the surface), others have weak intrinsic magnetic fields, and the remainder have no detectable intrinsic magnetic fields. One of the tasks of planetary interiors studies is to find a universal theory which can provide an explanation of the observations.

The first planetary magnetic field measured was that of the earth. Later, the Zeeman effect was observed in the solar spectrum and used to show that the sun also has a substantial general surface field amplitude of the same order of magnitude as the earth's. Following these observations, analysis of the properties of the ratio bursts from Jupiter revealed that this planet too has a strong intrinsic field. Further investigations with spacecraft have shown that no detectable field is present on Venus and possibly none on Mars, but that a weak field is definitely present on Mercury. No large-scale intrinsic field has been detected on the moon although weak local fields are observed. Saturn has a rather strong magnetic field of approximately the same surface strength as the earth's. No reliable data are yet available for the other solar system bodies, although there are reports of radio bursts from Uranus which may be consistent with the presence of a magnetosphere around that planet.[1] A summary of available magnetic field data for planets is given in Table 6-1. The field for each planet is represented in terms of an *eccentric dipole,* which is defined as a dipole with a total magnetic moment $\mathbf{M}$ (in this chapter only, we use the customary symbol M to denote the magnitude of a magnetic moment; elsewhere, the symbol M is reserved for mass), displaced from the center of the planet by a distance r in a direction given by a specified latitude and longitude. If the equivalent dipole is oriented so that field lines emerge in the southern hemisphere and reenter the planet in the northern hemisphere, as is true for the earth, it is considered to have negative polarity (the polarity can also be defined

Table 6-1. Properties of Planetary Magnetic Fields

(The spherical coordinates r, θ, ϕ give the location of the eccentric dipole with respect to the center of the planet; r is the distance of the equivalent dipole from the planet's center, θ is the colatitude (polar angle) of its location, and ϕ is the longitude with respect to the defined zero meridian. The tilt gives the angle between $\mathbf{M}$ and the z-axis.)

PLANET	M (gauss/cm³)	POLARITY	r (km)	θ (°)	ϕ (°)	TILT (°)
Mercury	2.4×10^{22}	—	500	?	?	2.3
Venus	$<4. \times 10^{21}$					
Earth	7.98×10^{25}	—	462	71.7	147.8	11.4
Moon*	$<4.4 \times 10^{13}$					
Mars**	2.5×10^{22} (?)					
Jupiter	1.5×10^{30}	+	7000	85	180	10
Saturn	4.6×10^{28}	+	2000 (?)	0 (?)		0.8

* The upper limit pertains to the moon's global magnetic field. Local regions on the moon have remanent fields as high as several $\times 10^{-3}$ gauss.
** The Martian dipole moment is deduced from measurements by Soviet spacecraft, but the interpretation of these measurements is in dispute (see reference).
Sources of data:
   Mercury: Whang, Y. C. Magnetospheric magnetic field of Mercury. *J. Geophys. Res.* **82**: 1024–1030 (1977).
   Venus: Russell, C. T., Elphic, R. C., and Slavin, J. A. Limits on the possible intrinsic magnetic field of Venus. *J. Geophys. Res.* **85**: 8319–8322 (1980).
   Earth: Zharkov, V. N., and Trubitsyn, V. P. *Physics of Planetary Interiors.* Tucson: Pachart, 1978. See also Rikitake, T. *Electromagnetism and the Earth's Interior.* Amsterdam: Elsevier, 1966.
   Moon: Dyal, P., Parkin, C. W., and Daily, W. D. Magnetism and the interior of the moon. *Rev. Geophys. and Spa. Phys.* **12**: 23–70 (1974).
   Mars: Slavin, J. A., and Holzer, R. E. The solar wind interaction with Mars revisited. *J. Geophys. Res.* **87**: 10285–10296 (1982).
   Jupiter: Smith, E. J., Davis, L., Jr., and Jones, D. E. Jupiter's magnetic field and magnetosphere. In *Jupiter* (T. Gehrels, ed.), Tucson: University of Arizona Press, 1976.
   Saturn: Smith, E. J., Davis, L., Jr., Jones, D. E., Coleman, P. J., Jr., Colburn, D. S., Dyal, P., and Sonett, C. P. Saturn's magnetosphere and its interaction with the solar wind. *J. Geophys. Res.* **85**: 5655–5674 (1980).

as the sign of the dot product of $\mathbf{M}$ and the planet's angular momentum vector). The angle by which the equivalent dipole is tilted to the rotation axis (the z-axis) is also given.

The earth's magnetic field was the first planetary field studied, and it is still the only one for which a significant quantity of information about secular variation is available. The formalism for the quantitative description of the earth's field is also used for other planetary fields; it is mathematically identical to the formalism already introduced for planetary gravitational fields in Chapter 4. If the magnetic field $\mathbf{B}$ is measured in a region containing no currents, as is essentially true for the surface and lower atmosphere of a planet, then it is given by the gradient of a scalar potential:

$$\mathbf{B} = -\nabla V_m, \tag{6-1}$$

where the magnetic scalar potential V_m satisfies Laplace's equation (Eq. 4-4). The expansion for V_m in spherical harmonics is therefore identical to Eq. 4-5. This expansion is valid in a spherical shell which excludes all currents. As in Chapter 4, we set the coefficients $\alpha_{lm} = 0$; these terms represent the the contribution to the magnetic potential from currents outside the spherical shell, and are not included in a description of the planet's intrinsic magnetic field, although they are needed for a description of its magnetic response to an external perturbation. The magnetic potential is then written in the form

$$V_m = a \sum_{l=1}^{\infty} \sum_{m=0}^{l} (a/r)^{l+1}[g_l^m \cos (m\phi) + h_l^m \sin (m\phi)]P_l^m(\cos \theta),$$

$$(6\text{-}2)$$

where the g_l^m and h_l^m, the magnetic analogs of the gravitational zonal and tesseral harmonics, are known as *Gauss coefficients*. As before, a is a normalizing radius which is usually taken to be the average planetary radius. There are the three lowest-order terms with $l = 1$; these three terms simply give the magnitude and orientation of an equivalent magnetic dipole. The higher-order terms correspond to higher magnetic multipoles. The farther away we are from the central region containing currents, i.e. the farther we are from the region of field generation, the more the dipole terms tend to dominate. However, there is no general prediction by any theory that the field will be purely dipolar.

Representation of a planetary magnetic field with a large number of Gauss coefficients can be carried out to great precision, but it is then difficult to physically grasp the field geometry. A more approximate, intuitive approach truncates the expansion of Eq. 6-2 at the $l = 1$ (dipole) terms, but allows the origin of the spherical coordinate system to differ from the center of the planet. This introduces three more free parameters—the vector components of the displacement. The resulting model is known as an *eccentric dipole*. The magnetic field is described by seven parameters: the total magnetic moment of the planet, M, the orientation of the equivalent dipole, and its offset. The field at a vector displacement $\mathbf{r}'$ from the dipole (which is *not* at the center of the planet, recall) is then given by

$$\mathbf{B}(\mathbf{r}') = [3\mathbf{n}(\mathbf{n} \cdot \mathbf{M}) - \mathbf{M}]/r'^3, \qquad (6\text{-}3)$$

where **n** is a unit vector in the **r**′ direction. Table 6-1 gives some of the parameters of eccentric dipole representations of planetary magnetic fields.

What property or properties do the magnetic planets have in common, and how do they differ from the nonmagnetic planets? A comprehensive theory of planetary magnetism is needed to answer this question, but so far such a synthesis has been lacking. Although theories do exist and are able to give a list of ingredients necessary to produce a magnetic planet, so far they have not had great predictive power. It seems so far that the existence of magnetism may depend on a great many parameters, most of which are observationally inaccessible.

There are two principal methods by which a planetary body may retain or generate an intrinsic magnetic field. The field may be a primordial one which has been retained up to the present time, or it may be continually regenerated by a dynamo process. There is no doubt that a dynamo of some sort is responsible for the earth's field. The principal evidence for this is the existence of reversals in the paleoremanent magnetization of minerals which were extruded and cooled in midoceanic rift zones, which indicates that the earth's field has reversed itself numerous times over the history of the planet, with a time between reversals which is highly variable, but which has typical values on the order of 10^6 years. Thus the earth's field has not been preserved intact from the time of the earth's formation. An analogous phenomenon is observed in connection with the 22-year period for the reversal of the solar magnetic field. We are therefore dealing with a dynamical process of magnetic field generation in the case of both bodies.

It is so far unclear whether there is any body in the solar system with a global, primordial ("frozen-in") field. There are two mechanisms by which a body could in principle retain an early field. If the planet contains a large amount of a ferromagnetic or ferrimagnetic substance which starts off at a high temperature and then cools below the Curie temperature (more precisely, below a range of blocking temperatures which are on the order of the Curie temperature) in the presence of a strong initial field, the primordial field will be retained by the ordering of the magnetic domains. Unless the material is reheated above the blocking temperatures, the planet will maintain a permanent magnetic moment. It seems unlikely that this mechanism could be important in any solar system object. First, only the terrestrial planets contain significant amounts of iron relative to their total mass. Second, the calculated temperatures within these planets, for plausible thermal evolution models (see below), are always well above the blocking temperatures, which are on the order of 10^3 °K. It is in principle possible that blocking temperatures might increase rapidly with pressure, but these temperatures are basically defined by interactions between adjacent magnetic elements, which in turn depend upon the lattice spacing. The lattice spacing, however, changes very little over the pressure range relevant to the terrestrial planets.

The other mechanism for retaining a primordial field relies upon the long diffusion times for magnetic fields to escape from a conductive medium. Consider a conducting fluid with electrical conductivity σ_c. In gaussian units, σ_c has dimensions of seconds^{-1}, so that one can form a quantity having the dimensions of a diffusion coefficient,

$$\eta = c^2/4\pi\sigma_c, \qquad (6\text{-}4)$$

where c is the speed of light. In fact, η *is* a diffusion coefficient, and represents the diffusion of the lines of force of a magnetic field through a conducting medium. This process is described by the equation

$$\partial\mathbf{B}/\partial t = \nabla \times (\mathbf{v} \times \mathbf{B}) + \eta\nabla^2\mathbf{B}, \qquad (6\text{-}5)$$

which describes the time rate of change of the magnetic field $\mathbf{B}$ as viewed by an observer at rest. If $\eta = 0$, corresponding to $\sigma_c = \infty$, then the magnetic lines of force are said to be "frozen" to the conducting medium, and if the fluid is at rest ($\mathbf{v} = 0$) the field remains constant. On the other hand, if $\mathbf{v} = 0$ but η is finite, the field decays away with a characteristic time constant which is given by

$$\tau \sim d^2/\eta, \qquad (6\text{-}6)$$

where d is a characteristic dimension of the system. For the metallic iron cores of the terrestrial planets, a typical value of the electrical conductivity is $\sigma_c \sim 10^{15}$ s^{-1} in Gaussian electrostatic units, or about 10^3 ohm^{-1}cm^{-1} (the conductivity in the mantle is orders of magnitude lower). Taking the core dimensions to be on the order of 1000 km, the characteristic time for diffusion of the magnetic field out of the core is $\tau \sim 10^4$ years. Even if the core dimensions are taken to be much larger and the conductivity even higher, primordial fields will in general disappear over the lifetime of the solar system.

What process can regenerate magnetic fields in planetary interiors? According to Eq. 6-5, the irreversible decay of a magnetic field by ohmic dissipation can be counteracted by the first term on the right, which depends on the distribution of fluid velocities in the conducting medium. For a suitable arrangement of fluid velocities, it may be possible to cause an initially small field to grow. The importance of the first term on the right of Eq. 6-5, relative to the second, is measured by the so-called *magnetic Reynolds number*

$$R_m = vd/\eta \qquad (6\text{-}7)$$

If $R_m \gg 1$, the magnetic field lines behave as if they were "frozen" to the medium, while if $R_m \ll 1$, the field lines diffuse rapidly through the medium. So far, investigations of the possibility of dynamo action have been largely limited to the study of the regenerative properties of specified fluid velocity distributions (so-called kinematic dynamo theories). These studies have shown that a suitable combination of differential rotation and convection can amplify magnetic fields. Differential rotation causes the magnetic field lines to "wind up," while turbulent convection with a sufficient degree of helicity can rotate the "wound-up" field lines in such a way that they align with an initial large-scale dipolar field and therefore amplify it as they diffuse outward. The requirement for this process to occur is that the *dynamo number, N,* be of order unity. The dynamo number is expressed as the product of two magnetic Reynolds numbers,

$$N = R_{m,d.r.} \cdot R_{m.conv.}, \qquad (6\text{-}8)$$

where $R_{m,d.r.}$ and $R_{m,conv.}$ are, respectively, the magnetic Reynolds numbers which are obtained using Eq. 6-7 with v taken to be equal to the order of magnitude of a typical velocity of differential rotation and a typical convection velocity respectively.[2]

As is evident, the theory of planetary dynamos is not in altogether satisfactory shape. In particular, the theory cannot specify how the appropriate dimensionless numbers are to be calculated from directly observable parameters. However, some useful qualitative results emerge from the theory. First, if a planet is to manufacture its own magnetic field, it must have a conducting region in its interior. Next, this region must be in a state of differential rotation and must also be convecting, which necessarily requires it to behave as a liquid. Since differential rotation is required, it seems that rotation of the planet as a whole is important for the existence of a dynamo, and it was commonly believed that very slowly rotating planets such as Mercury and Venus would not be found to possess intrinsic magnetic fields. Summarizing the requirements for a planet to possess a dynamo, we might say that (a) the planet must have a metallic inner region; (b) this inner region must be liquid and convecting; (c) the region must rotate differentially. A corollary of these requirements is that the planet must have an interior energy source to sustain the convection and differential rotation. One would thus not expect to observe dynamo action in small, primordial, undifferentiated bodies without iron cores or substantial interior energy sources.

Assuming that all of the necessary ingredients for a self-sustaining dynamo are present, it is possible to derive a scaling relation for the expected order of magnitude of the planetary magnetic moment. We assume that the Coriolis force per unit volume acting on a conductive convecting element of the magnetohydrodynamic dynamo is in approximate balance with the magnetic force per unit volume. Equating the two forces implies that

$$2\omega \times \mathbf{v} \sim (\nabla \times \mathbf{B}) \times \mathbf{B}/4\pi\rho, \tag{6-9}$$

where ρ is the mass density and ω is the planetary rotation rate. To order of magnitude, we replace $\nabla \times \mathbf{B}$ with B/a_c, where a_c is the core radius. Thus the field in the dynamo region scales as

$$B^2 \sim \rho a_c \omega v. \tag{6-10}$$

The velocity of differential rotation v is then assumed to scale proportionally to the rotational velocity ωa_c, and the planetary magnetic moment M scales as Ba_c^3, which leads to the scaling law[3]

$$M \propto \rho^{1/2}\omega a_c^4. \tag{6-11}$$

Figure 6-1 shows results of the application of this scaling law to the planets for which magnetic field measurements are available.[4]

The qualitative conclusions of dynamo theory should permit us to infer certain properties about the interior of a planet simply from an observation of the presence or absence of an intrinsic magnetic field. However, attempts to use dynamo theory to synthesize observations of planetary magnetism have not been entirely successful. Mercury possesses a weak but distinct global dipolar magnetic field which is strong enough to stand off the solar magnetic field and create a magnetosphere, whose geometry is shown in Fig. 6-2.[5] The value of the Mercurian dipole moment is in reasonable agreement with the scaling relation (Eq. 6-11). On the other hand, Venus, which like Mercury has a very slow rotation rate, has no detectable global magnetic field, a fact on which both Soviet and American space scientists agree. A magnetic field of equal intensity to Mercury's

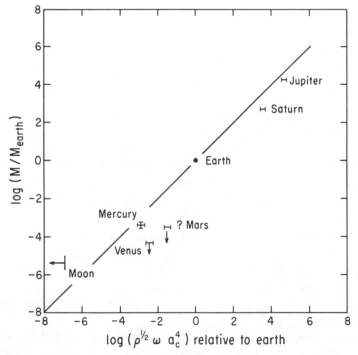

Fig. 6-1. Predicted value of magnetic moment M from scaling law 6-11. [From Russell, *et al.* (1980).]

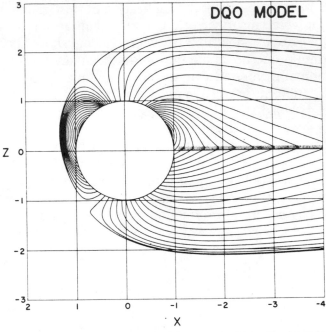

Fig. 6-2. Dipole-quadrupole-octupole model for Mercury's intrinsic magnetic field. The figure shows field lines and the magnetopause in the noon-midnight meridian plane. Unit of distance is one Mercury radius. [From Whang (1977); copyright American Geophysical Union.]

would have been readily detectable. The upper limit for a Martian magnetic moment also disagrees with the scaling law (Eq. 6-11).

It would not be fair to state that the existence of a magnetic field on Mercury and the nonexistence of a field on Venus represents a refutation of magnetic dynamo theory. As we have already stressed, the connection between the parameters needed to calculate the dynamo number and the parameters which are actually accessible to observation is extremely indirect. About all that we can conclude from this discussion is that there are probably important differences in the thermal and chemical properties of the deep interiors of the terrestrial planets, which are not incorporated into current models to the degree necessary for an adequate understanding of the possibilities for magnetic field generation.

The predicted geometry and time dependence of the field depends on the value of the dynamo number. Oscillatory fields, such as the earth's, can be produced for a number of values of the dynamo number. The oscillation frequency is on the order $10/\tau$, according to models.

So far, observations of magnetic fields on other planets have not been carried out over a sufficient time baseline to permit a detection of possible time dependence in the magnetic field geometry. Observations show that the absolute value of the earth's dipole moment is currently decreasing, thus confirming the nonstationary character of the earth's field.

Comprehensive observations of other planetary magnetic fields have so far only been made at Jupiter and Saturn. The Jovian field was observed well before the space age; it was initially detected because of its role in the trapping of energetic particles in the vicinity of Jupiter. Observations of synchrotron radio emission from trapped relativistic particles permitted a determination of the field's geometry, approximate surface value, and rotation rate. While these quantities have since been measured much more accurately *in situ* from spacecraft, the existence of a baseline for Jupiter radio observations now approaching thirty years in length will eventually permit conclusions to be drawn about the possible variability of Jupiter's field. So far, no clear variations in the field geometry have been detected above the scatter in the observations.

From a theoretical point of view, if the Jovian field varies, the characteristic period should be on the order of, or possibly as much as a factor ten smaller than, the time for the field to diffuse through the planet's metallic inner regions. Field transport may be governed by turbulent diffusion in Jupiter's interior, because of convective energy transport (see Chapter 8), and the estimated oscillation period for the field is on the order of 10^4 years. With the present baseline, a measurement accuracy of the field parameters approaching 10^{-3} would be needed to detect such a change.

As Table 6-1 shows, Jupiter's equivalent dipole is tilted by about ten degrees to the planet's rotation axis. Other properties of Jupiter's field are also analogous to those of the earth's field, although the field itself is much more intense, reaching a maximum value of about 14 gauss at the surface, vs. about 0.5 gauss at the earth's surface. The Jovian magnetic quadrupole and octupole moments, which are determined from the Gauss coefficients with $l = 2$ and $l = 3$ respectively, are about 20% and 15% of the dipole moment. This may be contrasted with the corresponding ratios for the earth, about 13% and

9%. Some researchers have used this comparison to infer that the active regions of the internal dynamo are closer to the planet's surface in Jupiter than they are in the earth, although this inference assumes that the Jovian and terrestrial dynamos are otherwise similar.

The tilt of Jupiter's magnetic moment with respect to the rotation axis has made it possible to use earth-based radio observations to measure the planet's rotation period with great accuracy. Relativistic electrons which are trapped in Jupiter's strong magnetic field emit strongly polarized radio signals at wavelengths on the order of a few tens of centimeters. The polarization of the signals is determined by the orientation of the field lines in the emitting region. Because of the tilt of the field geometry, the polarization "rocks" back and forth as the planet rotates, permitting an accurate determination of the period of the field rotation. The period which is thus measured, $9^h 55^m 29.7^s$ (see Table 1-1), is presumably the period at which the Jovian dynamo regions rotate on average. It is similar to the period at which cloud features rotate in the polar regions, but is about five minutes longer than the rotation period of equatorial cloud features.[6]

Saturn's magnetic field is distinctly different from any other field seen in the solar system. The magnetic moment is nearly aligned with the rotation axis. The maximum surface field is about one-third of a gauss. Although some spacecraft observations indicate that Saturn's equivalent dipole is displaced by about 2000 km northward along the rotation axis, this result has not been confirmed by other measurements. All observations agree that the dipole is tilted by about 1° to the rotation axis. The Saturn field is much more closely dipolar than Jupiter's field: observations indicate that the quadrupole and octupole terms in the field expansion are quite small. Again, some investigators have interpreted this circumstance as an indication that the dynamo regions in Saturn are confined to a relatively small region near the center of the planet. Theoretical models to explain the small tilt angle of the magnetic dipole are discussed in Chapter 8.

If we are observing internal magnetohydrodynamic dynamos in the four planets with observable magnetic fields, then it must be remembered that the fields which are produced by these dynamos are nonstationary in time. The field geometries which we currently observe may or may not be highly typical. Thus it may be dangerous to speculate in great detail on the relationship between specific field configurations and interior structure. It is probably significant, however, that all four planets have largely dipolar fields, and that the dipole moments tend to be aligned or antialigned with the rotation axis.

REFERENCES

1. Brown, L. W. Possible radio emission from Uranus at 0.5 MHz. *Astrophys. J.* **207:** L209–L212 (1976).
2. Levy, E. H. Generation of planetary magnetic fields. *Ann. Rev. Earth and Plan. Sci.* **4:** 159–185 (1976).
3. Busse, F. H. Generation of planetary magnetism by convection. *Phys. Earth and Plan. Interiors* **12:** 350–358 (1976).
4. Russell, C. T., Elphic, R. C., and Slavin, J. A. Limits on the possible intrinsic magnetic field of Venus. *J. Geophys. Res.* **85:** 8319–8332 (1980).
5. Whang, Y. C. Magnetospheric magnetic field of Mercury. *J. Geophys. Res.* **82:** 1024–1030 (1977).
6. Donivan, F. F. and Carr, T. D. Jupiter's decametric rotation period. *Astrophys. J.* **157:** L65–L68 (1969).

7
The Earth As a Paradigm

Since this book is about the interiors of other planets, we will not discuss the interior of the earth in detail. The principal purpose of this chapter is to convey an understanding of the role of the earth as a standard in comparative planetology, and to illustrate the major techniques which tell us about the earth's interior. Some of these techniques are now also used for the study of other planets, while others have not yet been applied anywhere else. An understanding of the way that these techniques are applied to the earth helps to clarify the technical problems which have so far limited their use in the study of other planets.

Finally, although the slogan is hackneyed by now, it is still true that the study of other planets helps to tell us about the earth. In the next chapter, we will look at some objects whose basic parameters differ from those of the earth. But first, we review the determination of these parameters for the earth.

SEISMOLOGY

Seismology is the source of the most detailed information about planetary interiors. A disturbance within or on a planet, either naturally-occurring or artificially induced, causes elastic waves to be radiated throughout the volume of the planet. These waves can be detected at surface stations, and with enough observations of the propagation times to various stations around the earth, the propagation paths within the earth can be reconstructed. This information can then be further analyzed to yield a profile of pressure, density, and elastic moduli within the planet. Unfortunately, with the exception of some fragmentary data for the moon, we have no seismic information about any planet other than the earth. However, the data for the earth, which are very detailed by this point, can be used as a starting point for generalizations about the other planets. Ultimately,

when data for other planets become available, their interpretation
will rest heavily on comparisons with the earth.

Modern seismological models of the earth need not assume complete spherical symmetry, but for our purposes we will consider a
spherical planet, so that all variables which describe its interior state,
such as pressure, density, temperature, elastic moduli, and chemical
composition, are functions only of the radial distance from the
earth's center, r. The task of seismology is to find these functions by
processing the primary experimental information which is available
to us, the travel-time curve. Suppose for simplicity that a disturbance
takes place at point E (for *epicenter*) on the surface of a planet (Fig.
7-1). For simplicity, assume for the moment that there are no discontinuities within this planet. Provided that the wavelength of the
elastic waves which are radiated by this disturbance are short compared with the scales over which properties of the earth's interior
change, the propagation of these waves can be described by the
ordinary equations of geometrical optics. A solid medium can sustain
two types of volume waves: longitudinal pressure (P) waves, which
are ordinary sound waves in which the direction of displacement of
the medium is parallel to the direction of propagation, and transverse
shear (S) waves, in which the medium is displaced in a direction

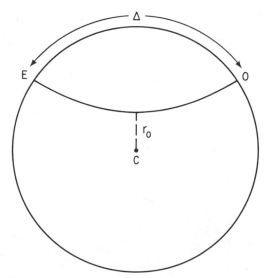

Fig. 7-1. A seismic ray path in a hypothetical planet with no discontinuities.

orthogonal to the direction of propagation. S waves can be resolved into two mutually perpendicular polarizations; in an isotropic medium these polarizations travel with the same velocity.

Now if we set up an observing station at point O on the planet's surface, seismic waves from E will arrive at this point after a propagation time interval T. A particular type of wave (either P or S) will travel from E to O along some curved path within the planet's interior, with the path in general being different for P and S waves. From general considerations of spherical symmetry, it is obvious that a given ray path must lie in the plane containing the points E, C, and O, where C is the center of the planet. Let Δ be the angle ECO, i.e. the angular separation of the source and the observation station. If enough observing stations and/or sources are available, the angle can be varied through a large range, and one can obtain a curve $T(\Delta)$. Separate curves of this type can be obtained for P and S waves (Fig. 7-2). Once the $T(\Delta)$ relation is known, three observations at separate stations of T from a given surface event will suffice to determine the location of the epicenter. But when this relation is initially unknown, and it is also unknown whether or not seismic disturbances are confined to the vicinity of a planet's surface, it is clearly a difficult task to map out the $T(\Delta)$ relation unless artificial disturbances can be provided at known locations. This is a major problem for extraterrestrial seismology.

Suppose the ray path joining E and O has a closest-approach distance to the center of the earth equal to r_0. Now $T(\Delta)$ is an integral

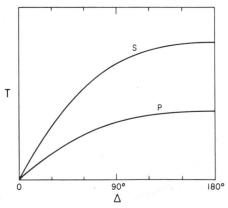

Fig. 7-2. Travel times for S- and P-waves versus angular separation in a hypothetical planet with no discontinuities.

measure of the range of propagation velocities along the ray path, and it samples all of these velocities from the surface ($r = a$) down to $r = r_0$. As Δ increases to $180°$, we sample all radial distances within the planet down to the center ($r_0 = 0$). It is thus not surprising that if we have a complete curve for $T(\Delta)$ with Δ ranging from 0 to $180°$, this curve can be exactly inverted to yield a corresponding curve for the elastic wave propagation velocities $v(r)$. The mathematical procedure for carrying out such an inversion relies on Snell's law from geometrical optics, adapted to a spherical geometry, and on the properties of the Abel integral transform pair,

$$F(w) = \int_0^w f(y) \, dy/(w - y)^{1/2} \tag{7-1}$$

$$f(y) = \pi^{-1} \int_0^y F'(w) \, dw/(y - w)^{1/2}, \tag{7-2}$$

where f and F are any sufficiently well-behaved functions. With appropriate manipulations, the relation for $\Delta(T)$ can be cast in the form of Eq. 7-2, where the integrand depends on the seismic wave velocity distribution $v(r)$. The latter function is then found by applying the inverse of Eq. 7-2, Eq. 7-1. A more detailed discussion of the procedure can be found in any standard geophysics text.[1,2]

Suppose that we now have available the P and S wave velocity profiles, $v_P(r)$ and $v_S(r)$. These can be related to thermodynamic properties of the medium through the formulas

$$v_P^2 = \left(K + \frac{4}{3}\mu\right)/\rho \tag{7-3}$$

and

$$v_S^2 = \mu/\rho, \tag{7-4}$$

where μ is the shear modulus of the medium, ρ is its mass density, and

$$K = \rho(\partial P/\partial\rho)_S \tag{7-5}$$

is the adiabatic modulus of incompressibility (P is the pressure and S is the entropy). By taking a suitable linear combination of $v_P(r)$ and

$v_S(r)$, we easily obtain the quantities μ/ρ and $(\partial P/\partial\rho)_S$, as a function of r. Liquid zones within the planet are readily detectable because μ vanishes within these regions: only P waves can propagate in liquids and gases. Now, as we have noted in Chapter 3, the pressure in a planetary interior can normally be written in the form $P = P_0 + P_T$, where P_0 is the pressure at zero temperature, P_T is a correction term for finite temperature which is approximately linear in the temperature T, and $P_T \ll P_0$. Thus, if this inequality is satisfied and if the planetary interior is *chemically homogeneous* as well, we can approximately write

$$\Phi \equiv (\partial P/\partial\rho)_S \simeq dP/d\rho, \tag{7-6}$$

where $dP/d\rho$ is the actual variation of pressure with density within the planet. We now assume that the planet is in hydrostatic equilibrium. It is convenient to write the equation of hydrostatic equilibrium (see Eq. 4-42) in the form

$$dP/dr = -g\rho, \tag{7-7}$$

where

$$g = (G/r^2) \int_0^r 4\pi x^2 \rho \, dx \tag{7-8}$$

is the gravitational acceleration as a function of r. Dividing Eq. 7-7 by ρ, differentiating the result with respect to r, and using Eq. 7-6, we find

$$r^{-2} \frac{d}{dr} \left[r^2 \rho^{-1} \Phi \frac{d\rho}{dr} \right] = -4\pi G\rho. \tag{7-9}$$

Since $\Phi(r)$ is a known function, this second-order differential equation can be integrated to obtain the density distribution $\rho(r)$, if we can supply two boundary conditions. In the case of our hypothetical planet with no discontinuities, this task is quite simple. First, we require that $\rho(r)$ be well-behaved near the center of the planet, which means that as $r \to 0$ we must be able to write

$$\rho(r) = \rho_0 - cr^2, \tag{7-10}$$

where ρ_0 is the central density and c is a constant. Substituting Eq. 7-10 in Eq. 7-9, and taking the limit as $r \rightarrow 0$, we find

$$c = 2\pi G \rho_0^2/(3\Phi_0), \qquad (7\text{-}11)$$

where Φ_0 is the central value of Φ. Now pick a provisional value of ρ_0. Starting the solution at the center with the help of Eqs. 7-10 and 7-11, we integrate Eq. 7-9 outward until the surface is reached at $r = a$. The total mass of the configuration must equal the observed planetary mass M; if it does not, we continue adjusting ρ_0 until agreement is obtained. The surface density $\rho(a)$ emerges as a byproduct of the solution.

Knowing $\rho(r)$, one can then integrate the equation $dP = \Phi \, d\rho$ from the surface ($P = 0$) to the center to obtain a $P(\rho)$ relation for the planet's interior.

The situation is not so simple in the real earth. The earth's interior differs in three important ways from the ideal model which we have been considering. (1) There are several distinct discontinuities within the earth, generally involving a change both in density and in chemical composition. (2) The "surface density," $\rho(a)$, is not really directly known for the earth, because the accessible part of the earth, the crust, is a product of partial melting and differentiation in the upper mantle, and its density differs substantially from the density of the upper mantle, which lies a variable distance (about 30 km, on average) below it. (3) Substantial compositional gradients and deviations from adiabaticity probably exist in the upper mantle and elsewhere, which would vitiate the use of Eq. 7-6. Since these three problems will be encountered in any planet which is large enough to have undergone substantial differentiation (which includes essentially all of the planets and most of the major satellites), it is worthwhile to briefly sketch how they are dealt with in the case of the earth.

The most significant discontinuity in the earth occurs at a mean radius of about 3480 km, and is manifested by its strong influence on the propagation of seismic waves. Such smooth $T(\Delta)$ curves as those shown in Fig. 7-2 do not extend all the way to $\Delta = 180°$. Instead, the curves terminate in the earth at about $\Delta = 105°$. Figure 7-3 illustrates the role of a dense core in forming a "shadow zone." For a core with a substantially lower seismic wave velocity than that of the overlying mantle, the situation is entirely analogous to the optical problem of light propagating at an angle through an interface between a layer

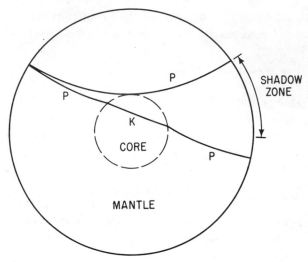

Fig. 7-3. Formation of a seismic shadow zone in a planet with an iron core.

with a low index of refraction (analogous to the mantle) and a layer with a high index of refraction (analogous to the core): the light is bent farther into the medium with the higher index. At the mantle-core interface, the P-wave velocity drops from about 13.7 km/s to about 8 km/s, and S-waves do not penetrate the interface. Ray paths for P-waves which are refracted once into the core are denoted PKP; such paths have a different $T(\Delta)$ relation than pure P ray paths (which terminate at the edge of the shadow zone).

Yet another discontinuity exists within the earth's core. Figure 3-17, which is based upon a very recent earth model (the PREM, or Preliminary Reference Earth Model[1]), shows an abrupt density jump from about 12.2 to about 12.8 g/cm^3 as one proceeds inward into the core. This jump occurs at a radius of about 1220 km. At the same time, analysis of the propagation of seismic waves through the core shows that a fraction of the P-wave energy which propagates thrugh the outer core is transformed to S waves within the inner core. Thus, the inner core is solid. Shock measurements on iron at 2 to 2.5 megabars (see Chapter 3) indicate that pure iron may melt at temperatures on the order of 5000° at these pressures. The discontinuity in the core may correspond to such a transition. However, the lighter component, which appears to be alloyed with the core iron, could cause a significant depression of the melting point below the value for pure iron. If the discontinuity corresponds to melting in such a two-component system, the chemical composition must also change across the discontinuity, according to the Gibbs phase rule. It is not known what the lighter component in the core may be; popular choices (on cosmochemical grounds—see Tables 2-1 and 2-2) include sulfur, oxygen, and silicon.

For a two-layer planetary model, such as a simplified earth model with separately continuous core and mantle, Eq. 7-9 must be solved separately in the core and mantle. We start at the center with a provisional ρ_0 and integrate outwards to the core-mantle boundary. The solution must then be restarted on the mantle side of this boundary, but this time both the starting density and its initial gradient must be guessed. Thus the solution depends on these two parameters plus ρ_0. The mass constrains one of these parameters, but two more constraints are needed. If $\rho(a)$ is known, it provides one more constraint. The final constraint can be obtained from the moment of inertia factor, C/Ma^2 (see Chapter 4 for a discussion of how this can be determined), if this is available, and thus $\rho(r)$ and then $P(r)$ can be determined in both the core and mantle.

In the real earth, and presumably in any real planet, we do not actually know the appropriate value to use for the surface density $\rho(a)$. The outermost tens of kilometers of the earth form the crust, a layer with distinct chemistry and density from the upper mantle. The density of the crust is variable but is typically about 2.5 g/cm³. The appropriate value for $\rho(a)$, however, would be the density of the upper mantle just below the crust. This density can be determined by indirect means, such as analysis of magmas, and is found to be about 3.3 g/cm³.

Since the upper mantle has been differentially processed by formation of the crust, there is of course no reason to assume that its composition is constant. Therefore approximation 7-6 is almost certainly invalid in the upper mantle. It is also suspected that the predicted phase transitions in mantle rocks (see Chapter 3) will have associated variations in chemical composition. The Fe/Mg ratio may vary through the mantle, possibly increasing in the lower mantle. Even if approximation 7-6 is piecewise valid in the mantle, the introduction of more interfaces obviously would require even more integral constraints on the interior model. In a subsequent section of this chapter, we discuss integral constraints which are provided by free oscillation data.

Average P- and S-wave velocities actually *increase* with radius between $r = 6151$ km and $r = 6291$ km, whereas elsewhere they always decrease monotonically away from the center of the earth. This region in the upper mantle is called the *low velocity zone,* or LVZ. The existence of a minimum in seismic velocities in this region is attributed to a change in the mechanical properties of the rock due to proximity to melting temperatures. Rock is more plastic in this region than it is in the layers extending from $r = 6291$ km to $r = 6371$ km (surface). The region above the LVZ, which includes the crust, and which is substantially less deformable, is called the *lithosphere.* In the lithosphere, the maximum deviatoric stresses supported by rocks are comparable to the hydrostatic (or lithostatic) pressures, and the material cannot be closely approximated by a state of hydrostatic equilibrium. As we will discuss presently, the existence of the lithosphere and its underlying LVZ appears to be a crucial element in a uniquely terrestrial phenomenon, plate tectonics.

We have briefly reviewed in this section a major technique for studying planetary interiors, but one which has so far only been applied in a comprehensive way to the earth. What prospects are there for applying it to other planets? First of all, a sparse network of seismic stations is of limited utility. We need to be able to construct complete $T(\Delta)$ curves for all of the principal seismic wave paths, which means that we need a widespread array of observing stations and/or sources at known positions. On the earth, nature (via plate tectonics) supplies the latter, while man supplies the former. As we shall see, certain planets can be expected to be seismically active, but others may be quite inactive. Moreover, even when detailed $T(\Delta)$ curves have been obtained, we have seen that it is not always possible to derive a unique interior model from them. A substantial amount of ancillary information concerning the planetary gravity field, surface chemistry, and other indirect constraints on interior structure, is required for full exploitation of the $T(\Delta)$ curves. We are still in an early phase of planetary exploration; seismic exploration of planetary interiors will eventually be fruitful when adequate initial studies have been carried out.

THE FIGURE OF THE EARTH

In Chapter 4 we discussed a number of characteristics of a planet's external gravity field. We will now discuss them further in the light of the seismic data on the interior of the earth. Although seismic data on other planets are lacking, data on their gravity fields, combined with what we know about the earth, may provide some indirect clues to their interior elastic properties.

Recall that contours on the geoid, which reflect the distribution of mass within the earth, are mostly uncorrelated with topography. This lack of correlation is due to the fact that the earth's surface layers are largely isostatically compensated: In effect, low-density continental blocks and mountain ranges "float" in the denser mantle. This buoyancy could be understood if we were to assume that the mantle is a fluid, and that crustal units float in it like ice cubes in water. The mantle transmits seismic S waves and therefore is a solid, at least for high-frequency disturbances such as seismic waves. However, it is possible for a substance to act as a solid over short time intervals and to act as a highly viscous fluid over long time intervals. The process can typically occur in a solid where successive crystal planes repetitively shift their positions with respect to each other by one unit cell length, gradually acting to relieve an applied stress. Although this process can in principle occur in a solid at any temperature, it is most rapid at high temperatures, where the lattice particles have a higher probability of probing nearby lattice states with lower strain energies.

As we have seen, the upper mantle of the earth exhibits the LVZ, where seismic velocities display an unusual minimum. The current interpretation of this zone is that temperatures within it are extremely close to temperatures at which a major liquid component would first appear (the so-called *solidus*). In the vicinity of the LVZ, the mantle, although it is still largely solid (at least to seismic waves), is able to slowly yield to imposed stresses. This plastic zone of the earth is known as the *asthenosphere*—the earth's "weak" layer. The existence of the asthenosphere is therefore, in all probability, closely related to the unusually smooth character of the earth's external gravity field.

When the earth's gravitational harmonic coefficient J_2 is combined with the value of the precession constant, the dimensionless polar moment of inertia is found to be $C/Ma^2 = 0.3308$, compared with 0.4000 for a Maclaurin spheroid. Even in a homogeneous-composition earth, we would expect C/Ma^2 to be smaller than 0.4 because of self-compression effects. However, the expected central pressure of such a homogeneous earth would only compress a rock-iron mixture by a few tens of percent, and thus C/Ma^2 would still be much larger than its observed value. We conclude that a comparison of the earth's C/Ma^2 with its interior structure deduced from seismic data shows that a dense core is clearly revealed by the external gravity field. The alert reader may have detected some circularity here: have we not used C/Ma^2 as a constraint to derive the seismic profile? In fact this is no longer necessary. The detection of free oscillation modes in the earth has provided an abundance of integral constraints for seismic models.

FREE OSCILLATIONS

Consider a seismic wave with wavelength λ. For $\lambda \ll a$, the wave propagation can be described by geometrical optics. But now suppose that λ is allowed to become comparable to the dimensions of the earth. The wave nature of the disturbance now becomes an essential feature. Large regions within the earth oscillate in phase. In general, P and S waves become intermingled, except for purely radial oscillations of the earth. Such large-scale displacements can be conveniently described as a superposition of excitations of the normal modes of a giant, spherically symmetric, elastic-compressive oscillator. Each of the normal modes has an associated eigenfrequency. If the system is initially caused to vibrate in one of these modes, and is then neither driven nor damped, it continues to vibrate forever with the corresponding eigenfrequency—a so-called free oscillation.

In a spherical geometry, the eigenvalue equation for the normal modes of free oscillation is separable into radial and angular functions. The angular functions are just the spherical harmonics $Y_{lm}(\theta,\phi)$ introduced in Chapter 4. There is a separate series of normal modes associated with each different value of l and m. Each member of such a series can be denoted by a radial quantum number n. Thus a given normal mode of free oscillation is specified by three quantum numbers: n, l, and m. In the exactly spherical case, the oscillation periods are degenerate with respect to m. That is, for equal values of n and l and different m, the spatial behavior of normal modes is different but the periods are the same. Free oscillations can be further divided into oscillations which involve no change in volume (purely torsional oscillations) and those which involve both shear and dilatation (spheroidal oscillations). There is a separate n, l, m series for each class.

The spectrum of eigenfrequencies for free oscillations is quite analogous to an atomic energy spectrum. Frequencies are discrete and well-separated for fundamental modes, and tend to become higher and closer together as the quantum numbers increase. Since the value of each eigenfrequency depends differently on the distribution of elastic moduli and gravity within the earth, an observational value for each eigenfrequency corresponds to a separate integral constraint on an earth model. Such integral constraints, which now number on the order of 10^3, are used to great advantage in refining seismic interior models.[2,3]

Free oscillations have not been observed for any planet other than the earth. The technical reason is quite obvious: only extremely powerful seismic disturbances are capable of exciting planet-scale normal modes, and sensitive seismometers must then be used to record these low-frequency oscillations, which typically have periods of tens of minutes. Planets with very high levels of internal activity are probably the best candidates for application of this technique. The sun, which has an active convection zone in its outer layers, appears to exhibit low-amplitude nonradial global oscillations in its visible atmosphere. Modes with high values of n and with l ranging from 0 to 5 have been identified. Free oscillation modes involving shear stresses are naturally absent in a gaseous, inviscid body, but otherwise solar oscillations are the stellar analog of free oscillations in planets, and are detected by means of the doppler shifts which they produce in spectral lines produced in the solar atmosphere. Models of the solar interior can be tested by means of such observations.[4,5]

Possibilities for remotely observing free oscillations in solid planets appear to be quite remote, as the expected displacements are far too small. But in the Jovian planets, behavior analogous to the solar case may perhaps occur.

Terrestrial free oscillations are of current interest for planetary interiors for another reason. Free oscillations of the earth gradually damp away, and their rate of damping provides us with an opportunity to measure the parameter Q (see Eq. 4-72) in the earth's interior. The Q which is obtained from the damping of free oscillations is not necessarily closely related to the tidal Q. Periods of tidal distortions are tens of hours or greater, and tidal amplitudes are much larger than free-oscillation amplitudes. Strictly speaking, Q's derived from free oscillations are different for distortions involving pure shear and pure compression respectively, but Q's for the latter are very large and may be neglected. We can speculate on the best way to extrapolate the earth's interior Q distribution, as a function of the earth's seismically-derived interior properties, to tidal Q's in other solid planets. In the earth's crust, Q is on the order of 500. It declines to values on the order of 100 in the vicinity of the LVZ, gradually rises to a maximum on the order of 1000 in the deeper mantle, and then declines again to about 100 near the surface of the core. The liquid outer core probably has an extremely large Q.[6]

The key point to note is that although the above values of Q are in the range that we adopted in discussing tidal friction phenomena in Chapter 4, the effective value of Q for the slowing of the earth's rotation due to lunar tides is on the order of 15, a value much smaller than any of those given above. Thus it appears that the elastic properties of rocks in the earth's interior do not actually determine its tidal friction, which is dominated by some other effect. The standard resolution to this paradox is that the turbulent flow of the oceans on the earth's surface, which may have no other planetary analog, is the source of the earth's unusually low tidal Q. But even if we assume that the seismic Q's for the earth's interior are the appropriate ones to use for other solid planets, we still note that these quantities vary by an order of magnitude. It appears that the proximity of a rock to partial melting may cause its effective Q to decrease drastically. The presence of liquid water may also cause rocks to become substantially more dissipative.

ELECTROMAGNETIC SOUNDING

As we have seen in connection with gravitational potentials, the response of a planet to a perturbation can provide information about its interior structure. The response of the earth to external magnetic disturbances offers yet another opportunity to apply this general technique. So far the moon is the only other planetary body whose response to electromagnetic disturbances has been studied in detail. We will first discuss the technique in very general terms, and then consider the prospects for its application to other planets.

Let us consider a very idealized problem. A uniform sphere of radius a with electrical conductivity σ_c is immersed in a constant, uniform external magnetic field $\mathbf{H}_0$. Now suppose that the external field varies sinusoidally with angular frequency ω, so that the imposed magnetic field is now

$$\mathbf{H} = \mathbf{H}_0 \exp(i\omega t), \qquad (7\text{-}12)$$

where t is the time. This varying magnetic field causes time-varying electrical currents within the sphere, which in turn produce an induced time-varying external magnetic field

$$\mathbf{H}' = \mathbf{H}_0' \exp(i\omega t), \qquad (7\text{-}13)$$

where $\mathbf{H}_0'$ is independent of time, such that the total external magnetic field is

$$\mathbf{H}_{\text{ext}} = \mathbf{H} + \mathbf{H}'. \qquad (7\text{-}14)$$

We are interested in the spatial dependence of $\mathbf{H}'$ and its amplitude and phase relation to the imposed field $\mathbf{H}_0$, for these quantities clearly must be related to the basic physical parameters of the sphere, its radius and conductivity. The solution is expressed as follows. Define a "skin depth" $\delta = c/(2\pi\sigma_c\omega)^{1/2}$, and form the dimensionless ratio δ/a. The skin depth is essentially the depth within the sphere to which the time-varying external field penetrates before it is masked by the induced currents. Clearly the character of the solution differs radically between the case for $\delta/a \ll 1$ and $\delta/a \gg 1$. The situation which corresponds most closely to real planets is $\delta/a \ll 1$, where one finds[7]

$$\mathbf{H}' = -\frac{1}{2}\,[3(\mathbf{n}\cdot\mathbf{H}_0)\mathbf{n} - \mathbf{H}_0]\,(a/r)^3 \exp(i\omega t + i\phi), \qquad (7\text{-}15)$$

where $\mathbf{r}$ is the radius vector from the center of the sphere, $\mathbf{n}$ is a unit vector in the direction of $\mathbf{r}$, and ϕ is the phase offset of $\mathbf{H}'$ from $\mathbf{H}$ given by

$$\phi = -\frac{3\delta}{2a}. \qquad (7\text{-}16)$$

The phase shift is negative; that is, the induced external field slightly lags the applied field in phase.

In this example, the fluctuating external field causes the sphere to respond with a dipolar induced field. The induced field is essentially antiparallel to the applied field at the sphere's magnetic poles and parallel at the magnetic equator (see Fig. 7-4). By measuring the magnitude of the induced field and its phase offset in time from the external field, one can in principle determine the parameters of the sphere, δ and a. The same general approach can be applied when the conductivity is some function of r and when ω decreases to the point that δ is comparable to a, although the solution of course differs from Eqs. 7-15 and 7-16.

Many other complications must also be taken into account when this method is applied to the earth. The solar wind, a plasma which flows radially outward from the sun with a velocity on the order of 400 km/s at the earth's orbit, is the principal source of a fluctuating external magnetic field at the earth. Magnetic field lines embedded in the solar wind plasma are carried past the earth but do not interact directly with the solid planet because the earth's intrinsic field creates a region about the earth, the *magnetosphere*, which repels the surrounding interplanetary medium and magnetic fields. Nevertheless, disturbances in the interplanetary plasma are felt at the surface of the earth because they can temporarily compress the magnetosphere, causing the earth's intrinsic field lines to move closer together and increase the field strength in the vicinity of the earth. Fourier analysis of such disturbances reveals many periods, but dominant among them is the earth's diurnal rotation period. Measurement of the earth's magnetic response to such disturbances can then be used to

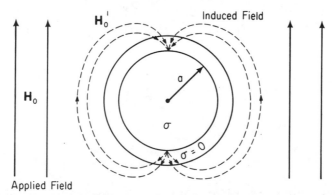

Fig. 7-4. Response of a planet with a conducting core of radius a to an external oscillating magnetic field.

infer a model for the distribution of σ_c in the earth's outer layers. Longer-period disturbances have greater associated electromagnetic skin depths, and thus measurement of the earth's response at a variety of frequencies can in principle be used to infer a multiparameter distribution of σ_c.

Application of this method to the earth gives the following results. In first approximation, the earth responds to an external field as if the earth were a uniform conducting sphere, but one with a radius substantially smaller than the actual mean radius of the earth. A typical model would have $\sigma_c \sim 10^{-4}$ (ohm $\cdot$ cm)$^{-1}$ (essentially a nonconductor) in its surface layers, followed by a nearly stepwise rise to about 10^{-2} (ohm $\cdot$ cm)$^{-1}$ at a depth of $\sim 500-1000$ km below the surface, and then a gradual rise in deeper layers. The earth therefore responds as if it were a conducting sphere with a radius about $500-1000$ km smaller than the earth's actual radius. The derived conductivity distribution in the earth can then in principle be combined with formulas such as Eq. 5-40 to infer a temperature distribution. Thus, the rapid rise of conductivity in the upper mantle suggests that temperatures attain some critical activation temperature for conduction electrons at the point where the conductivity increases rapidly. Unfortunately, such critical temperatures can depend very sensitively on the precise composition and thermodynamic environment of the sample.

Apart from the broad characteristics outlined above, there is rather little in common between models for the distribution of electrical conductivity in the earth's mantle.[8] The principal source of the disagreement may be the differing ways that complicating factors, such as the response of the oceans and the ionosphere, are handled in various models. We are presented with a much simpler situation in many potential applications of electromagnetic sounding in the solar system, where an atmosphereless body with no intrinsic magnetic field is exposed directly to the solar wind.

PLATE TECTONICS

The discovery of plate tectonics represents one of the unifying advances of terrestrial geology. It ties together, within a single conceptual framework, a wide range of geophysical and geological phenomena. However, since this book is about the planets rather than the earth, we will be concerned about plate tectonics only to the extent that it is a useful concept in comparative planetology. It must be initially admitted that so far we see no clear evidence of a mechanism precisely analogous to terrestrial plate tectonics on any other planetary body. The important questions which need to be answered are therefore as follows: What are the essential physical processes within the earth's interior which lead to plate tectonics? What does the apparent partial or complete absence of plate tectonics on a given planet tell us about processes within its interior?

If we could answer both of these questions, we would essentially be at the point of extending plate tectonics as a synthesizing concept to all of the planets, rather than to the earth alone. However, only partial answers are available at present, and the questions therefore pose one of the continuing tasks in the study of planetary interiors.

By "plate tectonics" is meant the following terrestrial phenomenon. The lithosphere, the mechanically rigid layer of the earth above the asthenosphere, is divided into a relatively small number (on the order of 10) of individual sections of varying size. These sections *(plates),* which are geometrically like portions of a thin spherical shell, act individually like rigid bodies which move over the surface of the deeper layers of the earth, in effect sliding upon the asthenosphere. A plate can rotate with one degree of freedom and translate with two degrees of freedom, being constrained to move upon the surface of the earth. The area of an individual plate is not in general conserved in the course of evolution of the earth's surface. It can be reduced in area if it slides under an adjacent plate (is *subducted*), but since the total surface area of the earth remains roughly constant, a corresponding amount of area must be added to a plate elsewhere. Area is usually added to plates along mid-oceanic ridges, where new material in the form of newly-erupted magma is "grafted" onto the edges.

Empirically, it is found that the relative velocities of the terrestrial plates are somewhat variable, but are of the order of $1-10$ cm/year. Since the lateral dimensions of plates are on the order of 10^3 km $= 10^8$ cm, it follows that a typical plate lifetime is $\sim 10^8$ years, short compared with the age of the earth. There is ample geological evidence to support this conclusion. With the exception of a few continental shield areas, most of the earth's surface has apparently been "reprocessed" many times, with the reprocessing taking the form of subduction of a plate into the mantle, where it loses its identity, followed by extrusion of upper mantle material along mid-oceanic ridges to complete the cycle.

What drives this process? There is still no universal consensus on a precise mechanism. However, it seems likely that mantle currents are involved. The mantle is solid, but may nevertheless be transporting energy from the interior of the earth by convection provided that it has a sufficiently low effective viscosity. If solid-state convection is actually occurring in the earth's upper mantle, the effective Rayleigh number (Eq. 5-46) must be large compared with 10^3. Unfortunately, many of the critical parameters which appear in the expression for the Rayleigh number are not well known. This is particularly true for the viscosity, which depends on temperature approximately as $\exp(-A/T)$.

There are problems with the concept that convective motions in the mantle are responsible for the shifting positions of plates. The most serious of these is the fact that the LVZ, the region where temperatures are closest to partial melting temperatures, is much thinner than a typical plate's horizontal extent. Thus, unless the convective cells can penetrate much deeper, into the mechanicaly rigid upper mantle, they must have extreme aspect ratios, a result not predicted by standard theories of convection.

For our purposes, we will adopt the following working hypotheses. First, it is clear that plate tectonics is a consequence of upper mantle motions in a solid but slowly deformable medium, and that these motions are probably related to the transport of energy from the interior of the planet. Such solid-state convection may even play a crucial role in establishing the earth's present interior temperature distribution. Therefore we would expect the existence of a substantial source of interior heating to be a vital ingredient for the production of plate tectonics in a planet. Relatively high temperatures in the mantle, leading to the possibility of such convection, may also be vital. Second, we have seen that terrestrial plate tectonics is a remarkably efficient process for resurfacing the planet. Given the present size of plates and their present average velocities, most of the earth's surface is relatively young compared with the planet's age of 4.6×10^9 years. With these hypotheses, we can seek to correlate some gross characteristics of planetary surfaces and heat flow data with possible models of planetary interiors. But, as was already mentioned, not enough is known about the fundamental process to permit a complete understanding of all the available data.[9]

CHEMICAL COMPOSITION

We do not, of course, have a direct determination of the overall chemical composition of the earth, but it is necessary to assume a bulk chemical composition for the purpose of computing interior models of planets, and the earth plays a key role as a reference object for this purpose.

Indirect evidence for the earth's bulk composition comes from comparison of seismic models for the pressure-density profile with theoretical and experimental pressure-density relations for various substances (which are, however, nonunique), and from analysis of surface minerals, some of which originate from deep subcrustal layers. The picture which emerges in this fashion is at least grossly consistent with an overall solar composition for the earth, minus, of course, such volatile components as the "ices" H_2O, CH_4, and the abundant gases H_2 and He.

The composition of the crust and of more recent magmas can be interpreted in terms of a plausible primordial composition for an initially homogeneous mantle. Certain constituents will appear upon partial melting of the primordial and then evolved mantle, and these constituents "float" upward to join the crust. Theoretical reconstruction of this process is complicated and not without controversy, but can in principle lead us back to the primordial composition of the mantle. A commonly accepted formula for such primordial material, known technically as "pyrolite," includes the components SiO_2 (about 45% by mass), MgO (38%), FeO (8%), Al_2O_3 (4%), and CaO (3%).[10] Assume that the bulk composition of the mantle is given by this recipe, and that the core is almost pure Fe, with a small admixture of some additional component. According to seismic earth models, the core's mass is very nearly one half of the mantle's mass. For a solar distribution of the elements in a nonvolatile sample, we would expect the mass ratio of pure iron to the compounds SiO_2, MgO, Al_2O_3, and CaO to be likewise very close to one half (between 0.4 and 0.5, to be more precise). Of course, some iron is retained in the mantle, about 10–15% of all the earth's iron, according to the pyrolite model, so that the earth may actually have somewhat more than its cosmic share of iron. However, correction for the non-iron component in the core again reduces the discrepancy. A reasonable estimate of the earth's total iron fraction by mass would be about 0.38 ± 0.02.

Sulfur is an abundant element in solar composition material. This element may possibly be somewhat underrepresented in mantle material, and it has been speculated that sulfur is the principal alloying agent in the core, responsible for the core's slight density deficit relative to pure iron. Laboratory experiments on the melting of Fe-FeS mixtures have shown eutectic behavior: The initial melting point of the mixture (solidus) is always lower than the melting points of the pure endmembers.[11] Thus estimates of core temperatures which are based upon the melting point of pure iron at high pressures could be too high.

The bulk composition of the earth is, of course, independent of processes of differentiation subsequent to the formation of the planet, but it will depend in detail upon chemical processes during and prior to formation. A major objective of interior modeling is to allow conclusions to be drawn about the earliest phases of planetary formation. Comparison of the planets, with the earth as the fundamental reference, is very helpful for this purpose.

REFERENCES

1. Dziewonski, A. M. and Anderson, D. L. Preliminary reference earth model. *Phys. Earth and Plan. Int.* **25:** 297–356 (1981).
2. Stacey, F. D. *Physics of the Earth.* New York: Wiley & Sons, 1969.
3. Garland, G. D. *Introduction to Geophysics.* Philadelphia: Saunders, 1971.
4. Christensen-Dalsgaard, J. and Gough, D. On the interpretation of 5-minute oscillations in the solar spectrum. *Mon. Not. R. A. S.* **198:** 141–171 (1982).
5. Scherrer, P. H., Wilcox, J. M., Christensen-Dalsgaard, J., and Gough, D. Observation of low-degree 5-minute modes of solar oscillation. *Nature* **297:** 312–313 (1982).
6. Zharkov, V. N. and Trubitsyn, V. P. *Physics of Planetary Interiors.* Tucson: Pachart, 1978.
7. Landau, L. D. and Lifshitz, E. M. *Electrodynamics of Continuous Media.* Reading, Mass.: Addison-Wesley, 1960, pp. 193–194.
8. Rikitake, T. *Electromagnetism and the Earth's Interior.* Amsterdam: Elsevier, 1966.
9. Hager, B. H. and O'Connell, R. J. Lithospheric thickening and subduction, plate motions and mantle convection. In *Physics of the Earth's Interior — Proc. Int. School of Physics "Enrico Fermi"* (A. M. Dziewonski, ed.), Course LXXVIII. North-Holland: Amsterdam, 1980.
10. Ringwood, A. E. Phase transformations and the constitution of the mantle. *Phys. Earth and Plan. Int.* **3:** 109–135 (1970).
11. Usselman, T. M. Experimental approach to the state of the core: Parts I and II. *Am. J. of Sci.* **275:** 278–303 (1975).

8
The Planets

It is convenient to divide the solar system into two regions: the region of the terrestrial planets, and the region of the Jovian planets. As Figs. 8-1a and 8-1b show, the two regions differ vastly in scale; they are separated by the asteroid belt, which comprises a large number of small solid bodies, most of which circle the sun between the orbits of Mars and Jupiter.

The most significant difference between the terrestrial and Jovian planets is in their mean composition. The Jovian planets are composed primarily of certain initial members of the periodic table, elements which have low atomic numbers and are cosmically abundant (see Chapter 2). Although there may be solid layers in the Jovian planets, these layers are not directly accessible to observation. Instead, we observe extensive atmospheres which are not readily distinguishable from the interior layers of the planet. In certain important respects, the Jovian planets are objects which occupy an intermediate position between stars and earthlike objects (which we tend to identify as true "planets"). The terrestrial planets, on the other hand, are vastly depleted in the cosmically abundant elements H, He, C, N, and O, and are mainly composed of the elements contained in substances which tend to condense into solid phases at relatively high temperatures on the order of $500 - 1500°K$, such as forsterite and enstatite. Their atmospheres are chemically distinct from their interiors, and do not comprise a significant mass fraction of the planet.

The terrestrial planets are much smaller in mass and size than the Jovian planets, occupy a much smaller volume in the solar system, and are much closer to the sun. These circumstances are clearly not accidental, but are a reflection of the relative cosmic distribution of the abundant and volatile elements H and He on the one hand, and the less abundant and volatile elements Mg, Si, and Fe on the other. The terrestrial planets are the minor, nonvolatile remains of amounts of cosmic material which may have initially rivaled the Jovian

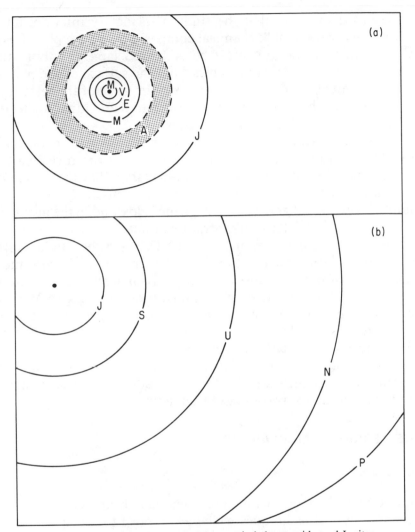

Fig. 8-1a. Orbits of Mercury, Venus, earth, Mars, main belt asteroids, and Jupiter.

Fig. 8-1b. On a compressed scale, the orbits of the outer planets Jupiter, Saturn, Uranus, Neptune, and Pluto.

planets in total mass. We do not know what became of the missing material, whether it was once incorporated in progenitors of the terrestrial planets, or whether it was lost at once before the terrestrial planets even formed. A reconstruction of the initial chemical state of the planets may help us to understand such processes.

The initial question that the study of planetary interiors must address concerns the bulk chemical composition of each planet. For the present, this question must be addressed largely without the benefit of direct analysis of samples from the surface of the planet, although limited data of this sort do exist for some of the terrestrial planets. We rely heavily on the use of terrestrial and lunar analogies in interpreting data from remote sensing.

Even if information about surface and/or atmospheric chemistry is available, such information still lies a long way from revealing the bulk composition of the planet. It is imperative that the degree of internal differentiation of the planet be understood. Clues can be found in the external gravity and magnetic fields, and in the heat flow and corresponding interior temperature profile.

The first seven chapters of this book discussed the fundamental tools and concepts which are available for probing planetary interiors. We turn now to a discussion of the individual planets, keeping in mind the ultimate objectives of the study, as outlined above. We will not find a fully consistent or unified picture of the development of the solar system as revealed by the interior structures of its major bodies. So far, most of the data have been obtained from earth-based observations or from exploratory flyby, orbiter, and probe missions. Although many regularities and common threads exist, at least as many contradictions and surprises have been found.

THE TERRESTRIAL PLANETS

The Moon

In the case of the moon we are able to bring to bear a number of the same geophysical techniques which are applied to the study of the earth's interior. The major new data base which has become available for the moon as a result of the space age includes analysis of surface chemistry from returned samples, and measurement of the interior electrical conductivity temperature profile from the moon's response to perturbations in its electromagnetic environment, leading indirectly to an inferred interior temperature profile. We also have more detailed measurements of the moon's gravitational figure and moments of inertia, as a result of analysis of perturbations to orbiting spacecraft and precise observations of the moon's motions using laser ranging to retroreflectors placed on the lunar surface. Finally, a sparse number of traditional geophysical measurements such as heat flow determinations and passive and active seismology have been carried out.

Prior to the first manned explorations and sample returns, there may have been a substantial number of scientists who held oversimplified views about the bulk composition of the moon and the structure of its interior. Because the surface is obviously heavily cratered and shows little evidence for recent alteration, one might have expected the moon to be a highly primitive body, perhaps reflecting the bulk composition of the undifferentiated earth. But the actual bulk composition of the moon, as deduced from measurements that we will now describe, differs markedly from the overall composition of the earth.

Composition of the Moon as Determined from Samples. Analysis of the first returned samples soon revealed that the moon was *not* a primitive, undifferentiated body. The composition of lunar samples differed both from the most primitive meteorites and from the presumed primordial terrestrial mantle material (sometimes called *pyrolite;* see Chapter 7). As in the case of the earth, one of the primary objectives of interior modeling of the moon is to reconstruct its bulk composition. And, as in the case of the earth, the composition of lunar surface samples does not straightforwardly tell us how to do this. The problem is, in fact, still unsolved in the sense that there is still some uncertainty about the precise origins of lunar surface rocks. We will describe a widely accepted model.

In general terms, the surface of the moon can be divided into two provinces. *Highland* regions, (sometimes called *terrae*) are saturated with craters, and are topographically rough and systematically higher in elevation than the smoother volcanic plains called *maria.* The lunar farside is essentially entirely composed of highlands regions, while the major maria are confined to the frontside. As we shall see, the structure of the lunar figure and gravitational field is also related to this distinction. Lunar highlands rocks tend to have greater ages (age being defined as the time since they last equilibrated and became closed systems) relative to mare samples. In fact, highlands samples are very ancient, with typical ages of 4×10^9 years or greater, while mare samples tend to be young (but still very old compared with most terrestrial rocks), with ages in the range 3.1 to 3.95 × 10⁹ years. Both highlands and mare samples are derived from rocks having an igneous history. However, many samples, particularly from the highlands, had been greatly modified by meteoroid impact-caused fragmentation subsequent to their solidification from magma.

The ancient highland samples cannot be primordial material because they have obviously crystallized from a melt. They tend to correspond to terrestrial igneous rocks in the *anorthosite-norite-troctolite* suite, which is produced during crystallization of a basaltic magma. Anorthosite is rich in a mineral called *plagioclase,* which is a solid solution of $NaAlSi_3O_8$ *(albite)* and $CaAl_2Si_2O_8$ *(anorthite).* The density is low (~ 2.7 g/cm³) compared with the moon's bulk density (~ 3.3 g/cm³). Norite and troctolite have more pyroxene and olivine, respectively, than anorthosite. It is generally believed that the initial lunar crust was composed of these materials, which are very rich in calcium and aluminum. Indeed, the bulk composition of the highlands crust is estimated to contain about 25% Al_2O_3. The primitive crustal material has been modified and redistributed by subsequent impact and melting events, but there is considerable petrologic evidence that such a primordial layer did exist. But if it existed, the implications for the moon are quite remarkable. Primitive terrestrial mantle material, or pyrolite, contains only about 4% Al_2O_3. If we assume that the initial moon had a similar composition (which is ultimately derived from solar abundances as modified by volatility; see Chapter 7), then a simple mass conservation argument tells us that the source region for the primitive lunar crust must have a mass not less than about seven times the mass of the crust.

Seismic evidence (discussed below) shows that the thickness of the lunar crust is on the order of 50 km. It follows that the molten region which produced this crust must have been at least hundreds of kilometers thick. A significant fraction of the entire moon must have been initially molten. The truly primordial layers of the moon, if these exist, would have to be deeper than several hundred kilometers, and never have been heated to solidus temperatures. We can understand the general origin of the composition of highlands rocks, then, if we postulate the existence of a primordial "magma ocean," whose legacy is the present highlands crust, an underlying upper mantle of residues from the formation of the crust, and a lower, never-melted mantle of pyrolitic composition. The composition of the differentiated upper mantle can be calculated from the assumption that it was once molten. Not only would aluminum-rich, lower-density material rise to the surface, but magnesium-rich, higher-density material would sink. Thus the upper mantle is expected to be primarily composed of olivine, while the primordial lower mantle is largely a mixture of olivine and pyroxene. A major difficulty with this picture, however, is that it is difficult to find a plausible mechanism for melting such a large fraction of the moon during its early history.

Theories for the origin of mare material are considerably more complicated. The maria are basaltic plains which differ in age, but are in general the most recent features on the moon. From the point of view of bulk composition, the maria are rather insignificant because they comprise only on the order of 1% of the mass of the crust. But they are of interest for our purposes because they give clues to the thermal state and composition of the deep lunar interior. Compared with highlands material, mare basalts are very enriched in FeO, and one class (the high-titanium basalts) is also very enriched in TiO_2. A fluid having the composition of mare basalts can be produced by partial melting of mantle material at pressures on the order of tens of kilobars. Thus there was still substantial partial melting in the middle mantle of the moon at the time that these basalts formed.

All of the petrologic evidence points toward a high-temperature origin for the moon. Abundant elements such as Mg, Al, Ca, and Si appear to have similar abundances in both the moon and the earth's mantle, but more volatile elements such as K, Na, and Cl are greatly depleted in the moon relative to the earth, suggesting that the moon's condensation sequence was completed at relatively high temperatures. There is no water whatsoever on the moon, either in free form or in hydrated minerals, as is consistent with this picture.

What has happened to the moon's iron? Reconstruction of the primitive composition of the lunar mantle from the analysis of samples suggests that there was on the order of 10–15% FeO present, a little more than the corresponding abundance in pyrolite. But since there is far more iron than this in a solar mixture (see Chapters 2 and 7), the iron complement of the lunar mantle must have separated early. It is sometimes stated that siderophile ("iron-loving") elements, such as Ni and Co, are depleted in the moon as a result of separation of iron from the mantle, but in fact the pressure regime under which this process occurred in the earth could not have been attained in the moon. In any case, the inferred abundance of Ni and Co in the lunar mantle is very similar to that in the earth's mantle. Abundances of other siderophiles show a more complex behavior, however.[1]

If the moon has a solar ratio of iron to rock, it follows that it should have a very substantial iron core, which must have differentiated early. In fact, the moon has no more than a miniscule core, which comprises at most only a few percent of the lunar mass. Evidence which limits the size of the lunar core is discussed in the following subsections. Thus the moon's iron abundance is on the order of 10%, and the moon is greatly depleted in this element.

Electromagnetic Response of the Moon.   In principle, the measurement of the moon's interior electrical conductivity profile proceeds in the same way as the corresponding terrestrial measurement. However, the moon's electromagnetic environment differs in significant ways from the earth's. As we have seen in Chapter 6, the moon possesses no significant intrinsic global magnetic field, and thus it is not isolated from the solar wind by the formation of a magnetosphere. The boundary of a magnetosphere is located at the point where the magnetic pressure becomes equal to the dynamic pressure of the solar wind. The earth's magnetospheric boundary is about ten earth radii distant in the solar direction.

The moon lies outside the terrestrial magnetosphere, except for the occasional times when its orbit carries it through the magnetotail, a long magnetic "shadow" of the earth. The ionized particles of the solar wind directly impact the lunar surface and are absorbed by it, creating a cavity on the leeward side of the moon. The supersonic solar wind expands into the cavity, compressing the magnetic field lines in this region. Thus the lunar situation is more complicated than the ideal geometry shown in Fig. 7-4. Nevertheless, the basic principles for determination of a conduction profile remain the same. Low-frequency disturbances penetrate deeply into the moon, while high-frequency disturbances are diagnostic of surface layers. The frequency range required to span the entire lunar interior depends of course on the average conductivity of the moon.

The response of the moon to an external electromagnetic disturbance is usually represented in terms of the *electromagnetic transfer* *function $A_i(v)$*, defined by

$$A_i(v) = [H_i(v) + H'_i(v)]/H_i(v), \qquad (8\text{-}1)$$

where $H_i(v)$ is the i-component of a Fourier component of the external field having time-frequency v, evaluated at the planetary surface, and $H'_i(v)$ is the same thing for the induced field.[2] The advantage of this representation is that the value of the transfer function does not depend upon the point at which it is measured; this is a consequence of the electromagnetic boundary conditions. Experimentally, the transfer function is determined by two different magnetometers. The first magnetometer is placed on the surface of the moon and simultaneously records the external magnetic disturbances and the moon's response to them. The second magnetometer is on board a lunar orbiter and also records both the external disturbances and the response, but the latter is weaker because of the greater distance from the lunar surface. Thus the two components can be resolved and the transfer function calculated.

The tangential transfer function A_{tan} contains information about the lunar conductivity profile. At high frequencies ($v \sim 10^{-3}$ Hz), A_{tan} significantly exceeds unity, but at low frequencies ($v \sim 10^{-5}$ Hz), it approaches unity, meaning that the induced field becomes very small compared with the forcing field (Fig. 8-2). It is the small difference between A_{tan} and unity at these frequencies which can reveal the existence of a highly conducting lunar core. If a conducting core is present, A_{tan} approaches unity more slowly than if a core is absent. Experimentally, the search for a lunar core by this technique is extremely difficult. Measurements of the response at frequencies 10^{-5} Hz requires time series of many days' duration. The data are inevitably gapped, because it is impossible to monitor a spacecraft without interruption for such periods of time. Nevertheless, meaningful limits have been placed on the maximum size of a metallic lunar core.

Although it is possible to carry out a formal inversion of lunar transfer function data to obtain an interior conductivity profile, the method most typically used to date assumes a model with a given set of parameters, which are then adjusted until a satisfactory fit to the data is obtained. Thus, one can assume a lunar interior consisting of,

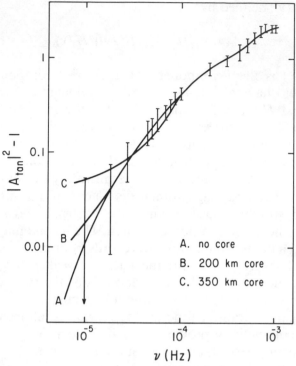

Fig. 8-2. Observed and calculated electromagnetic response of the moon as a function of inducing frequency. [From Sonett, C. P. Electromagnetic induction in the moon. *Rev. Geophys. and Space Phys.* **20**: 411–455 (1982). Copyright, American Geophysical Union.]

say, ten spherical layers with an electrical conductivity σ_j in the jth layer. The transfer function of such a model is then calculated, with the σ_j as free parameters. A model (generally nonunique) which fits the observed transfer function to within its error bars is considered satisfactory.

Figure 8-3 shows the result of a large number of attempts to fit the lunar transfer function with such conductivity models. The results indicate that, at comparable depths below the surface, the lunar conductivity is considerably lower than the earth's. We may attribute this result to a combination of two factors: (a) a paucity of metallic iron in the moon's mantle, and (b) lower temperatures in the moon's mantle compared with the earth, essentially because the moon's interior is a "stretched" version of the outermost ~ 150 km of the earth. Because of the low conductivity in the moon, it is possible to

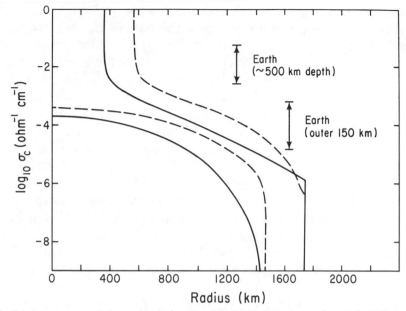

Fig. 8-3. Lunar conductivity profile. [After Hood, L. L., Herbert, F., and Sonett, C. P. The deep lunar electrical conductivity profile: structural and thermal inferences. *J. Geophys. Res.* **87**: 5311–5326 (1982). Data from Sonett, C. P., Smith, B. F., Colburn, D. S., Schubert, G., and Schwartz, K. The induced magnetic field of the moon: conductivity profiles and inferred temperature. *Proc. Lunar Sci. Conf.* **3**: 2309–2336 (1972) (solid curve); and Dyal, P., Parkin, C. W., and Daily, W. D. Structure of the lunar interior from magnetic field measurements. *Proc. Lunar Sci. Conf.* **7**: 3077–3095 (1976) (dashed curve).]

"see" rather deeply into the moon by this technique, and significant limits on the size of a lunar core with metallic conductivity can be set. As can be seen from Fig. 8-2, this limit is a radius of about 360 km. There could, in fact, be no metallic core at all.

In principle, the inferred distribution of electrical conductivity in the moon's interior can be used to derive a corresponding temperature profile. The temperature dependence of the electrical conductivity in minerals can usually be expressed in the general form

$$\sigma_c = \sum_{n=1}^{N} \sigma_{0,n} \exp\left(-E_n/kT\right), \qquad (8\text{-}2)$$

where N is a number which corresponds (at least approximately) to the number of distinct modes of charge transport, the $\sigma_{0,n}$ are constants, and E_n are corresponding constants, each of which is approximately equal to the activation energy of the nth mode. Given a composition, then, and a relation of the type in Eq. 8-2, the lunar conductivity profile yields a temperature profile. The problem with this procedure is that the exact composition of the lunar interior is not known, although it can be estimated from data on surface chemistry and from seismic data. Furthermore, the precise composition of the material can have important repercussions on the conductivity, because small admixtures of substances with low electron activation energies can, in certain temperature ranges, have a controlling effect on the conductivity (cf Eq. 8-2). Thus it is a difficult task to deduce accurate temperature profiles from conductivity data. The best that can be done is to *assume* a composition for the lunar interior, and then measure the temperature at which such a composition would have a conductivity equal to the observed one. The resulting temperature profile can then be compared with other lines of evidence, and the assumed composition can thus be checked for self-consistency.

A plausible model for the composition of the moon includes an outer zone which has been modified by processes of partial melting, differentiation, and crust formation, and an inner, primitive zone. According to seismic data (see below), the interface between the modified and unmodified parts of the lunar mantle may be located at a depth on the order of 300 – 500 km. There may even be no interface at all, but rather a region of gradually changing composition. Some estimates have used the model described above, with a dunite upper mantle composed predominantly of olivine, and a primitive lower mantle resembling the hypothetical terrestrial mantle material, pyrolite. With this composition distribution, the temperature profile inferred from the conductivity profile is approximately linear with depth and reaches a temperature of about 1700 °K at a depth of about 800 km. For the assumed composition of the lower mantle, at about this point the temperture reaches the *solidus,* which is the locus of the onset of initial partial melting in a given composition. Temperatures in current lunar models, as estimated from electromagnetic sounding, are substantially lower than initial estimates which did not properly account for the effect of "doping" by trace constituents in increasing the electrical conductivity. Current results seem to imply that there is a substantial amount of molten material in the lunar interior at depths on the order of 1000 km and greater. As we shall see, seismic data provide considerable support for this picture.

Finally, the global lunar heat flux can be deduced from the electromagnetic sounding data, as follows. The electrical conductivity distribution, together with an assumed composition, allows one to infer a temperature gradient in the lunar mantle. Knowing the thermal conductivity of the assumed composition, a heat flux then follows. The average surface heat flux which is deduced in this manner is found to be about $24-35$ erg/cm²/s.[3] This number is highly model-dependent, and somewhat exceeds the revised results from the two Apollo heat flow experiments.

Lunar Gravity Field and Tidal Response. The most basic quantities required for modeling the lunar interior are the mass, mean radius, and polar moment of inertia of the moon. All of these quantities are now known with considerable accuracy, as a result of precise measurements of doppler shifts from lunar orbiters and laser ranging data from McDonald Observatory in west Texas to four retroreflectors left on the moon's surface.

First consider the simplified case of a planet with a permanent equatorial bulge (such as the earth), which orbits the sun. If the rotation axis of the planet is inclined to the normal to its orbit about the sun, as is the case for the earth, then the equatorial bulge does not lie in the plane of the orbit and the sun will exert a torque on the planet. The components of the torque are proportional to differences in the principal moments of inertia of the planet, as follows from Eqs. 4-12 and 4-17–4-22. The axis of the planet's rotation then precesses under the influence of this torque, at a rate which is proportional to the so-called precession constant $H = (C - A)/C$ (see Eq. 4-23).

The motion of the moon is more complicated because of the commensurability between the moon's orbital and rotational periods, but is likewise affected by torques from the earth and the sun. In this case, the relevant motions of the moon are called *physical librations,* and correspond to a "rocking" of the moon about its equilibrium position with the axis of minimum moment of inertia pointing toward the earth. The rates of these motions depend upon the moments of inertia of the moon (analogous to the mass in a harmonic oscillator) and upon zonal and tesseral harmonic terms in the expansion of its gravitational potential (analogous to spring constants). Although torques on the moon are primarily produced by the second-degree harmonics, it has been found that third-degree harmonics of the lunar gravity potential contribute to an observable extent.

The problem of determining the lunar moments of inertia is addressed as follows. The physical librations of the moon are measured with great accuracy by sending a laser pulse from an observatory on the earth to a reflector on the moon, and measuring the round-trip time. With four reflectors, any small rotation of the moon about the earth-moon line is thus readily measured. Moreover, it is possible to use these measurements to observe the tidal flexing of the moon in response to the tidal stresses from the earth, which vary due to libration and changing earth-moon distance. At the same time, observations of doppler shifts of a spacecraft orbiting the moon provide information which can be used to derive the harmonic coefficients C_{lm} and S_{lm} of the moon's gravity expansion. As we have seen in Chapter 4, the second-degree coefficients are proportional to differences in the principal moments of inertia of the planet. Thus it is possible to combine all of these data in a simultaneous solution for the complete lunar gravity field, the lunar moments of inertia, and the lunar tidal response.

The results of the analysis are $A/Ma^2 = 0.3903 \pm 0.0023$ (earth-pointing axis), $B/Ma^2 = 0.3904 \pm 0.0023$ (orbital motion-pointing axis), and $C/Ma^2 = 0.3905 \pm 0.0023$ (polar axis), where M is the mass of the moon (equal to $1/81.3006$ of the earth's mass) and a is its mean radius, determined to be 1737.53 ± 0.03 km from orbiter data). The mean density of the moon is 3.3437 ± 0.0016 g/cm^3. The proximity of C/Ma^2 to the limiting value of 0.4 for a uniform-density sphere indicates that the moon can have only a small dense core. For example, the results correspond to a lunar interior model with a mantle density of 3.354 g/cm^3 and a core with a density of 8 g/cm^3 and a radius of 364 km, for a crustal density of 2.85 g/cm^3.[4]

Figure 8-4 shows an equipotential contour map of the moon, plotted in the same manner as the terrestrial geoid (see Chapter 4), except that all terms beginning with the second degree are included. Because the moon rotates very slowly, its expected equatorial bulge due to rotation is very small, and is in fact invisible on the scale of this figure. The equipotential map is generated from a sixteenth-degree fit to the lunar gravity obtained from data from four Apollo missions and five Lunar Orbiter missions.[5] Overall, the moon's equipotential surface is cigar-shaped, as expected, with the long axis of the cigar

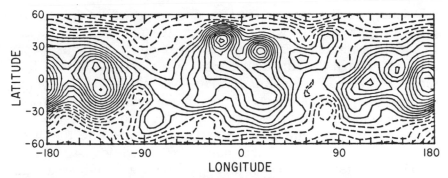

Fig. 8-4. Lunar gravitational equipotential contour plot. The contour interval is 50 meters. Dashed contours have negative elevation; the first solid contour is the zero contour. Zero longitude corresponds to the earth-facing side of the moon.

facing toward the earth: we see elevated equipotential contours on the front side of the moon near 0° longitude and on the far side near 180° longitude. Near the poles, at high latitudes, the contours tend to be negative, implying a compression along the rotation axis. Looking more closely at this map, correlations with well-known surface features are apparent. Pronounced gravity potential "highs," which are largely responsible for the gravitational elevation of the lunar frontside, are readily apparent on either side of the 0° meridian at about +37° latitude, −20° longitude, and about +25° latitude, +18° longitude. These features coincide with Mare Imbrium and Mare Serenitatis, two prominent circular basalt-filled basins ("maria"). A less-pronounced feature at about +18° latitude, +56° longitude coincides with Mare Crisium. This type of gravitational "high," which tends to be correlated with circular maria, is called a *mascon,* since it is probably caused by a concentration of uncompensated mass, possibly not very far below the lunar surface. Mascons are apparently the result of enormous ancient impacts on the surface of the moon, which hollowed out circular basins into which comparatively dense magma flowed and then solidified. They may also be caused by the upward flow of dense mantle material. The excess mass is thus not the impacting object itself, but is instead a result of geological processes which ensued as a result of the excavation, heating, and melting of the impact.

Do the mascons themselves determine the lunar figure, and thus determine which hemispheres shall point toward or away from the earth? First note that the lunar figure is not a result of hydrostatic adjustment to the moon's rotation and the earth's tides, for the parameter q associated with these effects (Chapter 4) is far smaller than the second-degree terms in the lunar potential. Thus it may seem reasonable that the actual lunar figure was established by a series of random major impacts. However, the situation is more complicated. Although many of the circular maria have associated mascons, most of them are much less pronounced features than the Serenitatis and Imbrium mascons, and one major feature, the Orientale Basin (at about $-15°$ latitude, $-96°$ longitude), which is a circular multiringed impact structure, even appears to have a slight *negative* mascon associated with it. Perhaps Orientale, being more recent and less basalt-flooded than the older frontside maria, was unable to establish a significant enhancement of density relative to surrounding material.

Because the origin of the gravitational figure coincides with the center of mass, the lunar farside must necessarily also have predominantly high equipotential contours. These, by and large, are not well correlated with features in the farside lunar crust (which is devoid of major maria). Interestingly, the *topographic* center of the moon does not coincide with the center of mass, but is offset by 2 km toward $-19°$ latitude, $194°$ longitude.[6] This result essentially implies that the average density of the farside hemisphere is slightly lower than the average density of the frontside hemisphere. One (nonunique) interpretation of this asymmetry is that the lunar crust is 10–20 km thicker on the lunar farside, but that the deeper density distribution is more symmetrical.

Compared with the geoid, the gravitational figure of the moon is very rough. This result can be understood qualitatively in terms of the strength of materials. Suppose a geologic process erects a surface feature of altitude h in a material of surface density ρ. If this feature is entirely uncompensated, then at its base there will be deviatoric stresses of order $\rho g h$, where g is the surface gravity. Now assume that the material yields to this deviatoric stress, and partially compensates when the stress exceeds a certain critical value σ_0. If this critical value is the same for the outer layers of all planets, it then follows that the maximum value of h will scale as $\sigma_0/\rho g$. Assume further that the horizontal scale of the surface feature of height h is of order a, i.e. on the same order as the planetary radius. The corresponding nonhydrostatic contribution to the gravity potential expansion will then scale as

$$C \sim h\rho a^2/M, \tag{8-3}$$

where M is the planetary mass and C is a low-degree harmonic coefficient. The fluctuation in the geoid height thus scales as

$$\delta r \sim aC \sim \sigma_0 a^3/gM. \tag{8-4}$$

If there is a universal value of σ_0, the roughness of planetary geoids should scale approximately inversely as the surface gravity. Now the surface gravity of the moon is about ⅙ of the earth's surface gravity, and its mean density is about 0.6 of the earth's, whence we predict that the moon's geoid heights should be about ten times larger than the earth's. In actuality, they are about five times larger, implying that the lunar geoid is a little smoother than it should be according to the scaling law. Such a modest discrepancy should not surprise us. The maximum stress which can be supported in the lithosphere of a planet must obviously depend upon the rate of accumulation and dissipation of such stresses, which in turn will depend upon physical conditions in the lithosphere.

On the basis of the above analysis, it thus appears that the moon's gravity field is actually no "rougher" than the earth's, once allowance is made for the differing gravitational accelerations. Some degree of isostatic compensation must therefore be occurring in the moon, particularly in view of the fact that the lunar topography is not generally correlated in detail with gravitational anomalies. Yet, in the case of certain of the circular maria, there is very good correlation, and so the compensation is undoubtedly less than in the earth. The answer may be that we are looking at very different types of geological processes in the earth and in the moon. The earth's surface is continually changing as a result of plate tectonics, and stresses accumulate and are relieved over relatively brief periods of time, compared with the case for the moon. Most likely, stresses within the moon were imposed very early in its history, and have been only very gradually relieved over subsequent time. The fact that these stresses are still comparable to those in the earth must indicate that relaxation rates are very slow in the moon, as would be expected if its outer layers are much colder than the earth's. This is consistent with evidence from electromagnetic sounding, as discussed above.

Laser ranging measurements have yielded other important parameters of the lunar interior. Because of the moon's libration and its eccentric orbit, there is a variable component of the earth's tidal stress field on the moon. Measurement of the maximum response of the moon to this variable stress, along with the phase shift of the response, yields the value of the moon's Love number and tidal Q. The Love number has not yet been determined to a useful accuracy, but the value of Q for disturbances with a period on the order of a month is on the order of 10. This is an astonishingly low value. Recall that in Chapter 4, it was argued that typical values of Q for solid planets are on the order of 100. The earth's tidal Q is of order 10, but presumably arises because of the effects of turbulent dissipation in shallow seas. The dissipative properties of the deep lunar interior to periodic disturbances are unusual, unexpected, and not well understood.[4]

Lunar Seismology. During the Apollo program, six seismographs were placed at various points on the lunar frontside. These instruments measured over 12,000 seismic events, including over 1700 large meteoroid impacts. However, despite the large number of observed events, relatively few had locations relative to the detectors such that the seismic ray paths penetrated to great depths within the moon. In fact, during the period of observation, only one weak farside event generated a ray path which passed close to the center of the moon, providing information about the possibility of a lunar core. The data base from these instruments will expand no further, for they were turned off in 1977 after eight years of operation. This data set comprises our only available seismic information on an extraterrestrial body. It is not complete enough to provide even the most basic datum that terrestrial seismology provides: information confirming or disproving the existence of a dense metallic core. Insufficient measurements have been made within the relatively small region in the moon's central regions where according to other lines of evidence a core could still exist.

The raw material for a seismological investigation is an arrival time, at a given station, of a disturbance which originated at an unknown point within the moon. In order to make sense of such an arrival time, it is necessary to correlate it with arrival times at other stations. One may proceed by a series of approximations. First we assume a constant seismic velocity for P and S waves within the lunar interior. This assumption is not unreasonable, because pressures within the moon are not high enough to cause significant compression of rocks. The central pressure in the moon is estimated to be around 50 kilobars. To place the number in perspective, note that this pressure is attained at a depth of about 150 km within the earth. The entire lunar interior can thus be regarded as an expanded version of the superficial layers of the earth. *If* no chemical discontinuities or gradients exist within the moon, one would therefore expect seismic signals to travel approximately in straight lines within its interior. Measurement of arrival times at four stations from a single disturbance then allows one to solve for the time and location of the disturbance, if the velocity is known. If both P and S waves arrive at the stations, their velocities can be inferred.

Initial analysis of lunar seismic data using the above technique showed that at separations greater than $\Delta \sim 70°$ (for disturbances with sources located on the lunar surface), the inferred velocities were slightly lower than those found from closer disturbances. This was the first indication that velocities were not constant within the moon's interior, but that velocities in fact *decreased* slightly below a certain depth, somewhat in analogy to the earth's low-velocity zone.[7] A possible explanation of this phenomenon is that there is a slight density inversion within the moon, corresponding to a layer at depths on the order of 300–500 km where the density reaches a minimum. This can be explained in terms of a substantial temperature gradient: if the temperature increases rapidly enough with depth, thermal expansion of rocks occurs more rapidly than volume decrease due to compression, leading to an inversion. However, thermal models of the lunar interior (see below) have shown that, with plausible values of the mantle conductivity, the temperature gradient is probably not large enough to produce an inversion. Alternately, it is possible to explain the decrease in velocity with depth by postulating a chemical gradient, such as a gradual increase in the $Fe/(Fe + Mg)$ ratio with depth.

Even a crude model of seismic velocities within the moon suffices to provide considerable information about the distribution of moon-quake sources. The picture which emerges differs substantially from the terrestrial case. Almost all terrestrial disturbances originate close to the earth's surface, but moonquakes are observed both deep within the moon and close to its surface. There appear to be four principal types of moonquakes. The first type is caused by impacting objects. The second type also originates close to the surface, but is apparently triggered by some sort of intrinsic tectonic mechanism related to the release of stress. These are the most powerful moonquakes. They are not particularly correlated with variable tidal stresses. The third type of moonquake, which is also shallow but quite weak, appears to be caused by variation in thermal stresses during the course of a lunar day. The fourth type is found only deep within the moon, at depths below about 800 km. The latter is generally of low amplitude, and has a highly reproducible signal. Such moonquakes tend to occur near perigee, and are evidently caused by variable tidal stresses on the moon from the earth. This interpretation is reinforced by the observed pattern for the location of deep moonquake sources within the moon. The sources tend to lie along deep seismic belts which show considerable bilateral symmetry with respect to the zero meridian.[8]

Figure 8-5 schematically shows the locations of seismic stations on the moon and locations of various sources of moonquakes. Also shown are seismically-derived layers in the lunar interior, which we will shortly describe. First note the pronounced asymmetry in the distribution of moonquake foci. Most of these appear to be located within the frontside hemisphere. We may interpret this distribution in two possible ways: Either the distribution is truly asymmetrical, or else there is an observational bias related to the fact that all seismic stations are located on the frontside. Now the time-varying tidal stress field, which is probably responsible for the deep moonquakes, is symmetric with respect to the front- and farside hemispheres (see Chapter 4), and so a real asymmetry would have to be related to a difference in the ability of the two hemispheres to support stress. While this possibility cannot be totally ruled out, it seems somewhat more likely that the lunar farside is simply not very seismically "visible" because the deep lunar interior is highly dissipative to seismic waves.

Figure 8-6 shows velocity profiles for P and S waves in the lunar mantle, as derived from inversion of available arrival times. The solution is found by assuming a layered model with constant velocities in each layer. The values of the velocities, along with source

FRONTSIDE

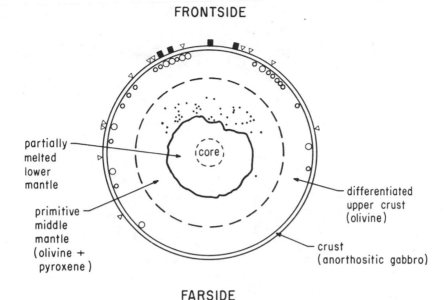

partially
melted
lower
mantle

primitive
middle
mantle
(olivine +
pyroxene)

core

differentiated
upper crust
(olivine)

crust
(anorthositic gabbro)

FARSIDE

Fig. 8-5. Seismic data for the lunar interior. Solid squares show locations of seismic detectors. Arrowheads show detected major meteoroid impacts. Circles are shallow moonquake sources, dots are deep sources. [From Nakamura, Y. Seismic velocity structure of the lunar mantle. *J. Geophys. Res.* **88**: 677–686 (1983). Copyright, American Geophysical Union.]

locations, are obtained by progressive minimization of arrival time residuals. Because of the relatively small number of stations and data points, one cannot distinguish between such a stepwise model and one in which the velocity distributions change more continuously. Note that S-waves do not appear to propagate in the moon at depths below about 1000 km. This is strong evidence for partial melting, and is consistent with the data from electromagnetic sounding discussed above. The layers of the lunar interior which are shown in Fig. 8-5 correspond to the velocity zones of Fig. 8-6. Outermost, we have the lunar crust, which is highly variable in thickness but on the order of 50 km thick. This layer comprises the pulverized rock layer (the *regolith*) of the outermost few hundred meters of the moon, the product of eons of "gardening" by meteoroid impacts, and a chemically differentiated layer produced by partial melting. The next layer is the upper mantle of the moon which extends from the base of the crust down to the end of the velocity inversion at about 500-km depth. The upper mantle may also be chemically altered by partial

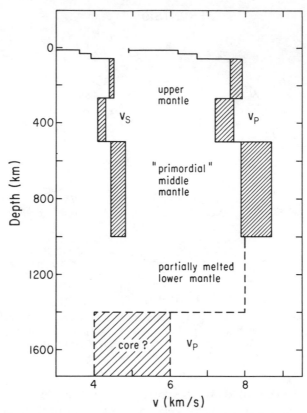

Fig. 8-6. Lunar seismic velocity profile. [After Nakamura, Y. Seismic velocity structure of the lunar mantle. *J. Geophys. Res.* **88**: 677–686 (1983). Copyright, American Geophysical Union.]

melting. From about 500-km depth to about 1000-km depth, we have the middle mantle, a region which may correspond to the most primitive (unmelted) layers in the moon. The *P*- and *S*-wave velocities in the upper and middle mantle are consistent with laboratory data on synthetic minerals consisting mostly of olivines with some pyroxenes. The olivines must be on the order of 80–85% forsterite. The composition is thus not grossly dissimilar to the presumed composition of the primitive terrestrial mantle.

Below 1000 km depth, we find the lower mantle, which is highly opaque to *S*-waves. This part of the moon appears to be the analog of the terrestrial asthenosphere. The analogy is made even more suggestive by noting that there is a fairly well-defined cutoff in the loci of deep-focus moonquakes at about 1000-km depth, as though the "lubrication" due to partial melting were able to preclude any accumulation of tidal stresses below this level.

Figure 8-6 shows a region near the center of the moon where P-wave velocities decrease drastically, by a factor on the order of two. Such a decrease would be analogous to the case in the earth, where P-wave velocities decrease by a similar factor from the mantle to the core. The seismic evidence for the existence of this region in the moon comes from an observation of a single weak meteoroid impact on the lunar farside. Unfortunately, such an event never occurred again during the period that the seismometers were active. For a homogeneous lunar interior, the expected time for the seismic disturbance to propagate from the lunar farside to the seismometers on the frontside was on the order of 8 minutes. However, the actual arrival at the most diametrically-opposed seismometer was delayed by some tens of seconds, as if the wave had traveled briefly through a medium with a very low P-wave velocity. The result could be interpreted as corresponding to propagation through an iron-rich core with a P-wave velocity equal to ~ 4 km/s and with a radius not exceeding 360 km.[7]

Observations of the propagation of seismic waves in the moon give information about the dissipative properties of the lunar interior. The seismic Q of the moon's outer layers is found to be very high, ranging from values on the order of 6000 in the lunar crust to about 1500 in the middle mantle.[8] Although the seismic Q in the earth's mantle is likewise high, values of Q in the earth's crust and asthenosphere are substantially lower, ranging from ~ 500 in the crust to ~ 100 in the low-velocity zone. The precise cause of the difference between the earth and moon is not known, but is probably related to the fact that the moon's outer layers are much farther from melting, and are very dry: the presence of water in minerals can greatly change their dissipative properties. Because the moon's outer layers are highly non-dissipative, records of moonquakes have a very distinctive appearance. Following the arrival of the first wavefronts from the disturbance, the signal continuously builds up as scattered (and relatively unattenuated) waves arrive. After reaching a maximum, the signal slowly decays as the moon reverberates.

We have already mentioned the unusually low tidal Q of the moon. How is this to be reconciled with its large seismic Q? First, we note that the two types of Q's are measured at very different frequencies. Seismic Q's pertain to oscillations with frequencies on the order of 5 Hz, while tidal disturbances are at frequencies $\sim 10^{-6}$ Hz. Furthermore, although the moon's seismic Q may be large in the outer layers, it may be much smaller in the deep interior, where temperatures are close to melting.

Attempts have been made to identify free oscillations of the moon, following large seismic disturbances, but no such oscillations have been found. Detection of these oscillations would provide additional integral constraints on lunar interior models, as in the case of terrestrial models (Chapter 7), and would help to make up for the paucity of travel-time data. It appears, however, that their detection may require seismographs of greater sensitivity.

Heat Flow and Interior Temperature Distribution. Table 5-1 gives the value $H_i = 17$ erg/cm²/s for the moon's average intrinsic surface heat flow. This number is the simple mean of values of 21 and 14 erg/cm²/s measured at Hadley Rille (Apollo 15) and Taurus Littrow (Apollo 17) respectively. These were the only direct measurements of heat flow (via the thermal gradient method) carried out on the lunar surface, and may or may not be representative of the global average.[9] If we assume that both the earth and moon are in a steady thermal state, with the energy radiated from the interior equal to the energy released by radioactive decay (possibly not an accurate assumption for any planet), then the luminosity per unit mass of each object gives an indication of its average concentration of radioactive species. As Table 5-1 shows, the specific luminosity of the moon is comparable to the earth's but somewhat higher, possibly indicating that the moon may be enriched in heat-producing materials relative to the earth. The corresponding temperature profile within the moon will depend on the distribution of these materials and on the thermal conductivity of the lunar mantle. Since indirect information about the temperature distribution is available from electromagnetic sounding and from seismic experiments, we may use this information to draw conclusions about the distribution of heat sources in the moon.

Tables 2-1 and 2-2 give mass abundances of the elements which contribute the principal heat-producing isotopes in planetary interiors. These elements are K, U, and Th. From the observed lunar luminosity/mass ratio, it is not unreasonable to assume that U and Th were present in approximately solar ratio to Si in the primordial moon. Because of its volatility, K is globally depleted relative to U and Th, and thus is not as important a contributor to the global heat flux of an equilibrium lunar thermal model (cf Fig. 5-1). Now it is a characteristic of U and Th that when rocks containing them are partially melted, these elements tend to concentrate in the initial liquid. This occurs because U and Th have large ionic radii and charges, and therefore "prefer" to leave a dense crystal lattice in favor of the liquid phase. We therefore expect virtually all planetary crusts to be enriched in U and Th at the expense of the underlying differentiated mantle. Given the initial ratio of U and Th to silicates, and given the observed present abundances of these species in the lunar crust, one can then infer limits on the depth of the differentiated mantle. If the underlying differentiated mantle was entirely depleted in U and Th, the observed enrichment of these elements in the lunar crust can be combined with an average crustal thickness of ~ 50 km to yield a differentiated mantle thickness of 300 – 600 km. Obviously, if the mantle was not fully depleted of its U and Th, then this thickness is a lower limit. This result, which is dependent only on the global heat flux to the extent that the latter tells us the overall abundance of U and Th in the moon, is consistent with seismic evidence and with petrologic evidence for the existence of a primordial deep magma ocean on the primitive moon.

A more model-sensitive, but nevertheless provocative, result emerges if we attempt to calculate the temperature profile within the lunar mantle. Two poorly known parameters enter into this calculation: the thermal conductivity of the lunar mantle and the distribution of heat sources. If *all* of the heat sources in the moon were concentrated in the crust, the interior of the moon would be isothermal in equilibrium (since the heat flux would be zero below the crust). Actually, a significant fraction of the heat sources remain in the deep, primitive mantle, and contribute on the order of 50% of the total heat flux. Experimental measurements of the thermal conductivity of a presumed forsterite mantle material give a value of $K \sim 3 \times 10^5$ erg/s/cm/°K at $T = 1000$ °K. This implies that temperatures reach the range 1100 – 1600 °K at a depth of 300 km, and significantly exceed solidus temperatures at depths ~ 800 – 1000 km.[9] Thus a simple thermal model produces a moon whose interior is too hot: it melts at layers which are solid according to the seismic data. The dilemma may be resolved by assuming that convection occurs within the outer solid layers of the moon. If solid-state convection can proceed with enough efficiency to depress the thermal gradient substantially below the conductive gradient, the heat flow data and seismic data can be reconciled. This hypothesis raises the interesting possibility that the moon's interior may still be active, although it is overlain with a thick thermal lithosphere which prevents active development of plate tectonics. The moon's nonequilibrium gravity field might also express density fluctuations associated with deep interior solid-state convection.

Mercury

Interior Structure and Evolution: Geological Evidence. By default, a large fraction of the available data bearing on the thermal evolution and internal structure of Mercury comes from photogeological studies carried out during 1974 and 1975 from the Mariner 10 spacecraft. Such evidence is necessarily highly indirect, and is not a satisfactory substitute for geophysical measurements such as heat flow, seismology, and detailed gravity mapping. Geological observations which could be relevant to the study of Mercury's interior involve the investigation of tectonic features caused by global changes in the size and shape of the planet.

Mercury's figure could have undergone substantial perturbations over the history of the planet. First consider an incompressible liquid planet with density ρ and mean radius a, rotating at an angular rate ω. The shape of the planet is given by Eq. 4-33. The planet is oblate, with an oblateness on the order of q (defined by Eq. 4-34). If primordial Mercury had a rotation period on the order of 20^h, then $q \approx 0.005$ initially. Because of tidal despinning (see Chapter 4), Mercury's present q is essentially zero, and thus Mercury's oblateness could have changed by a few $\times 10^{-3}$ since the planet's formation. Such a change may have had geological consequences, depending on the thermal state of Mercury at the time that the spindown occurred.

Furthermore, when compressibility and thermal expansion are taken into account, Mercury's cooling and the formation of a dense iron core would cause the mean density of the planet to change.

Such a deformation would be spherically symmetric and could in principle be distinguished from a change in oblateness by appropriate geological measurements.

Geological evidence indicates that Mercury has (a) probably formed a distinct iron core, and (b) undergone a significant change in radius since developing a lithosphere.[10] Although Mercury seems to have no exact analog to the lunar maria, extensive plains areas have been interpreted as being volcanic in origin, and may thus be evidence for significant melting in the primordial Mercurian mantle. Partial melting of a primordial silicate-iron mixture would then lead to formation of a differentiated iron-rich core. This evidence is quite indirect, however, and no definitive evidence of Mercurian volcanism has been found. In any case, Mercury's heavily cratered, moonlike appearance shows that if a core formed in this manner, the process must have occurred quite early in Mercury's history, since partial melting would have partially obscured the cratering record if it occurred subsequent to the period of heavy bombardment. This suggests that Mercury's core was present prior to about 4×10^9 years ago.

The evidence that Mercury has contracted consists of observations of a widespread system of arcuate scarps, which appear to be thrust faults (Figs. 8-7a and 8-7b). Such geological features are produced by compressive stresses in crustal material, which tend to cause one crustal block to ride up over another. This would be a consequence of reduction of the surface area of the planet. The fact that such scarps are found over a significant fraction of the observed surface area of Mercury, and are not compensated by normal faults resulting from tensional stresses, indicates that the reduction of surface area was probably global. Estimates from photogeological measurements suggest that the planet underwent a reduction in radius on the order of 1 to 2 km.[10] The most plausible mechanism for such a contraction would be the volume decrease which accompanies cooling. Geological evidence therefore suggests that the most recent thermal evolution of Mercury has involved substantial interior cooling. In order to determine whether this should have occurred prior to, subsequent to, or concomitantly with despinning, we must now consider the stress patterns that are established in a planet with a decreasing rotation rate.

Suppose that Mercury is originally liquid and rotating with a

Fig. 8-7a. Discovery scarp on Mercury (courtesy NASA); south is up.

primordial period $\simeq 20^h$. Now suppose that it is allowed to solidify into a hydrostatic equilibrium figure appropriate to such a rotation period. Thus, in the initial solid state, there are no stresses in the planet's mantle other than hydrostatic ones. As some unknown period of time goes by, the planet is despun by tidal friction to its present rotational resonance with its orbital period. If Mercury's tidal Q is constant and on the order of 100, this time interval will be on the

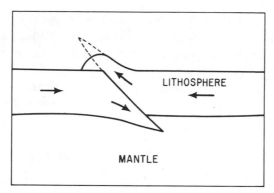

Fig. 8-7b. Hypothetical vertical cross-section of a Mercurian scarp.

order of 10^9 years. In the new rotation state, the oblateness in hydrostatic equilibrium must be much smaller than the initial value $\sim 10^{-3}$. But since the planet is solid, nonhydrostatic stresses will be established as it attempts to deform from this initial value of the oblateness to a more spherical configuration. Let f be the initial oblateness and f' be the final oblateness. In hydrostatic equilibrium $f \gg f' \simeq 0$. The stress tensor will be described using the same spherical coordinate system as was used to describe planetary gravity fields (Chapter 4). Thus $\sigma_{\theta\phi}$ denotes a stress (or force per unit area) acting on a unit surface which is normal to the θ-direction, with the force directed in the ϕ-direction, σ_{rr} denotes a stress acting in the radial direction on a surface normal to the radial, etc. By symmetry, in this problem all stresses σ_{ij} with $i \neq j$ vanish. For an incompressible elastic planet with shear modulus μ, one finds

$$f - f' = (\tfrac{5}{4})(q - q')/[1 + (\tfrac{19}{2})(a\mu/\rho GM)], \qquad (8\text{-}5)$$

(cf Eqs. 4-59 and 4-64) where $q' \simeq 0$ is the final value of the rotation parameter.[11] The corresponding surface stresses in the final configuration are given by

$$\sigma_{\theta\theta} = -(\tfrac{2}{5})\mu(f - f'), \qquad (8\text{-}6)$$

$$\sigma_{\phi\phi} = (\tfrac{1}{5})\mu(f - f')(1 - 3\cos 2\theta). \qquad (8\text{-}7)$$

Although only very crude information about Mercury's gravity field is available, it is sufficient to rule out the above model. Let us adopt $\mu \simeq 6.5 \times 10^{11}$ dyne/cm^2, a value comparable to that in the earth's mantle (as determined from seismic data). Then with $f = 6.25 \times 10^{-3}$ (appropriate to $q = 0.005$ for a Maclaurin spheroid) and with $q' = 0$, we find that $f' = 5.8 \times 10^{-3}$. Using the theory of Chapter 4, it is easily seen that the final model will have a $J_2 \sim f'$, whereas in hydrostatic equilibrium J_2 would be zero. Now the value determined from spacecraft flybys is $J_2 = (8 \pm 6) \times 10^{-5}$ (Table 4-1). This number is much larger than the hydrostatic equilibrium value of J_2 corresponding to the present rotation rate, but is considerably smaller than the J_2 which would correspond to the f' calculated above. Also, the stresses which are calculated from Eqs. 8-6 and 8-7 are on the order of a few hundred bar. Such stresses are somewhat too low to account for the formation of such features as the Mercurian scarps. Thus we need a model which produces considerably larger stresses in the lithosphere and a considerably smaller final equatorial bulge (to account for the observed limits on J_2).

In the opposite limit, we assume that Mercury did not crystallize simultaneously, but instead developed a thin solid lithosphere over a liquid layer as it was despinning. In the limit of a lithosphere thin compared with the planetary radius, the surface stress field is given by

$$\sigma_{\theta\theta} = -(5/12)(q - q')\mu[(1 + \sigma_p)/(5 + \sigma_p)](5 + 3 \cos 2\theta), \quad (8\text{-}8)$$

$$\sigma_{\phi\phi} = (5/12)(q - q')\mu[(1 + \sigma_p)/(5 + \sigma_p)](1 - 9 \cos 2\theta), \quad (8\text{-}9)$$

where σ_p is Poisson's ratio ($\simeq 1/3$ for rocks).[12,13] These stresses can be considerably larger than for the case of the entirely solid planet, because they are proportional to $q - q'$ rather than the much smaller difference $f - f'$. At the same time, since we have only a thin solid layer floating on a liquid planet, the shape of this layer must conform quite closely to hydrostatic equilibrium, and thus the expected final value of J_2 must be on the order of $f' \sim q'$, i.e. much smaller than $f \sim q$. This value may also be incompatible with observation, since Mercury's present $q' \sim 10^{-6}$, substantially smaller than its observed limits on J_2. A consistent story can then be formulated if we assume that Mercury initially had a molten mantle, thus differentiating an iron core, and then formed a substantial solid lithosphere over this molten mantle, during the period of time while the planet was being despun by solar tides. This epoch must have been substantially completed by the end of the last great bombardment of Mercury's surface by meteoroids ($\sim 4 \times 10^9$ years ago by analogy with the moon). Of course, the present value of Mercury's J_2 may have nothing to do with a primordial rotation state.

The reader will note that there are many analogies between the proposed history of Mercury's interior and that of the moon. However, the moon shows no clear evidence, either geological or gravitational, of having been spun down. Does Mercury? First let us examine the predicted fracture pattern for the surface of Mercury. In both limits (Eqs. 8-6 – 8-7 and Eqs. 8-8 – 8-9), $\sigma_{\theta\theta}$ is negative (tensional) for all θ, while $\sigma_{\phi\phi}$ is negative near the poles and becomes positive (compressive) toward the equator. Thus, near the poles, both stresses are tensional, which should lead to normal faulting, possibly leading to the production of graben (Fig. 8-8). Closer to the equator, east-west stresses are compressive but north-south stresses are tensional. This would tend to produce strike-slip faulting (Fig. 8-9). In neither case would we expect to see reverse faulting, which requires compressive stresses in both directions. Actually, it turns out that for the intermediate case of a lithosphere of nonnegligible thickness overlying a liquid mantle, the despun planet does develop an equatorial band with compressive stresses in both the θ- and ϕ-directions. But in no case does the predicted faulting pattern agree convincingly with the observed distribution of arcuate scarps. The essentially global distribution of these may instead imply that Mercury has contracted globally, and that this has probably occurred more recently than tidal despinning, erasing most traces of the earlier process.[13]

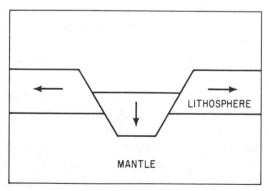

Fig. 8-8. Vertical cross-section of normal faulting, produced by purely tensional stress.

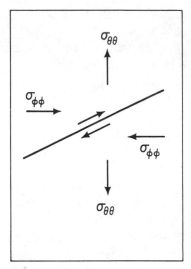

Fig. 8-9. Diagram of north-south tensional stresses, east-west compressive stresses.

Static Interior Models. Mercury's mean density of 5.44 g/cm^3 is almost the same as the earth's (5.51 g/cm^3), despite the fact that Mercury's mass is only $\frac{1}{18}$ of the earth's.[14] With such a small mass, central pressures in Mercury are low, on the order of a few hundred kilobars, and compression of minerals in Mercury's interior should be small. Therefore Mercury must differ substantially in composition from either the earth or the moon. With such a high density, "rock" must be a minor component of its interior composition. Iron seems a likely candidate for the major component because of its cosmic abundance (Tables 2-1 – 2-2) and high density. It is straightforward to deduce the required abundance of iron, since the corrections for compression are small. Thus, we take a typical silicate ("rock") density of 3.3 g/cm^3 and a density for γ-iron of 8.95 g/cm^3. Requiring the mixture of rock and iron to have an average density of 5.44 g/cm^3 gives the result that iron must comprise about 60% of the planet's mass. We found that iron comprised about 40% of the earth's mass (Chapter 7), and about 10% or less of the moon's mass (previous section). For a solar composition mixture of rock and iron, one would expect iron to comprise about 30 – 35% of the mass. Thus there is a strong, virtually model-independent indication that Mercury is greatly enriched in iron relative to solar composition. Mercury's superficial resemblance to the moon's heavily cratered surface therefore does not extend to its interior.

Geological evidence discussed in the previous section provides a circumstantial case that Mercury differentiated an iron core relatively early in its history. Static models fitted to Mercury's mean density by adjustment of the size of a central iron core overlain by a rock mantle find that the mantle must be about 600 km thick, or about ¼ of the planet's radius. At the base of the mantle, the pressure is about 70 kilobar, which for reasonable interior temperatures (see below) implies that the iron core is entirely in the γ-phase (Fig. 3-16). The central pressure is about 400–450 kilobar, which produces about a 20% compression of the iron relative to the density at the base of the mantle. Such a model has $C/Ma^2 = 0.325$, but this prediction will remain unverifiable unless Mercury's constant of precession can be determined.[15,16]

Thermal Structure of Mercury. As discussed in Chapter 5, large planets ($a \gg 10^3$ km) tend to retain their primordial heat, while small planets can cool substantially over the age of the solar system. Mercury is a smallish planet, and determination of the temperature distribution in its interior represents one of the major problems in understanding its internal structure. Mariner 10 discovered that Mercury sustains an intrinsic planetary magnetic field, and according to conventional concepts about the production of such fields, a liquid, convecting, metallic region must exist in the planet (Chapter 6). Detection of the field was a surprise because Mercury would not be expected to have such a region in its interior. Recall that the geological and gravitational evidence suggests that Mercury's mantle was initially hot enough to melt out an iron core from an iron-rich primordial condensate. In order to understand the existence of a dynamo at present, we must demonstrate that the core becomes liquid and remains liquid to the present, and that the core contains sufficient heat sources to remain convectively unstable to the present.

We will first make a rough calculation of the cooling history of a pure iron, Mercury-sized object. Start with Eq. 5-9, assuming that the only source of energy release is the decline in the thermal energy of the planet. The planet is assumed to be incompressible, so that the equation reduces to

$$\rho\epsilon = -\rho C_P \partial T/\partial t, \tag{8-10}$$

where C_P is the heat capacity per gram at constant pressure. The partial differential equation for the temperature T as a function of time t is given by Eq. 5-48, which we rewrite as

$$r^{-2} \frac{\partial}{\partial r} \left(r^2 K \frac{\partial T}{\partial r} \right) = \rho C_P \frac{\partial T}{\partial t}. \tag{8-11}$$

The temperature at the surface of the planet will be assumed to be held constant; for the purposes of this example it can be taken to be zero. Thus the solution will obey the boundary condition $T(r = a,t) = 0$. Assume that the full solution is given by a linear combination of solutions of the form

$$T_n(r,t) = f_n(r) \exp(-t/\tau_n), \tag{8-12}$$

where τ_n is the time constant for the decay of the nth mode. Substituting Eq. 8-12 in Eq. 8-11, we find

$$f_n(r) = T_{0n} \sin(kr)/(kr), \tag{8-13}$$

$$k = n\pi/a, \tag{8-14}$$

and

$$\tau_n = (\chi k^2)^{-1}, \tag{8-15}$$

where $\chi = K/\rho C_P$ is the thermal diffusivity and T_{0n} is some arbitrary initial central temperature for the nth mode. Note that the lowest-order mode with $n = 1$ will decay the most slowly, and therefore will tend to describe the eventual thermal state of the planet regardless of its initial temperature distribution. Let us estimate τ_1 for parameters appropriate to Mercury. The thermal diffusivity of iron is a function of temperature and density, but for present purposes can be taken to be $\chi \approx 0.056$ cm^2/s. Then, using $a = 2439$ km, we find $\tau_1 = 3.4 \times 10^9$ years. This result shows that for an arbitrary initial temperature distribution in Mercury, the temperature will decline to about $1/e$ of its initial value in about 3.4×10^9 years. If the melting temperature of the iron core is about 2000 °K near its center (see Fig. 3-16), then it would need to be formed at an initial temperature greater than 5000 °K in order to remain liquid to the present.

Several additional complications must be taken into account when a realistic thermal model of Mercury is calculated. The outermost 600 km of the pure iron planet must be replaced by a silicate mantle with about an order of magnitude lower thermal conductivity. A realistic initial temperature profile must be estimated, a distribution of radioactive heat sources must be assumed, and it is necessary to take into account the possibility of convection.

First assume that Mercury initially formed as a homogeneous mixture of iron and rock. If the cosmic ratio of U to Fe was preserved in the initial Mercury, then we find an initial U/Fe $= 13.5 \times 10^{-8}$ by mass (Chapter 2). The initial Th/U ratio is about 2.2. Regarding Mercury's bulk composition as an indication that volatile elements such as K are absent, we do not include K^{40} among these heat-producing species. It is then straightforward to compute ϵ_N for the primordial planet. The initial temperature profile is determined by accretional heating plus adiabatic compression of the accreted material, and may be on the order of 1000 °K. The model then heats up internally due to radioactive energy release while it cools at the surface. The time scale for significant melting of the iron to occur is on the order of 10^9 years, and is relatively insensitive to the precise values of adopted parameters. This can easily be seen from the following considerations. For the assumed initial concentrations of U and Th, we find a global radioactive heat production in Mercury of about 4.2×10^{19} erg/s. The heat of fusion required to melt the iron in Mercury amounts to 5.6×10^{35} erg. Thus it takes about 0.4×10^9 years merely for the radioactive species to supply the iron's heat of fusion. About 0.9×10^9 years of additional time is required to heat the primordial planet from ~ 1000 °K to 2000 °K. On the other hand, it is currently believed that the epoch of intense cratering of planetary surfaces in the solar system ceased about 4×10^9 years ago, and therefore that Mercury's heavily-cratered surface is generally about this old. Calculations indicate that segregation of an iron core in Mercury would be accompanied by a substantial *increase* in the radius of the planet (~ 10 km) due to decompression of the silicates.[16]

Shrinkage would only occur later, after the liquid core begins to cool. But we see no geological evidence of a major expansion of the lithospheric area subsequent to cessation of intense cratering; in fact, the opposite is true. Thus we are forced to conclude that Mercury's massive iron core was already essentially present as a distinct entity at the time of the planet's formation. It may be possible to salvage a homogeneous-accretion model for Mercury, but only by postulating chemically unlikely combinations, such as large amounts of sulfur in the iron to depress its melting temperature, or extreme enrichments of radioactive species.

Assuming an iron core to be initially present in Mercury, can it be maintained in a molten state? It is instructive to consider a steady-state model in which the radioactive heat sources are uniformly concentrated in the silicate mantle in an amount sufficient to raise the base of the mantle to the melting temperature of iron. This time we solve Eq. 8-11 again, but with the right-hand side replaced with $-\epsilon_N$. The boundary conditions are: (1) $T = T_1$ (the surface temperature) at $r = a$; (2) $dT/dr = 0$ at $r = r_c$ (the base of the mantle), because there are no heat sources in the core and therefore no heat flux emerges from it. The reader can verify that the following temperature profile is the appropriate solution[17]:

$$T = T_1 + r_c^3 \frac{\rho \epsilon_N}{3Ka} [0.5(a/r_c)^3(1 - x^2) - (x^{-1} - 1)], \quad (8\text{-}16)$$

where $x = r/a$. Because there are no heat sources in the core and the solution is a steady-state one, $T = $ const. for $r < r_c$. The mean surface temperature of Mercury is not an essential parameter as long as it is low; we shall adopt $T_1 = 400\,°K$. In order for the outermost core layers to be molten, we require $T(r_c) \approx 2100\,°K$. Equation 8-16 then requires $\epsilon_N = 1.4 \times 10^{-7}$ erg/g/s. When integrated over the mantle, this gives a total heat production of about 2×10^{19} erg/s, which is reasonably consistent with the heating that could be supplied by a solar abundance of U and Th relative to Fe, as modified by the decline in abundance due to decay over the lifetime of Mercury. Thus it would appear that it is marginally possible to keep the iron core partially molten in a steady state. However, we must now test solution 8-16 for instability to convection. For this purpose we compute the Rayleigh number Ra, given by Eq. 5-46. The most uncertain of the parameters which enters into the definition of Ra is $v = \eta/\rho$, the kinematic viscosity (which has the units cm^2/s and is a diffusion coefficient for microscopic momentum). As discussed in Chapter 5, v or η normally has a very strong temperature dependence. Rock is unable to plastically deform to any meaningful extent over the age of the solar system if $T \lesssim 1300$ °K. Thus in this temperature range, v can be regarded as essentially infinite. For temperatures somewhat higher than this value, terrestrial studies suggest the value $v \sim 10^{22}$ cm^2/s.

Above the layer where $T = 1300$ °K, Mercury's silicate mantle behaves as a rigid solid body and can be regarded as a lithosphere. Using the steady-state solution given above, one finds a lithospheric thickness of about 200 km, substantially greater than the corresponding lithospheric thickness for the earth (which is about 50 km). The thickness d which should be used in the expression for Ra is therefore about 400 km, the thickness of the mantle minus the lithosphere. The resulting Rayleigh number then turns out to be about an order of magnitude greater than the critical value, which means that Mercury's deep mantle must be undergoing solid-state convection. This result is not necessarily inconsistent with evidence from surface geology, since the thick lithosphere might prevent plate motions and thus there would be no manifestations of plate tectonics comparable to the terrestrial case.

Equation 5-47 gives the Nusselt number in terms of the Rayleigh number. For a Rayleigh number of the value given above, the Nusselt number turns out be be about 2, which means that the value of ϵ_N must be roughly doubled in order to maintain the same temperature profile in the presence of convection. Thus deep solid-state convection in Mercury, if present, would make it even more difficult to keep the core liquid.

In order to have a dynamo in the iron core, of course, this region must not only be liquid but also stirred by convection or some other mechanism. But, as we have already discussed, when an iron-silicate mixture is melted, there is a strong tendency for the radioactive atoms U and Th to partition into the silicate fraction. It is difficult to conceive of a plausible chemical process which would leave the core sufficiently enriched in these species so that convection could proceed. Indeed, this same problem arises in connection with the earth's magnetic field.

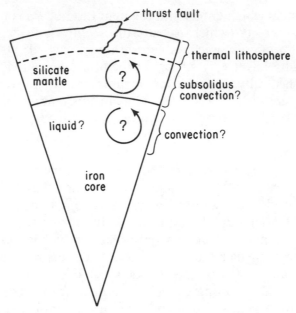

Fig. 8-10. Model of Mercury's interior.

More exotic mechanisms, such as dissipation associated with the tidal flexing of Mercury as it approaches and recedes from the sun on its eccentric orbit, do not appear promising. Using $Q = 100$ and the mechanical properties of the mantle mentioned above, Eq. 4-86 gives a value $\sim 10^{14} - 10^{15}$ erg/s for the energy deposited in Mercury by this mechanism. This number is about three orders of magnitude too small to be significant for the present thermal state of Mercury.

Figure 8-10 summarizes our current understanding of the internal state of Mercury. Since our information is entirely based upon (a) radar measurements of Mercury's spin state, and (b) results from an initial reconnaisance flyby mission which mapped only half the planet's surface and carried out initial observations of the magnetic field geometry, it is not surprising that there are many question marks on this figure.

Venus

Bulk Composition. Superficially, Venus closely resembles the earth. With a mass of 4.869×10^{27} g (81% of the earth) and a mean radius of

6051.5 km (95% of the earth), the mean density is 5.245 g/cm³ (95% of the earth) (see Table 1-1). Is Venus' bulk composition, in particular the iron/rock ratio, the same as the earth's? Is the small difference between the earth's mean density and Venus' significant, or can it be accounted for solely in terms of a greater degree of self-compression in the earth?

Although Venus' gravity field probably provides significant clues to the nature of its lithosphere and upper mantle (see below), it says nothing about the possible existence of a core. Like Mercury, Venus rotates at a very slow rate, and the rotational potential is not a significant perturbation on its interior structure. For Venus, the small parameter describing the rotational perturbation (see Chapter 4) is $q = 6.1 \times 10^{-8}$. This is only sufficient to create an equatorial bulge with a height of a few tens of centimeters. In order to measure the planet's response to the rotational perturbation and thus infer the existence of a core via Radau-Darwin theory, it would be necessary to detect this bulge in the presence of nonhydrostatic undulations in the planetary gravitational figure with an amplitude on the order of tens of meters. Unless a way can be found to observe Venus' precession constant, the value of C/Ma^2 will remain unknown.

There is a reasonably model-independent way to compare Venus with the earth. We adopt the hypothesis that Venus' interior is identical to the earth's in every respect, except that the total mass is different. From seismic inversions and other data, we have available pressure-density relations for the earth's core, mantle, and crust. These pressure-density relations can then be incorporated in the equation of hydrostatic equilibrium, which is integrated, starting at the center, through Venus' core, mantle, and crust. It is assumed that the mass fraction of each of these regions is the same in the earth and Venus.

The results of the integrations for Venus show that zero pressure, i.e. the surface of the planet, is reached at a radial distance about 40 km lower than the actual mean surface. The real Venus is therefore a little larger, and has a mean density on the order of 1–2% lower, than an earthlike Venus.[18,19] We would certainly conclude from this result that the general distribution of density in Venus is very close to that of the earth, and that Venus possesses an iron core and a mantle with a mass ratio much like the earth's.

Is the small discrepancy in the mean density significant? Because it is quite small, the value of the discrepancy is rather sensitive to the detailed way that the terrestrial pressure-density relation is adapted to Venus. Since a given pressure level falls deeper in Venus than in the earth because of lower gravitational acceleration, use of the unmodified terrestrial pressure-density relation is tantamount to assuming that a given temperature level in the earth falls deeper in Venus than in the earth. This makes the interior of Venus comparatively cooler than the earth's. Now thermal effects cannot account for the full 40-km radius difference, but if we assume that temperatures in Venus are equivalent level-by-level to those in the earth, and if allowance is made for a much higher lithospheric temperature in Venus (see below), an earth-like Venus model can be made to expand enough to reduce the radius discrepancy to about 20 km. Apparently the discrepancy is real, and cannot be made to go away altogether.

This tiny density discrepancy vexes planetary scientists for the following reasons. We have seen that the composition of Mercury is certainly different from the earth, and that when allowance is made for gravitational self-compression, the intrinsic mean density of Mercury is considerably larger than the earth's. We will find (see below) that the intrinsic mean density of Mars is lower than the earth's, and that the trend of decreasing intrinsic mean density with heliocentric distance continues to Jupiter (although it then reverses with the subsequent giant planets). The result that Venus seems to have a slightly *lower* intrinsic mean density than the earth contradicts this tidy trend (see Fig. 8-19), and seems to demand a special explanation.

As we have discussed in Chapter 2, a condensation sequence may provide a useful framework for understanding some aspects of the density trend in the terrestrial planets. The compositions of the terrestrial planets, as a function of heliocentric distance, may reflect differing levels of volatility and incomplete condensation from a primordial nebula, with the degree of condensation regulated by the prevailing nebular temperature at the corresponding distance from the sun. How does the condensation model deal with the apparently anomalous density of Venus?

The density of Venus could be decreased by $\sim 1-2\%$ either by removing a relatively abundant high-density constituent from an earthlike interior composition, or by adding a relatively abundant low-density constituent. The constituent in question must be one that is changing its concentration in the primordial nebula and in the protoplanet at the time and temperature corresponding to accretion. Finally, there must be a substantial difference in the amounts of this constituent incorporated in the earth and Venus respectively.

Sulfur is a candidate for this substance because it is almost as abundant as iron (Tables 2-1 and 2-2), and because it condenses from a primordial nebula at about 700 °K, a temperature which lies within a plausible range for conditions at the orbits of Venus and the earth. The sulfur reacts with iron to form FeS, and thus would tend to be incorporated in the planetary core.[20] It is for this reason that sulfur is a popular choice for the lighter element which alloys with iron in the earth's core, causing its density to be about 10% lower than the density of an iron-nickel alloy. Suppose that temperatures were never low enough at Venus' orbit for sulfur to condense as FeS before the primordial nebula was removed. Venus would then be essentially devoid of sulfur. Its core would be about 10% denser than the earth's, but of lower overall mass. Thus the overlying silicate mantle is not quite as compressed as before, and the bulk density of Venus could be reduced slightly, by an amount on the order of a percent. A possible problem with this hypothesis is that Venus is not devoid of sulfur. It is now known that H_2SO_4 (sulfuric acid) is a principal constituent of the dense clouds in Venus' atmosphere.[19,21] However, the mode of origin of the sulphuric acid droplets via interactions between the dense, hot atmosphere and minerals in the planetary surface is not well understood, and thus one cannot yet infer a bulk sulfur abundance. The sulfur seen in the atmosphere could be a surface phenomenon having little or nothing to do with abundances deep within the planet, but it is generally recognized as a datum unfavorable to equilibrium condensation theories for the origin of planetary compositions.

A diametrically-opposed argument can also be given to explain the anomalous density of Venus. Suppose that Venus accreted at a temperature *lower* than the earth. This assumption is not unreasonable if we assume that the main factor governing accretion temperatures was not the proximity to the sun (or the central part of the primordial nebula) but the energy deposited by accreted bodies on the surface of the primordial planet. Because Venus has a lower mass than the earth, it can be argued that it accreted at a lower temperature, and therefore incorporated substantially more FeO in its silicate mantle than did the earth. In this picture, Venus occupies an intermediate position between the earth (which has ~ 10–15% of all its iron in the mantle, in the form of FeO) and Mars (which has most of its iron oxidized and in the mantle in the form of FeO). In this alternative model, about 35% of all the iron in Venus would be present in the mantle, and the iron-nickel-sulfur core would make up about 23% of the planetary mass, compared with 32% for the earth's core.[19]

The very low upper limit on the maximum size of a Venusian magnetic dipole moment (Chapter 6) sheds little light on the likelihood of various alternative models of the interior. The problem is that there are too many possible explanations of the apparent absence of a dynamo. Scaling relations which allow for the slow rotation rate of Venus predict a substantially larger dipole moment than is observed. There should be an iron core of some size in Venus, but perhaps it is unable to convect. If sulfur is absent from Venus' core, the eutectic depression of the melting temperature in Fe-FeS mixtures would be absent, possibly allowing Venus' core to freeze entirely at temperatures where the earth's core is partially liquid. On the other hand, Venus' core may even be completely liquid. It is possible that the phase transition in the earth's core plays an essential role in liberating the energy which maintains convection. If Venus' core does not lie within the narrow band of temperatures which includes the core solidus and liquidus, this energy source may be absent.[20]

Thermal Structure. In first approximation, it is usually assumed that temperatures within Venus resemble those at comparable depths within the earth. There is little direct information which bears upon this problem, since no seismic data or electromagnetic sounding data are available. However, one important data set from Soviet landers tells us that Venus, like the earth, has a differentiated crust. According to measurements of abundances of radioactive isotopes in primitive meteorites, a solar mixture of rock would have today a mass ratio of uranium to silicon of about 0.09 parts per million (ppm). The thorium to uranium ratio would be about 3.4. This ratio of thorium to uranium tends to be preserved in rocks from the earth's surface, but the amount of thorium and uranium is greatly enhanced, typically by factors on the order of 10–100. We have already discussed the mechanism by which this occurs, which involves partial melting of the primordial mantle and concentration of large-radius species such as U and Th in the liquid fraction.

Several Soviet landers have been placed on the surface of Venus and have carried out rudimentary measurements of the density of the surface rocks and their levels of radioactivity (a U.S. probe which also sent back data from the surface of Venus was designed to study only the atmosphere). One of the Soviet landers was situated on uplands terrain which may correspond to some of the oldest crustal regions on Venus, and two were placed on the flanks of a large plateau (Beta Regio), which may be a more recent volcanic construct. The rocks were found to have a relatively low density ($\rho \simeq 2.7 - 2.9$ g/cm^3), as might be expected for igneous, differentiated material. Measurements of radioactivity showed that U and Th were present in amounts comparable to terrestrial crustal rocks, with a uranium abundance of about 2 ppm at the uplands site and about 0.5 ppm at the two "volcanic" sites. The Th/U ratio was similar to the earth's.[22] Thus we conclude that the thermal history of Venus must have included major episodes of mantle melting and differentiation, as seems to be the case for the earth and the moon as well.

Thermal evolution calculations for Venus confirm that substantial melting of the mantle should have occurred as a result of initial accretional heating and radioactive heat release. A model calculation of the present temperature profile in Venus is shown in Fig. 8-11.[23] Note that the planet is predicted to have a lithosphere, underlain by a partially molten region which would correspond to the terrestrial asthenosphere. Thus, according to this model, many of the ingredients for plate tectonics should be present on Venus. However, there is a major difference in the thermal structures of the outermost layers of the earth and Venus. This difference has to do with consideration of the density and composition of Venus' atmosphere, for which we now digress.

Atmosphere: Effect on Interior Structure. We are used to ignoring the atmosphere as a portion of the planet which contributes negligibly to bulk properties such as composition, mass, radius, heat flow, or temperature distribution. However, in certain planets, the structure of the atmosphere can have important repercussions on the planet's interior structure. Venus has a dense, hot atmosphere composed principally of CO_2. The surface temperature is about 730 °K and the surface pressure is about 90 bars.[24] The surface temperature is far above the equilibrium temperature with sunlight, T_s, and is appar-

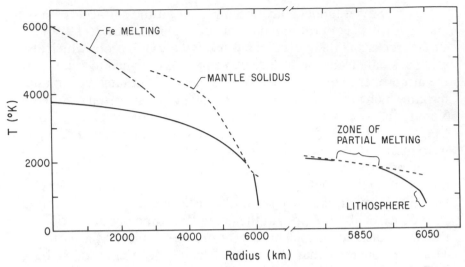

Fig. 8-11. Thermal model of Venus; outermost layers are shown on an expanded scale. [Data from Toksoz, M. N., Hsui, A. T., and Johnston, D. H. Thermal evolutions of the terrestrial planets. *The Moon and Planets* **18**: 281–320 (1978).]

ently caused by a substantial greenhouse effect (Chapter 5) attributable to the density of the CO_2 atmosphere and cloud layers. The origin of this greenhouse and the dense atmosphere seems to be related to an instability produced by substantial warming of the outer layers of a planet whose rocks contain substantial amounts of CO_2. The total amount of CO_2 on the earth and on Venus is roughly the same, but the CO_2 in the earth is primarily bound in carbonate rocks, such as limestone. If these rocks are heated sufficiently, the CO_2 is outgassed to the atmosphere, where it acts as an efficient infrared aborber. If enough CO_2 is present, the atmosphere begins to act as a "blanket" in the infrared while remaining transmissive to incoming visual-wavelength radiation, and the surface temperature rises further, thus accelerating the process of outgassing. The end point, which seems to have been reached on Venus, has all of the available CO_2 in a dense, hot atmosphere. The surface temperature is essentially limited only by the surface pressure of the atmosphere.[25]

Although Venus resembles the earth in a number of respects, its solid surface is at a much higher temperature. Recall that we have defined the lithosphere to be that part of the planet's outer layers which is at a temperature below about 1100–1300 °K, and which therefore acts as a rigid body over the age of the solar system and is unable to participate in mantle convection. Since the earth's surface temperature is about 300 °K, the terrestrial lithosphere spans a temperature range of 800–1000 °K. The thermal conductivity of silicates at a temperature of about 1000 °K is $\approx 3 \times 10^5$ ergs/cm/s/°K, and with an average heat flow of 62 ergs/cm^2/s (Chapter 5), we infer a terrestrial lithosphere thickness on the order of 40 km. After allowing for a temperature gradient which declines with depth due to concentration of heat sources in the crust, the earth's lithosphere may be somewhat thicker. But in any case, Venus' lithosphere only spans a temperature range of ~400–600 °K, and therefore if the temperature gradient in Venus is similar to that in the earth, which seems reasonable in view of the similar surface abundances of radioactive species, the Venus lithosphere is only about half as thick as the terrestrial one.

What kind of plate tectonics would we expect on Venus? Although mantle convection is apparently important for terrestrial plate tectonics, it is believed that the process of plate subduction (Chapter 7) may require a type of gravitational instability, whereby a cooling lithospheric plate shrinks enough to become denser than the underlying material, causing the edge of the plate to "dive" into the mantle. Possibly, Venus' lithosphere can never cool enough to produce this type of density inversion, and so no recycling occurs between the mantle and the crust.[26] These speculations cannot be resolved at present because available radar data on Venus topography are not really detailed enough to show features which would be diagnostic of plate tectonics. We can only say that there is so far no evidence for terrestrial-style tectonics.

In principle, the structure of Venus' gravity field can shed light on the properties of the lithosphere. If the lithosphere is thinner than the earth's, as predicted, then less nonhydrostatic deformation of the surface should occur. One might then expect Venus' gravity map to be somewhat smoother than the earth's, and like the earth's, to be little correlated with topography. However, very different properties are found.

Gravity Field. Figure 8-12 shows a sixth-degree and order equipotential contour map of Venus.[27] It is qualitatively very similar to the earth's equipotential map plotted to the same degree and order and

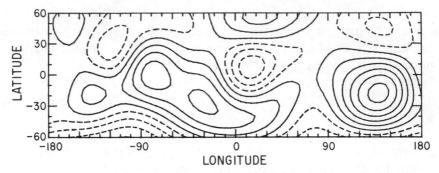

Fig. 8-12. Venus gravitational equipotential contour plot. The contour interval is 20 meters. Dashed contours have negative elevation; the first solid contour is the zero contour.

with the rotational response removed (Fig. 4-2). Like the terrestrial geoid, the Venus map shows excursions with maximum amplitudes on the order of several tens of meters. But on closer examination, significant differences appear. Recall that the terrestrial geoid shows little or no correlation with major topographic features such as continents and mountain ranges. However, "continents" *are* somewhat discernible on the Venus equipotential map. Figure 8-13 shows a shaded relief map of radar altitude contours on Venus (imagine that the light is coming from above and to the right). Some of the major high points and low points from the map of Fig. 8-12 are indicated on the relief map for comparison. Unlike the earth, Venus has only a few clearly identifiable "continents." The major elevated regions on Venus are Ishtar Terra at ~65°N. latitude, ~0° longitude, which includes the highest point on Venus, Mount Maxwell, Beta Regio at ~30° N. latitude, ~80° longitude, which may be a volcanic construct, and a large, elongated "continent" running roughly east-to-west along the equator from about 60° longitude to about 210° longitude (Aphrodite Terra). Note that a "high" of the gravitational potential is found on or near each of these elevated regions.

The gravitational expansion coefficients can be used to derive the orientations of the principal axes of the planet. The C-axis (axis of maximum moment of inertia) lies quite close to the rotation pole, being offset by an angle of about 5°. However, this result is quite uncertain, and considerably larger offsets are consistent with the error bars on the gravitational coefficients. The reason that the orientation of the C-axis remains uncertain is that Venus' equatorial

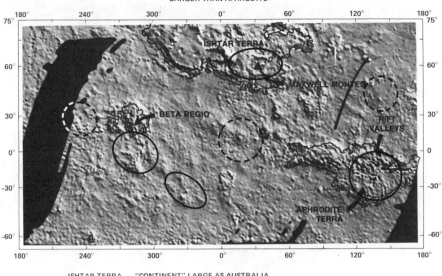

ISHTAR TERRA — "CONTINENT" LARGE AS AUSTRALIA
BETA REGIO — DOUBLE SHIELD VOLCANOS LARGER THAN HAWAII — MIDWAY CHAIN
MAXWELL MONTES — HIGHER THAN MT. EVEREST
APHRODITE TERRA — "CONTINENT" BIG AS HALF OF AFRICA

Fig. 8-13. Major highs and lows from Fig. 8-12 are shown superimposed on a shaded relief map of Venus (courtesy NASA).

bulge has a height of only a few tens of meters, and it is therefore difficult to measure its precise orientation. It is not known whether this bulge is a "fossil" distortion from an earlier period of more rapid rotation. Such a bulge might be largely eradicated over the age of the solar system because of solid state creep, owing to the average high temperature of Venus' lithosphere. The A-axis (axis of minimum moment of inertia) passes through the surface at about 130° longitude and −2° latitude. Venus' figure is essentially triaxial, since the amplitude of the bulge at either end of the A-axis is comparable to the equatorial bulge. The A-axis appears to pass through Aphrodite Terra, which may in fact be the planetary mass feature which is responsible for this orientation of the principal axes. If the rotation state of Venus were locked in a rotational resonance with the earth (see Chapter 4) either Aphrodite Terra or its antipodal point would have to face the earth at each Venus-earth conjunction. This does not appear to be the case, and as discussed in Chapter 4, the *present* bulge in Venus' gravitational figure associated with the location of Aphrodite Terra does not appear to be big enough to provide a stable resonance lock.

An even more convincing correlation between gravitational *acceleration* and topography is shown in Fig. 8-14.[28] Figure 8-14a shows a contour map of the vertical acceleration at 200 km altitude above the mean surface of Venus, relative to the average spherically symmetric perturbation. This map is more detailed than the equipotential contour map because it is constructed on the basis of instantaneous accelerations of the spacecraft rather than on the basis of long-term perturbations. Figure 8-14b shows a gravitational contour map for Venus which is synthesized by assuming that all of the gravitational perturbations are produced by undulations in the surface topography. The equation which is used for such a synthesis is the analog of Eq. 4-27:

$$g_l^m = \frac{3(l+1)}{2l+1} \frac{GM}{a^2} (\rho_s/\rho_0)\delta r_{lm}, \tag{8-17}$$

where the g_l^m are spherical-harmonic components of the gravitational acceleration evaluated at the mean surface of the planet $(r = a)$, ρ_s is the assumed density of the surface layers (2.7 g/cm^3), and ρ_0 is the average density of Venus.

The agreement between Figs. 8-14a and 8-14b is quite striking in most cases. Figure 8-14b shows the gravitational accelerations that would be present if Venus were a precisely spherical body overlain by a thin surface layer with density ρ_s and a thickness corresponding to the observed surface topography. Accelerations shown in Fig. 8-14b

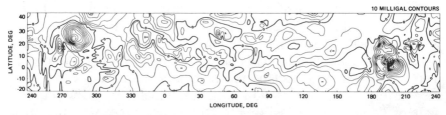

Fig. 8-14a. Venus vertical gravity at 200 km altitude derived from line-of-sight gravity measurements by Pioneer Venus Orbiter. [From Sjogren, *et al.* (1983). Copyright, American Geophysical Union.]

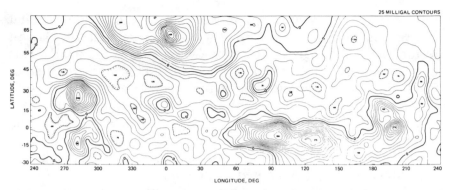

Fig. 8-14b. Venus vertical gravity at 200 km altitude, synthesized from surface topography using Eq. 8-17. [From Sjogren, *et al.* (1983). Copyright, American Geophysical Union.]

are typically about three times larger than the observed accelerations, except in the region of Aphrodite Terra, where they are about ten times larger. Thus the topography on Venus is isostatically compensated, but to a far lesser degree than on the earth. This result is somewhat unexpected, and is not entirely consistent with the picture of a thin lithosphere which we discussed above. A thinner lithosphere than the earth's should be able to yield more under the stress of an overlying "continent," and in equilibrium one would expect more isostatic compensation to be present than in the earth. Perhaps Venus is a very dynamic planet which forms surface features more rapidly than they can be compensated by solid-state creep processes. More detailed geological investigations of the ages and characteristics of the surface will be needed in order to investigate this possibility.

Mars

Interior Structure and Evolution: Observational Data. After the moon, Mars is the terrestrial body whose interior structure is best constrained by an observational data base. The mean density is 3.96 g/cm^3 for a mean radius of 3389 km (Table 1-1). Unlike the case for Mercury and Venus, for Mars there exists the possibility of inferring the dimensionless moment of inertia C/Ma^2 from hydrostatic equilibrium theory. This is because the value of J_2 (or $-C_{20}$) is clearly dominated by the response to rotation (Table 4-1). Unlike Mercury and Venus, Mars rotates relatively rapidly (rotation period $= 24^h 37^m$), because it has not been significantly despun by tidal interactions with the sun. A formal application of the Radau-Darwin formula (Eq. 4-54) gives the result $C/Ma^2 = 0.377 \pm 0.001$. This number is sufficiently close to the limiting uniform-sphere value of 0.400 that the nonuniqueness of the Radau-Darwin relationship between C/Ma^2 and $\Lambda_{2,0}$ is unimportant. However, we must still worry about whether the Martian J_2 represents solely a hydrostatic equilibrium response to rotation. If this number is contaminated by a nonhydrostatic contribution, a corresponding correction must be applied before $\Lambda_{2,0}$ is computed by dividing J_2 by q.

Other geophysical data are scanty. Although rudimentary seismic measurements were attempted by a Viking surface lander experiment, the sensor was not sufficiently well coupled to the solid surface to definitely reveal any intrinsic vibrations other than wind noise.[29] Other experiments on the Viking landers carried out important surface chemistry studies of relevance to interior structure, but the Viking program was basically oriented toward a search for living organisms on Mars (which ultimately proved unsuccessful).

As discussed in Chapter 6, it is not certain whether Mars possesses an intrinsic self-sustained magnetic field. Even accepting a controversial Soviet measurement as a positive detection, the Martian dipole moment falls somewhat below the value suggested by a simple scaling law. We will return to possible implications that this may have for a dynamo in Mars after discussing interior models.

The geology of Mars reveals several phenomena which have bearing on the structure of the interior. The familiar ubiquitous reddish appearance of the soil appears to be primarily caused by a high abundance of Fe_2O_3 in the surface minerals (about 18% by mass, according to Viking lander measurements).[30] The surface of Mars is rusty! As we shall see, interior models predict that the high iron abundance in the outer layers continues deep into the mantle.

There is little doubt that the Martian mantle has undergone deep igneous differentiation. Unlike Mercury and the moon, volcanos are a prominent feature of the Martian landscape. In particular, a region of Mars known as Tharsis ($\sim 0°$ latitude, $\sim -110°$ longitude) is dominated by a volcanic plateau containing several enormous volcanos. The characteristics of these volcanos have important implications for the structure and composition of the Martian mantle. The first important feature of these volcanos is that they are *shields,* which means that they have convex, gently-sloping surfaces (Fig. 8-15). Such a volcano is produced by the eruption of very low-viscosity lavas, which flow rapidly over large distances before solidifying at a small angle to the horizontal. The viscosity of terrestrial lavas varies greatly, but typically lies in the range 10^2 to 10^5 poise. In contrast, the Martian lavas may have a viscosity on the order of 10 poise. Such a low viscosity would be found in a lava which is produced by partial melting of exceptionally iron-rich rock, producing an FeO abundance in the lava which could be as large as 25% by mass.[31] Thus the physical properties of the Martian lavas, combined with the large volume which they represent, provide strong evidence that Mars has a mantle which is enriched in iron relative to the earth.

How deep within Mars do the magmas originate? If they are produced by partial melting in rock at a temperature of ~1100 °K, then they will be ~10% less dense than the surrounding solid material and will therefore rise to the surface. A shield will be built up to progressively higher altitudes as more and more magma erupts, flows, and solidifies. The process ultimately terminates when the hydrostatic head of the column of molten material equals the lithostatic pressure on its base. Thus, if the relative density contrast is 0.1, and the magma source region is a distance y below the mean planetary surface, then the magma column can be raised a distance $0.1y$ above the mean surface (Fig. 8-15). The shield volcanos in the Tharsis ridge area have elevations on the order of 20 km, so their source

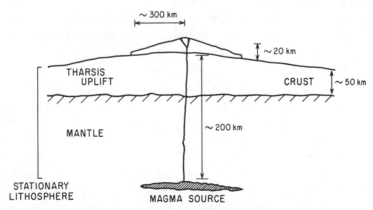

Fig. 8-15. Schematic vertical cross-section of the surface layers of Mars in the vicinity of the Tharsis uplift, showing a typical shield volcano. Vertical scale is exaggerated by a factor of about three.

region is 200 km deep, confirming that the iron-rich lavas have their origin within the Martian mantle.

The Tharsis volcanos are much larger than any comparable feature on the earth. Their diameters are typically ~ 500–600 km, not including the Tharsis uplift itself. One of the larger shield volcano constructs on the earth is the island of Hawaii, whose diameter is only ~ 100 km. How is Mars, a small planet, able to produce such enormous volcanos compared with the earth? First note that the island of Hawaii is only the most recent of the Hawaiian Islands. A chain of volcanic islands and seamounts extends from Hawaii toward the northwest, with age increasing in that direction also. It is as if the magma source for these volcanos has been gradually moving under the crust, toward the southeast. More likely, the lithospheric plate which contains the Hawaiian Islands has been sliding over a more or less fixed magma source in the mantle. This relative motion prevents the shields from growing to great size. In contrast, it appears that the Martian volcanos are in place over a magma source (possibly now extinct) which has never moved substantially with respect to them. There is evidence of a partially molten layer in Mars at ~ 200 km depth: this layer may correspond to the Martian asthenosphere. Because Mars is a small planet compared with the earth, its outer layers may be cooler, and therefore Mars' lithosphere may be thicker than the earth's, which may prevent terrestrial-type plate motions. Subsequently, we will show how these conjectures are confirmed by models of the interior.

The age of the Martian volcanos is not known. Relative age-dating techniques can be applied by counting crater densities on the flanks of the volcanos and comparing them with more heavily cratered regions elsewhere on Mars. These techniques indicate that the large Tharsis features are among the youngest on Mars, and there is some indication that Mars has remained volcanically active longer than the moon. However, an absolute age, which would be useful for calibrating thermal models of the Martian interior, cannot be reliably inferred at present.

There is no evidence for terrestrial-type plate tectonics on Mars. Tensional faulting seems to be a characteristic feature of Martian tectonics, in contrast to Mercury. Starting to the east of Tharsis, and extending for about four thousand km in an east-west direction along the equator, lies a large and complex canyon system known as Valles

Marineris. The origin of this feature is unclear, but it may have been related to a global expansion of Mars, which caused the crust to split in this region. It is therefore important to determine whether thermal models of the interior evolution of Mars are able to account for such an event.

The condensation-sequence theory for the origin of planetary bulk compositions (Chapter 2) meets an interesting challenge in Mars. According to this theory, Mars, which formed at a greater distance from the sun than the other terrestrial planets, should have a composition which reflects condensation at generally lower temperatures. Other than data on FeO abundance discussed above, geological observations may show evidence for substantial amounts of water, another volatile material. Superficially, Mars is an extremely dry place. Only small amounts of water are present in the Martian atmosphere and in the polar caps. However, certain geological features, such as craters with lobate scarps and eroded channels, seem to be best explained in terms of a layer of permafrost (H_2O frost) in the Martian crust. The total amount of this water, and whether any of it remains today, is unknown.

Gravity Field. Figure 8-16 shows the areoid, or gravity equipotential contour plot for Mars.[32] In this plot, the large contribution from the J_2 term has been removed, so that the excursions all represent departures from hydrostatic equilibrium. The excursions are very large, and are highly correlated with topography. Note the pronounced elevation of the contours in a region which coincides with

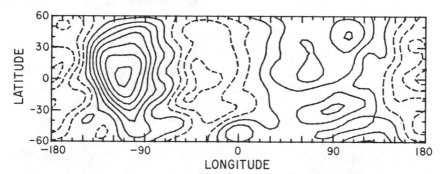

Fig. 8-16. The areoid, or Mars gravitational equipotential contour plot. The contour interval is 200 meters. Dashed contours have negative elevation; the first solid contour is the zero contour.

the Tharsis uplift. Indeed, to a first approximation, we could reproduce the areoid with a model of a spherical planet and an extra mass point at about the position of Tharsis. Areoid heights exceed 1 km over the Tharsis ridge, indicating that this feature is not highly compensated isostatically. The simple scaling relation which we derived from the expected variation in geoid height, Eq. 8-4, does not predict such large-amplitude variation for Mars. According to the scaling relation, the global undulations of the areoid should be typically only about one-third of the amplitude of the lunar undulations, while they are actually about two times greater. The Martian lithosphere supports greater stresses than the lunar lithosphere. Other than the equatorial bulge produced by rotation, the Martian gravitational figure is dominated by a single "mascon," the Tharsis ridge. However, this feature is not analogous to a lunar mascon, for it corresponds to an actual topographic high.

In hydrostatic equilibrium, the difference between the equatorial and polar radii of Mars would be about 18 km. This figure is not vastly greater than the nonhydrostatic variation of the areoid radius (~ 1 km), which suggests that the measured value J_2 for Mars may not represent a pure hydrostatic response to rotation, but may be partly "contaminated" by a nonhydrostatic component at the level of a few percent. This is an important issue because the value of C/Ma^2 for Mars is deduced, via Radau-Darwin theory, by assuming that the value of J_2 represents a purely hydrostatic response to rotation. If J_2 is not purely hydrostatic, then we must somehow correct it by an appropriate amount before applying the Radau-Darwin formula.

This situation has a terrestrial analog. Knowing the distribution of density in the earth from seismic models, one can predict the value of J_2 that the earth would have in precise hydrostatic equilibrium. This number is about 1% smaller than the measured value of J_2, and at the level of accuracy of the seismic models and the hydrostatic equilibrium theory, the discrepancy is significant. The discrepancy has been attributed by some investigators to a fossil equatorial bulge produced by the earth's response to an earlier, more rapid rotation rate. But other investigators believe that the excess oblateness is merely caused by a planet's tendency to rotate about the axis of maximum moment of inertia (see the proof in Chapter 4).[33]

For Mars we are faced with a similar problem of evaluating the hydrostatic response at an earlier epoch, presumably before Tharsis was formed, but because of the lack of seismic data, we do not know the "true" value of J_2. Instead, the following assumptions are made. First, we assume that Mars developed its present equatorial bulge at a time when it was able to reach a state closer to hydrostatic equilibrium. It is not clear whether there was ever such a time, but if it occurred, it may have been at the time of extensive partial melting of the mantle and volcanic activity about 2×10^9 years ago, according to thermal models (see below). At this time, Mars had the same rotation rate as at present because of negligible tidal slowing. The planet then became rigid and able to support stresses. Next, magma from all around the planet concentrated in the Tharsis region, erupted onto the surface, and formed the Tharsis bulge, a symmetric dome-shaped feature. The rigidity of the underlying lithosphere prevented any substantial isostatic compensation, so that the extra mass now sits atop the lithosphere without deforming it. Similarly, there was no deformation of the lithosphere during the process of draining the magma from elsewhere within the planet. Under these assumptions, some of which are debatable, it is possible to derive a Tharsis-removed value of J_2 through a simple procedure.[34,35] Let the hydrostatic component of Mars' figure have corresponding principal moments of inertia A^0, B^0, and C^0. If C^0 is the polar moment of inertia, then $A^0 = B^0$. Let the observed moments of inertia be $A = A^0 + A'$, $B = B^0 + B'$, etc., where A', B', and C' are purely nonhydrostatic contributions. The C axis coincides with the original rotation axis. The A^0-axis is indeterminate of course, because the initial planet is axially symmetric, but the A axis (minimum moment of inertia) must pass through Tharsis if the latter is entirely responsible for the nonhydrostatic component of the planetary figure. From Chapter 4 and the gravity data for Mars, we have

$$(C_{22}{}^2 + S_{22}{}^2)^{1/2} = (B - A)/(4\ Ma^2) = (B' - A')/ (4\ Ma^2) = 0.000063. \tag{8-18}$$

If the nonhydrostatic component of J_2 is J_2', then we must have

$$J_2' = [C' - 0.5\ (A' + B')]/Ma^2. \tag{8-19}$$

Because the Tharsis bulge is assumed to have a symmetrical shape (or to be negligible in horizontal extent), we have $C' = B'$. Thus Tharsis contributes $J_2' = (1.26 \pm 0.05) \times 10^{-4}$ to J_2, leading to a corrected (hydrostatic) value $J_2 = (1.829 \pm 0.012) \times 10^{-3}$. The Radau-Darwin approximation then gives $C/Ma^2 = 0.365$, the "undistorted" dimensionless moment of inertia. As we shall now see, the precise value of this quantity turns out to be critical for determining the chemical composition of the Martian interior.

Interior Structure: Static Models. The central pressure of Mars does not exceed about 400 kilobars, a pressure which is reached at a depth of about 1000 km in the earth. Thus material in the Martian interior is not greatly compressed. Using this fact, and the constraint on the interior density profile provided by C/Ma^2, it is possible to conclude that the Martian mantle is denser than the earth's. We will now present an approximate model which illustrates this point.[36]

We have already concluded that the geological evidence shows that Mars at some time underwent deep mantle melting and differentiation, and that an iron-nickel or perhaps, Fe-FeS + Ni core formed. Assume that the interior model can then be represented by a central core of constant density ρ_c and a mantle of density ρ_m. Let $\langle \rho \rangle$ be the average density of the entire planet, equal to 3.96 g/cm³. An elementary calculation then shows that the dimensionless moment of inertia is related to these quantities by the equation

$$C/Ma^2 = \frac{0.4}{\langle \rho \rangle} [\rho_m + (\langle \rho \rangle - \rho_m)^{5/3}/(\rho_c - \rho_m)^{2/3}]. \qquad (8\text{-}20)$$

Table 8-1 gives some solutions to this equation. If we take the older, uncorrected value $C/Ma^2 = 0.377$, then a core density of 6 g/cm³ implies a mantle density of about 3.65 g/cm³, while a core density of 8 g/cm³ implies a mantle density of 3.7 g/cm³. The corrected value $C/Ma^2 = 0.365$ gives $\rho_m = 3.54$ g/cm³ for $\rho_c = 8$ g/cm³, and $\rho_m = 3.42$ g/cm³ for $\rho_c = 6$ g/cm³. We conclude that the mantle density is not very sensitive to the assumed core density, but it is quite sensitive to the value of C/Ma^2.

The two values of the core density given above correspond to two limiting hypotheses about the composition of the Martian core. If the core is similar to the earth's, being composed largely of iron with small amounts of nickel and some lighter element such as sulfur, then its average density at a pressure of a few hundred kilobars would be

Table 8-1. Solutions of Eq. 8-20

$\rho_c = 6$ g/cm³		$\rho_c = 8$ g/cm³	
ρ_m (g/cm³)	C/Ma^2	ρ_m (g/cm³)	C/Ma^2
3.2	0.3554	3.2	0.3457
3.3	0.3594	3.3	0.3513
3.4	0.3638	3.4	0.3573
3.5	0.3686	3.5	0.3637
3.6	0.3739	3.6	0.3705
3.7	0.3799	3.7	0.3778
3.8	0.3867	3.8	0.3857
3.9	0.3945	3.9	0.3943

about 8 g/cm³. On the other hand, if the core formed under conditions close to the eutectic temperature and composition of an Fe-FeS mixture, then $\rho_c \simeq 6$ g/cm³. In the latter case, the core radius would be about 500 km greater than the radius of the 8 g/cm³-core. This may some day be measured by a suitable seismic experiment, but until such a measurement is carried out, there is no real observational constraint on the possible composition of the Martian core.

Because of the absence of constraining data, the composition of Mars' core is a controversial matter. According to one school of thought, Mars is enriched in volatiles which were able to condense due to Mars' distance from the primordial sun. A cosmic or near-cosmic complement of sulfur was therefore incorporated in the solid condensate, primarily in the form of FeS, most of which now resides in the Martian core.[20] However, another school of thought argues that the abundances of argon isotopes in the Martian atmosphere imply that the planet is actually globally depleted in volatiles, and is closer in bulk composition to the moon than it is to the earth.[37] As we have shown, the deduction of the mantle density of Mars does not depend upon which of these schools of thought is correct, although thermal models turn out to be more sensitive to the controversy.

Allowing for compressibility of the mantle and core material, and then correcting back to the zero-pressure density of the mantle, more detailed models[1] find an uncompressed mantle density of about 3.55 g/cm³, using the revised value of C/Ma^2. For comparison, the uncompressed density of the earth's mantle is calculated to be 3.34 g/cm³. Since a variety of assumed compositions for the Martian core give essentially the same result for the mantle density, the difference between Mars and the earth is probably significant. What is responsible for the difference? As we have seen, there is geological evidence that the Martian mantle is rich in iron. Thus the most straightforward and chemically plausible procedure for producing a higher density is to suppose that the Martian mantle consists of (a) pyrolite, the proposed primitive mantle composition for the earth (see Chapter 7), and (b) FeO in an amount sufficient to raise the density to the required value. Pyrolite contains about 8% FeO. If we instead assume that the Martian mantle contains about twice as much ($\sim 16\%$) FeO, with the other proportions in pyrolite remaining the same, then the mantle density comes out about right. Moreover, when a mantle of this composition is melted, the magmas which are produced are iron-rich, with viscosities in the right range to explain the properties of the Martian shield volcanos.

Nevertheless, Mars as a whole appears to contain less iron than the earth. Most models indicate that iron in the Martian core comprises about 15% of the mass of the planet. As a whole, the planet is about 27% iron by mass. These results are not very sensitive to the assumed sulfur content of the core. With a silicon abundance of about 16% by mass, Mars apparently has essentially the same Fe/Si mass ratio as the primordial solar abundances (Tables 2-1 and 2-2). From this, one would conclude that Mars continues a trend which starts with Mercury. This trend corresponds to a decreasing iron/rock ratio with increasing heliocentric distance, or equivalently, to a decreasing uncompressed average planetary density (we have already discussed some bothersome problems raised in this regard by Venus). A possible explanation for the trend is that the iron component was everywhere completely condensed when all of the terrestrial planets began to form from planetesimals, but that the rock component was incompletely condensed in Mercury's region, somewhat more condensed in the region of Venus and the earth, and entirely condensed in the region of Mars.

Thermal Structure. The first task of thermal modeling of Mars is to explain the presence of a dense core. The existence of such a core is implied by the value of C/Ma^2, as discussed above. If Mars initially formed out of accreted cold planetesimals, and if this process occurred so slowly that little accretional heating occurred (see Chapter 5), then core formation could have proceeded only as a result of gradual radioactive heating from the decay of uranium, thorium, and potassium isotopes. Eventually partial melting would occur, leading to gravitational separation of iron from silicates. Let us first perform a computation to see if Mars could have developed a core in this manner.

We begin with the full equation of heat conduction,

$$r^{-2} \frac{\partial}{\partial r} \left(r^2 K \frac{\partial T}{\partial r} \right) = \rho C_P \frac{\partial T}{\partial t} - \epsilon_N. \tag{8-21}$$

Assume that the initial body is homogeneous, with constant thermal conductivity K, heat capacity C_P, and energy generation rate ϵ_N. The body is initially at a constant temperature T_1, and at all subsequent times the surface remains at T_1. The solution will then be a linear combination of the two solutions given earlier in the section on Mercury: Eqs. 8-12 and 8-16, which are respectively the solutions with $\epsilon_N = 0$ and $\partial T/\partial t = 0$. Since no core has been formed yet, we adopt Eq. 8-16 in the limit r_c (= core radius) $\rightarrow 0$. The solution to Eq. 8-21 is then given by

$$T = T_1 + \frac{\rho \epsilon_N}{6K} (a^2 - r^2) + \sum_{n=1}^{\infty} T_{0n} \frac{\sin (kr)}{kr} \exp (-t/\tau_n), \tag{8-22}$$

where the T_{0n} must be determined from the initial temperature distribution. We will assume that the initial temperature is uniform and equal to T_1. The sum of the last two terms on the right of Eq. 8-22 must therefore be zero at $t = 0$. Multiplying the resulting expression by $kr \sin (k'r)$ and integrating over r from 0 to a, we find

$$T_{0n} = -2\pi n \frac{\rho \epsilon_N}{6K} a^2 \int_0^1 dx \, (x - x^3) \sin (\pi n x). \tag{8-23}$$

It is convenient to express the solution in terms of a dimensionless temperature T', time t', and radius x. We let $T = T' T_u$, $t = t' t_u$, and $r = ax$, where

$$T_u = \rho \epsilon_N a^2 / 6K \tag{8-24}$$

is the unit of temperature and

$$t_u = \rho C_P a^2 / K = a^2 / \chi \qquad (8\text{-}25)$$

is the unit of time.

Figure 8-17 shows $T'(x)$ for various values of t'. Here we have set $T_1 = 0$. The actual temperature can be obtained by multiplying T' by T_u and adding T_1. This general solution can be applied to any homogeneous planetary body with a uniform distribution of heat sources by computing the appropriate values of T_u and t_u. For $t \ll t_u$, the temperature is essentially constant everywhere within the planet except close to the surface, and it rises linearly with time. In the asymptotic limit $t \gg t_u$, a steady-state situation is reached, with $T' = 1 - x^2$.

We have already noted that over planetary scales greater than about 1000 km, $t < t_u$ over the age of the solar system. Thus, for example, Mercury's silicate mantle could conceivably have reached a steady state over the age of the planet, but the entire planet could not. In the case of Mars, we adopt $K = 3 \times 10^5$ erg/cm/s/°K, $\rho = 4$ g/cm³, $C_P = 1.2 \times 10^7$ erg/g/°K, and for ϵ_N we take a present average value

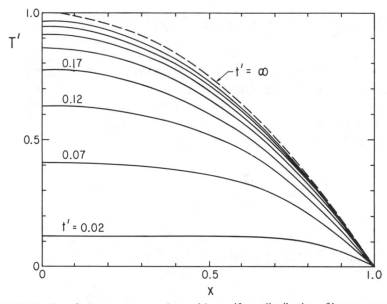

Fig. 8-17. Heating of a homogeneous sphere with a uniform distribution of heat sources.

5.3×10^{-8} erg/g/s (see Table 5-1), which is very similar to the value in primitive meteorites. We then find $T_u = 13400$ °K and $t_u = 577 \times 10^9$ years. The solutions corresponding to the present Mars would have $t' = 0.008$ and a nearly uniform interior temperature of $T' = 0.048$, or $T = T_1 + 640$ °K. Assuming $T_1 \simeq 300$ °K, the planet could not have heated up enough to melt out an iron core over the age of the solar system.

Naturally, some of the parameters used in the above calculation are somewhat uncertain. The most uncertain is the actual concentration of radioactive species. We have used the concentration which corresponds to the estimated *current* concentration of radioactive species in the earth assuming steady-state heat flow, which happens to nearly coincide with the present concentration in a solar composition mix. In the early solar system, the concentration would have been higher, and it might have been possible for radioactive heating to increase the Martian interior temperature from an initial temperature T_1 to perhaps $T \simeq T_1 + 1000$ °K. If Mars could achieve a central temperature of ~ 2000 °K, it would be marginally possible to melt a eutectic mixture of Fe and FeS (see Fig. 3-16). But in this case, core formation would have occurred only very recently. The associated volcanism would likewise be a very recent phenomenon. Although we do not know the absolute ages of the Martian volcanoes, most photogeologists do not consider them to be this recent. From observed crater densities, they may be as much as 2×10^9 years old.

The problem of an energy source for core formation is compounded when we consider available data for the abundances of radioactive species on the surface of Mars. Gamma-ray spectrometry carried out by Mars 5, a Soviet orbiter, has shown that the K/U ratio on Mars is about 2200, or about a factor of twenty lower than the *present* cosmic ratio[22] (remember that Table 2-1 gives the *primordial* abundances of U and Th). Interestingly, a similar depletion of K is observed on the moon. This result complicates our simple picture of Mars as a body assembled from low-temperature condensates. It also reduces the amount of heat which can be generated by decay of K^{40} (see Fig. 5-1).

All of the above results lead to the following picture of the thermal evolution of Mars. We must assume that sufficient heat sources are present to provide for core formation and partial melting of the silicate mantle, leading to volcanism midway through the evolution of Mars. It does not appear to be possible to do this with an assumed (present) cosmic abundance of U equal to about 15 ppb relative to rock plus iron. If one assumes that the present U abundance in the planet as a whole is about 28 ppb, with Th/U = 3.6 and K/U = 2200 globally, then globally $\epsilon_N = 21.5 \times 10^{-8}$ erg/g/s when Mars first forms, declining to 5.3×10^{-8} erg/g/s at present. This might provide a large enough heat source to account for the geological observations. But even so, early formation of a Martian core seems to require a high initial T_1, say 1300 °K. This assumption is somewhat *ad hoc,* but it insures that Fe-FeS melting ensues immediately with the addition of a small amount of accretional heating. The latter amounts to about 500 °K for an accretion time of 10^5 years. Since the Martian core is small masswise, not a large amount of extra energy is available from core formation. This comprises about 200 °K of additional heating, in all.[38]

Current thermal models of Mars postulate early core formation rather than deriving it from fundamental considerations. The mantle then continues to increase in temperature as a result of radioactive heating. Melting of the mantle and resulting volcanism would occur when mantle temperatures reach about 2000 °K, which may have occurred at a planetary age of about $2-3 \times 10^9$ years. The Tharsis bulge presumably dates back to this era. Temperatures in Mars have generally tended to increase with time, according to models. There has been substantial thermal expansion as a result, with the planetary radius increasing by up to 10 km over the age of the planet. This may have been responsible for the appearance of the large equatorial canyon system.

Some investigators believe that solid state convection has been an important process in the Martian mantle, and may be continuing today. However, it is not manifested in terrestrial-style plate tectonics because of the great extent of the Martian lithosphere. Adopting the conventional definition of the lithosphere, such that it has a base at a temperature of about 1300 °K, current thermal models predict a lithospheric thickness of about 200 km. This seems consistent with the geological evidence for a relatively stationary lithosphere.

Since thermal models invoke an initial Mars which was hot enough to melt out an Fe-FeS core, and since interior temperatures tend to remain high or increase further with time, it seems difficult to avoid the conclusion that the core is still molten. But if this is so, why does

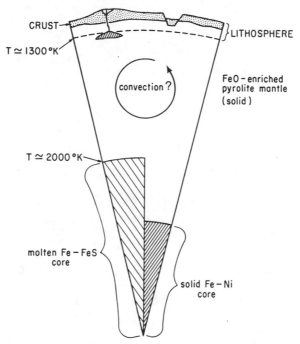

Fig. 8-18. Model of Mars' interior. Two alternative versions of the core are shown.

Mars lack an appreciable magnetic moment (Chapter 6)? All of the ingredients for a dynamo seem to be present; we must conclude that dynamo theories seem to have little predictive power when applied to the terrestrial planets. It is of course possible that Mars contains very little sulfur in its core, as is suggested by alternative models. In this case, Mars may have an iron-nickel core with a melting point on the order of 3000 °K instead of 2000 °K. The planet could then be hot enough inside to produce the Tharsis volcanos and still have a solid core incapable of dynamo action. However, in solving the problem of the magnetic field, this hypothesis creates another problem: how does the core get formed if the iron does not melt? A number of basic geophysical measurements, such as heat flow and determination of the seismic properties of the core, would help to resolve these issues.

Figure 8-18 shows a "typical" interior model of Mars, which is based upon the hypothesis of low-temperature condensation. As discussed above, there exist alternative models in which more volatile substances such as S and K are depleted.

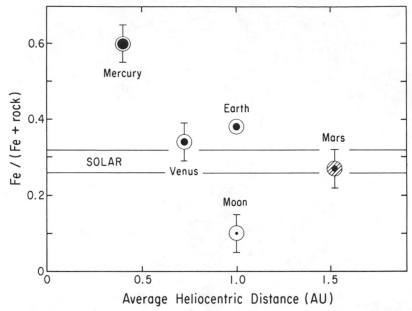

Fig. 8-19. Iron content of terrestrial planets (as derived from models) as a function of heliocentric distance. Solid shading within each symbol shows proportional size of the iron core (upper limit for the moon). Shading for the Mars symbol indicates a substantial quantity of iron in the mantle.

To conclude this section on the terrestrial planets, Fig. 8-19 illustrates the observed trend in the iron/rock ratio as a function of the heliocentric distances of the terrestrial planets.

THE JOVIAN PLANETS

Much of the discussion of the interiors of the terrestrial planets employed extensions of geophysics. Where geophysical data were lacking, geological observations were used to infer important characteristics of the planetary interior.

An entirely different approach is employed for the study of the Jovian planets. First, there is no geology in the conventional sense. We see no solid surfaces, but only deep atmospheres with multiple cloud layers. There is no mineralogy, although the main planetary constituents probably undergo high-pressure phase transformations like minerals. The chemistry of the Jovian planets is dominated by the volatile and most abundant complement of the cosmic mixture, the complement that is largely absent in the terrestrial planets. Historically, research on the interiors of the terrestrial planets has been carried out by geophysicists applying their talents to more exotic objects than the earth. Conversely, much of the research on Jovian planet interiors has been carried out by astrophysicists who use extensions of the techniques which are applied to stellar structure.

Although the Jovian planets are much more massive and less dense than the terrestrial planets, and are therefore probably composed of a suite of lighter elements than are found in rock, it is not necessarily obvious that they are composed of matter which resembles solar composition. In order to understand why Jupiter and Saturn are believed to consist largely of hydrogen and helium, we must now return to some fundamental results which were discussed in Chapters 3 and 4.

We write the equation of hydrostatic equilibirum (Eq. 4-24) as a dimensional relationship:

$$P/a \sim (GM/a^2)(M/a^3), \qquad (8\text{-}26)$$

where P is the central pressure of a planet with a typical dimension (radius) a and mass M, and G is the gravitational constant. Thus the central pressure of a body in hydrostatic equilibrium should be of order GM^2/a^4. The central pressure of the earth is about 3 megabars, so according to this relation Jupiter's central pressure would be about 20 megabars, Saturn's about 3.5 megabars, and Uranus' and Neptune's about 3 megabars. These pressures are not high enough to carry us into the asymptotic regime for the pressure-density relation discussed in Chapter 3, but they are high enough to ensure that a substantial degree of compression occurs within these planets. The mean density of a self-gravitating body goes up because of this effect, and if the effect is large, it must be taken into account before an identification of chemical constituents can be made.

Consider the following thought experiment. We take a small amount of a solid material of composition X. The amount is so small that self-gravitation is negligible. The volume of the sample is simply equal to the mass of the sample divided by its uncompressed mean density. If the mass of the sample is then doubled, the volume also doubles. Mathematically, this is expressed as a dimensional relationship between the planetary mass and the radius:

$$M \propto a^3; \qquad (8\text{-}27)$$

valid for an incompressible planet. If the bulk modulus of the material is K_S, then the mass-radius relation of the material will follow Eq. 8-27 as long as the central pressure is much less than K_S. But if sufficient material is added to the planet, the matter at the center will begin to compress substantially, and the mass-radius relation will begin to deviate from Eq. 8-27. The radius will increase more slowly than the 1/3-power of the mass, and if the material is sufficiently "squashy," the radius may even become stationary and then begin to *decrease* with mass. Such behavior, which may seem strange to those used to thinking about objects in the terrestrial-planet mass range, is in fact predicted by the asymptotic pressure-density relation which was presented in Chapter 3.

Suppose that the pressure-density relation for a substance can be represented within a given pressure range by a power law:

$$P = K\rho^{1+1/n}, \qquad (8\text{-}28)$$

where K and n are respectively called the polytropic constant and polytropic index. At very low pressures, the mass density must be essentially independent of the pressure, and therefore $n \simeq 0$ as $P \rightarrow 0$. But in the opposite asymptotic limit of very high pressures, the pressure-density relation is given by Eq. 3-11, and in the limit $r_e \rightarrow 0$, it corresponds exactly to Eq. 8-28 with $n = 5/3$. We now write two dimensional relations for the central pressure, in this limit:

$$P \propto GM^2/a^4 \propto K(M/a^3)^{1+1/n}. \qquad (8\text{-}29)$$

For $n = 5/3$, this requires that $M \propto a^{-3}$. Thus, as more mass is added, the object *shrinks*. The material is so compressible that the average density must increase rapidly in order to maintain hydrostatic equilibrium as mass is added. The remarkable objects that behave in this manner are called *white dwarfs,* and represent an extension of the planetary sequence to masses a thousand times greater. In all of this discussion, of course, we are dealing with objects which are so cold that thermal effects play a small role in their pressure equation of state. Thus the sun is not described by such physics.

If the radius increases with mass in the planetary mass range and decreases with mass in the stellar mass range, it follows that there must be maximum radius. Let us derive this radius for various chemical compositions. First consider the value that the polytropic index must assume at the maximum radius. Note that the mass M cancels out of Eq. 8-29 for $n = 1$, leaving the result $a^2 \propto K/G$. Thus the polytrope of index one has a radius which is independent of mass. With the addition of more mass to the object, it compresses by precisely the amount required to leave the radius unchanged. The effective polytropic index must closely approximate unity near the maximum of the radius vs. mass curve.

The special case of $n = 1$ is one of the few polytropic relations for which the equation of hydrostatic equilibrium can be solved analytically, giving a density distribution

$$\rho = \rho_c \sin (kr)/(kr), \tag{8-30}$$

where

$$k = (2\pi G/K)^{1/2}, \tag{8-31}$$

ρ_c is the central density, and the radius is $a = \pi/k$. One then finds the central pressure $P_c = 0.393 \ GM^2/R^4$, and the relation between the central density ρ_c and the average density $\langle \rho \rangle$: $\rho_c = 3.29 \ \langle \rho \rangle$.

The actual pressure-density relation is of course never precisely represented by a polytropic relation, but we will assume that near the maximum in the radius, one has

$$1 + 1/n_{\text{eff}} = d \ln P/d \ln \rho = 2, \tag{8-32}$$

where n_{eff} is the effective polytropic index defined by Eq. 8-32, evaluated at the object's central density. To evaluate this quantity, we use Eq. 3-11, noting that $d \ln P/d \ln \rho = -(\frac{1}{3}) \cdot d \ln P/d \ln r_e$. Requiring $n_{\text{eff}} = 1$, one finds

$$r_e = 0.5/(0.407 \ Z^{2/3} + 0.207), \tag{8-33}$$
$$P_c = 0.176 \ e^2/(2a_0{}^4r_e{}^5), \tag{8-34}$$

and

$$\rho_c = 3Am/(4\pi Z r_e{}^3 a_0{}^3), \tag{8-35}$$

where A is the atomic mass and m is an atomic mass unit. Combining Eqs. 8-33 – 8-35, we find for the maximum radius

$$a_{max} = 101430 \ (Z/A)(Z^{2/3} + 0.509)^{-1/2} \text{ km}. \qquad (8\text{-}36)$$

Although some of the assumptions used in deriving Eq. 8-36 are rather crude, the equation gives results which are in good agreement with calculations of the mass-radius curve for more detailed pressure-density relations.[39] For example, for hydrogen, $a_{max} = 82600$ km, for helium $a_{max} = 35000$ km, and for carbon $a_{max} = 26000$ km.

It is important to note that the value of a_{max} for hydrogen is considerably larger than the value for any other element. This is a consequence of two circumstances: (a) hydrogen has a value of Z/A which is about twice as large as the value for any other element; (b) the coulomb correction to the pressure for hydrogen is substantially smaller than the correction for any other element. Just as the maximum radius for hydrogen is well-separated from the maximum radius for any other element, the radius vs. mass curve for hydrogen lies well above the curves for other elements. Since the observed radii of Jupiter and Saturn lie well in excess of 35,000 km, which is the maximum radius for an object of any composition except hydrogen, it follows on fundamental grounds that these planets must be predominantly composed of hydrogen. This result depends only on the assumption that Jupiter and Saturn are at interior temperatures substantially smaller than an atomic unit of temperature $(3 \times 10^5 \ ^\circ\text{K})$.

Figure 8-20 shows theoretical radius-mass curves for several chemical compositions. Observational values for the four Jovian planets (J, S, U, N) are also shown. Before presenting a discussion of this diagram, we need to define what is meant by the planetary radius a. Unlike the terrestrial planets, the Jovian planets lack well-defined "surfaces." Instead, it is customary to define the surface to be that layer in the atmosphere which is at a mean pressure of 1 bar, although the actual atmosphere extends for hundreds of kilometers above and below this level. In addition, the 1-bar surface is highly oblate because of the rapid rotation of these planets. Therefore the parameter which is used to characterize the planetary radius a is usually taken to be the average 1-bar radius at the equator. It is this quantity which is plotted

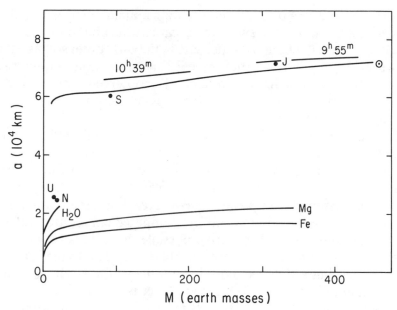

Fig. 8-20. Radius versus mass for the Jovian planets. Solid curves show theoretical relations for water, Mg, and iron at $T = 0$, and for solar composition with an adiabatic temperature distribution (⊙). Effect of rotation in Saturn and Jupiter is also shown.

in Fig. 8-20. Correspondingly, a correction for the effect of rotation must be made to the theoretical radius-mass curves before they can be quantitatively compared with the observational data.

The mass corresponding to the maximum possible radius for the various possible chemical constituents is in all cases substantially greater than any planetary mass in the solar system (for hydrogen it is about 1000 earth masses). Thus all of the planets lie on the ascending portions of their respective curves. Figure 8-20 shows $a(M)$ curves for cold spheres of magnesium and iron, computed using Thomas-Fermi-Dirac theory (Chapter 3). These curves are generally representative of matter similar in composition to the terrestrial planets, i.e. solar-composition matter from which the gaseous and volatile components have been removed. It is clear that none of the Jovian planets can have a composition of this type.

Next, Fig. 8-20 shows a short segment of a curve computed for cold spheres composed of H_2O, using an empirical pressure-density curve obtained from shock compression data on water (Chapter 3). The proximity of this curve to the points for Uranus and Neptune shows that H_2O would be a plausible choice for a major constituent of these planets, particularly in view of the great cosmic abundance of H_2O and its likely existence as a condensate in the early outer solar system. However, Fig. 8-20 shows that there is substantial nonuniqueness in Uranus and Neptune interior models. The observed radius could be reproduced by a variety of combinations of rock, water, methane, hydrogen, helium, etc.

The upper curves in Fig. 8-20 (marked $\odot$) are computed under the assumption of exact primordial solar composition as given in Tables 2-1 and 2-2. The proximity of Jupiter and Saturn to these curves provides a clear indication that these planets do not differ greatly from this assumed composition. However, the small differences between the observed points and the computed curves are probably significant, and are used to derive possible differences between Jupiter and Saturn's composition and the sun's. The longer curve is calculated using a detailed liquid-state theory for H-He mixtures. The temperature distribution in the planet is assumed to follow an adiabat starting at $T = 140 \, °K$ at $P = 1$ bar, and the transition from molecular to metallic hydrogen is assumed to be continuous. However, effects of rotation are not included in this curve, and as a result Jupiter lies above it while Saturn falls below it. The two segments marked "$10^h \, 39^m$" and "$9^h \, 55^m$" indicate the appropriate equatorial radius-mass curves for planets which rotate with these periods (corresponding to Saturn and Jupiter, respectively). If the temperatures in the model planets were reduced from the assumed adiabatic values to zero, producing a pressure reduction on the order of 10%, the computed curves for rotating planets would again be lowered to the vicinity of the longest curve (corresponding to the nonrotating case). The point is that it is not hard to conclude that Jupiter and Saturn are close to solar composition, but a deduction of their detailed composition requires careful inclusion of the effects of temperature and rotation.

Jupiter

Atmosphere. Jupiter's atmosphere, unlike terrestrial planet atmospheres, has a composition which is approximately representative of the bulk composition of the planet. There is no clear demarkation between the interior and the atmosphere. Thus the atmospheric chemistry of Jupiter plays a role analogous to that of the surface mineralogy in terrestrial planets in constraining the composition of the interior. Like chemical abundances in the surface layers of a terrestrial planet, the abundances in the atmosphere of Jupiter are in general not equal to the bulk abundances in the planet, because of chemical reactions which can occur at various points between the center and the surface of the planet. To illustrate this point, we will now consider a simplified model of the Jovian atmosphere.

Suppose that the lower Jovian atmosphere resembles the lower atmosphere of the earth, the so-called *troposphere*. In this layer, the infrared opacity is sufficiently great and the heat flux from below is sufficiently large that the atmosphere becomes convectively unstable. As discussed in Chapter 5, the temperature-pressure relation in such an atmosphere closely follows an adiabat, and we may adopt an adiabatic relationship for the purpose of computing a numerical model. The atmosphere departs from this relationship in the uppermost layers, where the gas density becomes small enough for the infrared photons to have a significant probability of escaping directly into space. The temperature in the atmosphere tends to become constant in the vicinity of this region. The top of the troposphere, the *tropopause,* is the layer at which the temperature gradient falls to zero. Convection ceases at a point somewhat below the tropopause.

Suppose further that the Jovian troposphere has a composition which is identical to primordial solar composition. Since it is convectively unstable, one might suppose that it is well mixed and would therefore have a uniform composition throughout. Measurement of the composition at the top of the troposphere would therefore suffice to determine the composition throughout the region of the planet included by the troposphere. In fact, this cannot be so, because as a parcel of solar-composition gas is carried through an adiabatic decompression from high temperature to low temperature, certain species will condense from the gas phase, forming liquid and solid particulate matter which will tend to remain near the layer at which it condensed. Being denser than the gas, such particles will "rain out" in the atmosphere until they reach thermodynamic conditions where they can redissolve in the gas phase.

We have already discussed the sequence in which various precipitates will appear in a cooling gas of solar composition. Chapter 2 noted that, under nebular pressures, the first precipitates which appear are the refractory compounds, then iron-nickel condenses, then the silicates, and then the "ices"—H_2O, NH_3, and CH_4. This same general sequence is preserved when we consider an adiabatic transformation of such a gas, although the temperatures of condensation and detailed chemistry of the condensates are altered by the much higher pressures. In this model, then, the present-day Jovian atmosphere serves as a kind of static recapitulation of the temporal and spatial condensation sequence of the primordial solar nebula.

Figure 8-21 shows a theoretical temperature-pressure plot of the Jovian atmosphere, along with similar curves for Saturn and Uranus/Neptune (which will be discussed in subsequent sections).[40] The part of the curve at pressures above a few bars is based upon actual spacecraft data from radio-occultation and infrared sounding measurements.[41] The temperature structure above the 100-millibar level appears to be spatially and temporally variable. At deeper layers it follows an adiabat with the boundary condition $T = 165\ °K$ at $P = 1$ bar. Jupiter thus has a troposphere similar to the earth's. The Jovian troposphere appears to start at a pressure of about 0.4 bar. We do not know how deeply into the planet it extends, although some theoretical ideas on this topic are presented below.

As a parcel of gas from deep in Jupiter's atmosphere is carried toward the surface by convection, it decompresses and cools, and condensates form. The dashed lines in Fig. 8-21 show the approximate loci of first appearance of several important condensates in a gas of solar composition. For example, liquid water first condenses out of the gas at a pressure of several bars. The adiabat then passes through a complicated region where phases of solid and liquid water with dissolved ammonia (not shown) can exist. At somewhat lower pressures, H_2S gas, if it is present, can react with ammonia to form NH_4SH clouds, and then, at a pressure somewhat less than a bar, the remaining NH_3 appears as a solid condensate. The Jovian atmosphere never gets cold enough to precipitate CH_4. Thus, above the upper ammonia haze layer, one would expect to see solar proportions of uncondensed species such as H, He, Ne, Ar, and CH_4, but such major components as N, O, and S would be absent or greatly depleted. According to this model, how deep would one need to penetrate in order to observe such major elements as Mg, Si, and Fe? Mg and Si would be present as silicate "cloud" layers at a temperature on

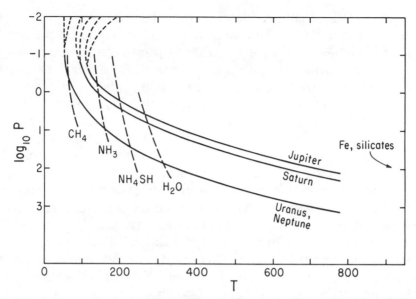

Fig. 8-21. Solid curves, pressure (in bars) versus temperature (in °K) in adiabatic models of the outer layers of the Jovian planets, assuming solar composition. Dashed curves intercepting the solid curves show the onset of condensation of major atmospheric constituents. To the right of these curves, the indicated constituents are completely dissolved in the gas phase (after Weidenschilling and Lewis, 1973).

the order of 1500 °K, and Fe would appear at still deeper layers. It would be impossible to ever confirm the existence of such layers, but as we shall see, it is doubtful that they exist.

The condensation-layer model of the Jovian atmosphere has had great influence on planetary science, because it successfully explains much of what is observed. Remote measurements of the composition of the Jovian atmosphere are based upon the spectral absorption of sunlight which passes through the upper atmosphere and then is reflected back into space.[42] Very little of such sunlight penetrates much below the main Jovian cloud deck, and so species which are not present as vapor above this layer are not in general detected. Since the main cloud deck is presumed to be the NH_4SH layer together with deeper clouds composed of water and ammonia, one should principally see absorption caused by H_2, CH_4, and NH_3. H_2S and H_2O should be largely invisible, together with many other molecular species which are present at higher temperatures. The observations are generally consistent with this picture (with some notable exceptions to be mentioned later). There is some difficulty involved in deriving abundances from remote spectroscopic observations because the path lengths depend upon the opacity and scattering properties of the atmosphere at the relevant wavelengths, which of course differ from one absorption feature to another. Therefore it is desirable to obtain "ground truth" measurements of Jovian atmospheric abundances by means of an atmospheric entry probe. For reasons discussed above, such abundance determinations will be most useful if they can be carried out at least to depths where H_2O is completely dissolved into the gas phase ($P \sim 10$ bar).

Available remote sensing measurements of Jovian atmospheric abundances give the following results. H_2 is the predominant species, followed by He. If the total mass of H_2 and He in the atmosphere is 1, then He comprises 0.20 ± 0.05 of this mass.[1] The corresponding He/H number ratio is 0.063.[43] This result is in good agreement with the primordial abundances given in Tables 2-1 and 2-2. A number of measurements[44,45] indicate that C/H is enhanced by a factor of two to three relative to solar composition; results typically fall in the range C/H (number ratio) = (8 to 17) $\times$ 10^{-4}. On the other hand, the N/H ratio appears to be approximately solar (to within a factor 2) at the 1-bar pressure level. Determination of the atmospheric H_2O abundance is quite difficult because it is necessary to probe to depths below the water condensation level, as discussed above. Measurements to date indicate that the oxygen/hydrogen ratio is lower than the solar ratio by about a factor of 30 at the deepest levels probed.[45] It is not clear whether this result reflects the fact that a substantial fraction of the water is still condensed at this level in the atmosphere, perhaps owing to lateral inhomogeneities caused by Jovian weather, or whether oxygen is truly depleted in the Jovian atmosphere.

As discussed in Chapter 2, the sun is depleted in deuterium, although its primordial D/H number fraction is estimated at about 2×10^{-5} (Table 2-1). The observed D/H value in the Jovian atmosphere is approximately consistent with this number, or perhaps up to a factor of two larger.[45] Since deuterium is rapidly destroyed by the nuclear reaction D^2 (p,γ) He^3 at temperatures greater than about 5×10^5 °K, this result implies that Jovian material has never been heated to such high temperatures.

Some of the detections of species in the Jovian atmosphere are inconsistent with the unmodified condensation layer model. For example, the molecules PH_3 (phosphine) and GeH_4 (germane) are present in the observable atmosphere, although according to the model of equilibrium chemistry in an adiabatically stratified atmosphere they should not be seen. The observed abundance of PH_3 in the 1–4-bar region corresponds to a number ratio $P/H \simeq 3 \times 10^{-7}$, in good agreement with the solar ratio. On the other hand, the GeH_4 abundance gives $Ge/H \simeq 3.5 \times 10^{-10}$, about a factor of ten smaller than the solar ratio.[45] A trace amount of CO (a few parts in 10^9) is also observed, but according to equilibrium chemistry calculations for the pressure and temperature conditions in Jupiter's accessible atmosphere, the available carbon prefers to form CH_4 so overwhelmingly that the predicted CO abundance would be orders of magnitude smaller than the observed amount.[46]

How do we explain the presence of these anomalous species in the Jovian atmosphere? A parcel of atmosphere which undergoes an adiabatic transformation requires a finite time to come to thermodynamic equilibrium with its surroundings. At the very low temperatures characteristic of the upper Jovian troposphere, equilibration times may become long compared with the typical time for circulation of an element of gas between observable layers and deeper, hotter levels. Thus the anomalous species may be "frozen-in" products of an equilibrium which is established at deeper layers. Continuing the supposition that the Jovian atmosphere is approximately of solar composition and is adiabatically stratified, we then ask at what temperature level the anomalous species were equilibrated. The answer to this question then gives a (model-dependent) lower limit to the extent of the Jovian troposphere.

If the deep Jovian atmosphere contains a solar abundance of oxygen, which may possibly be untrue, then phosphine is thermodynamically stable only at temperatures greater than about 800 °K. Higher in the atmosphere than the 800 °K level, PH_3 reacts with H_2O to form P_4O_6; the latter compound dissolves in the water clouds at about the 300 °K level.[47] Thus the presence of PH_3 in the atmosphere may indicate that the Jovian troposphere extends at least down past the 800 °K level, and that material from this level is brought rapidly into the observable atmosphere by convection before it can establish thermodynamic equilibrium. Similar arguments would apply to the observed abundances of GeH_4 and CO: they are quenched products of equilibration in high-temperature layers deep in the Jovian atmosphere. However, some of these arguments depend upon the assumption that a solar complement of oxygen is present in the deep layers to react with carbon and phosphorus.[48]

Nonequilibrium species such as C_2H_2, C_2H_6, and HCN have also been observed in the spectrum of Jupiter.[49] These molecules are thought to be produced by nonequilibrium chemistry in the extreme upper atmosphere and are thus not directly relevant to the interior structure of Jupiter.

Observations of the emission of the Jovian disk at radio wavelengths place additional constraints on the extent of the troposphere. A gas of solar composition is highly transparent to microwaves ranging from ~ 1 cm to $\sim 10^2$ cm wavelength. Ammonia is the principal absorber in this band, but with an absorption cross-section that declines rapidly with increasing wavelength. Thus a microwave signal of longer wavelength penetrates somewhat deeper into the Jovian atmosphere than a signal of shorter wavelength (Fig. 8-22).

Figure 8-23 shows the microwave emission spectrum of Jupiter,[50] expressed in terms of an equivalent disc temperature. This temperature corresponds to the temperature of a blackbody emitter with the same angular extent which emits power at the same rate in a given wavelength interval. In an atmosphere with a microwave opacity that increases with wavelength, the disc temperature at a given wavelength is approximately the same as the actual atmospheric temperature at the level where the path-integrated opacity, or optical depth, is unity. Thus Fig. 8-23 implies that, if the opacity is indeed determined

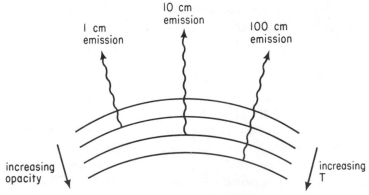

Fig. 8-22. Diagram showing that the intensity of microwave emission from a Jovian planet at increasing wavelengths gives information about the temperature at progressively deeper layers.

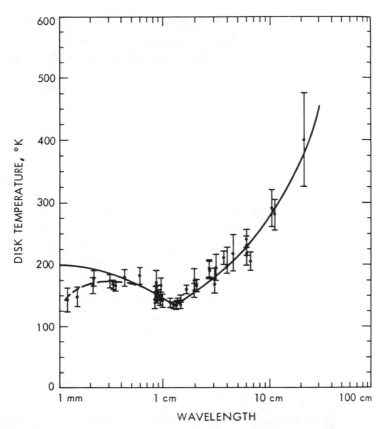

Fig. 8-23. Observations of Jupiter's microwave emission, compared with a theoretical model based upon a deep convective atmosphere with a solar abundance of ammonia. [Used by permission, Berge, G. L., and Gulkis, S. Earth-based observations of Jupiter: millimeter to meter wavelengths. In *Jupiter: Studies of the interior, atmosphere, magnetosphere, and satellites* (T. Gehrels, ed.). University of Arizona Press: Tucson, copyright 1976.]

by the ammonia, and if the ammonia mixing ratio is constant with depth, then the temperature of the Jovian atmosphere increases with depth, approaching values ~ 500 °K. The microwave emission spectrum can be reproduced well by assuming that the atmosphere is adiabatically stratified and that the NH_3 abundance is approximately solar. Signal-to-noise problems prevent detection of emission from layers deeper than that corresponding to a brightness temperature of ~ 500 °K. The microwave data imply that the Jovian troposphere extends at least to the 500 °K level.

Predominance of Hydrogen in Jupiter's Interior. We have found that Jupiter has a radius close to the maximum radius for a cold, hydrogen-rich body. Could this be a coincidence, such that Jupiter is actually composed of higher-Z material at an elevated temperature? How high would the temperature need to be in order for this to be possible? We adopt a typical interior pressure for Jupiter of 20 megabars, and equate this number to the approximate thermal pressure of a solid derived in Chapter 3:

$$P_T \simeq 3\gamma n_i kT \simeq 20 \text{ Mbar,} \qquad (8\text{-}37)$$

where n_i is the number density of heavy particles, given by ρ/Am_p^*. This formula will also be valid to order of magnitude in the ideal gas phase. Taking $\rho = 1$ g/cm^3 as a typical interior density, we find that Eq. 8-37 implies $T \sim A \cdot 1.3 \times 10^5$ °K. For a typical silicate ($A \sim 30$), interior temperatures of millions of degrees are implied. With a large temperature gradient between the center and surface, convection would ensue. All of the planet's deuterium would be quickly consumed. In any case, there is no plausible mechanism for producing and maintaining such an absurdly high internal temperature.

If Jupiter is primarily composed of hydrogen, and if the hydrogen is cold in the sense that average internal temperatures are substantially smaller than 10^5 °K, then the proximity of Jupiter to the maximum possible radius for a hydrogenic body implies that its internal structure should bear a resemblance to an $n = 1$ polytrope. This hypothesis can be tested by considering Jupiter's external gravity field. We noted in Eq. 4-37 that to lowest order in the rotational parameter q, the zonal harmonic J_2 is given by

$$J_2 = \Lambda_{2,0} \, q \qquad (8\text{-}38)$$

for a planet in hydrostatic equilibrium. The response coefficient $\Lambda_{2,0}$ for a polytrope of index $n = 0$ (which is equivalent to a Maclaurin spheroid or uniform-density liquid object) can be calculated exactly and is equal to 0.5. It is also possible to calculate $\Lambda_{2,0}$ exactly for the

---

* A = atomic weight; m_p = proton mass.

$n = 1$ polytrope; the result is[51]

$$\Lambda_{2,0} = \left(\frac{5}{\pi^2} - \frac{1}{3}\right) = 0.173. \qquad (8\text{-}39)$$

Substituting $q = 0.0888$ for Jupiter, Eq. 8-38 gives $J_2 = 0.015$, compared with the observed value $J_2 = 0.01473$. The $n = 1$ polytrope thus gives a valid lowest-order approximation to the interior structure of Jupiter. The corresponding pressure-density relation is $P = K\rho^2$ with $K \simeq 2$ megabars for ρ in g/cm^3. The central density of a polytropic model of Jupiter is about 4.8 g/cm^3. These results are consistent with the assumption that the interior of Jupiter has a pressure-density relation similar to that of hydrogen with a small admixture of helium.

The foregoing qualitative considerations have shown that Jupiter is predominantly composed of hydrogen. A more accurate treatment is necessary in order to determine whether Jupiter has a bulk composition that corresponds precisely to solar abundances. The composition of the atmosphere suggests that it does not. In order to quantitatively address the problem, it is necessary to (a) allow for the effects of a finite interior temperature, (b) allow for the effects of small proportions of nonhydrogen components on a pressure-density relation which is dominated by hydrogen, and (c) make some statements about transport mechanisms within the planet and their relation to atmospheric abundances.

In the terrestrial planets, one may neglect the interior temperature distribution to first approximation when deriving interior abundances because the thermal perturbation to the pressure is small compared with the total pressure when $A \simeq 30$ and $T \sim$ few $\times 10^3$ °K (cf Eq. 8-37). For Jupiter, we will take a typical interior temperature to be $\sim 10^4$ °K (a derivation of this number will be given below), and take $A \sim 1$. Then $P_T \simeq 1.5$ megabars for $\rho \sim 1$ g/cm^3. Since a typical interior pressure is ~ 15 megabars, we conclude that the thermal pressure comprises $\sim 10\%$ of the total pressure. This perturbation to the pressure is of the same order of magnitude as the perturbation due to the presence of nonhydrogen components in a solar composition mixture, and therefore cannot be neglected.

Thermal Structure. The phase diagram of hydrogen was discussed in Chapter 3, where it was noted that molecular hydrogen probably undergoes a transition to a pressure-ionized phase (metallic hydrogen) at a pressure on the order of 3 megabars, with an uncertainty in the critical pressure of about a factor of two. Continuing to use the polytropic model for Jupiter's interior, one finds that the critical pressure is attained at a radius of about 0.7 to 0.8 of the total radius of Jupiter. This radial distance also encloses about 0.7 to 0.8 of the total planetary mass. Thus the bulk of Jupiter is probably in the metallic hydrogen phase. This large metallic zone is enclosed by an extensive envelope of molecular hydrogen with a thickness on the order of 15,000–20,000 km.

No significant amount of energy can be transported by radiation in the metallic portion of Jupiter. Metals are opaque to photons because of their high density of free electrons. Photon frequencies are lower than the electron plasma frequency,

$$\omega_{p,e} = (4\pi n_e e^2/m_e)^{1/2}, \qquad (8\text{-}40)$$

(n_e = electron number density, e = electron charge, m_e = electron mass) over a broad range of wavelengths from the extreme radio and infrared band to hard ultraviolet radiation, which prevents propagation of photons in this wavelength range. Such a wavelength range also includes any thermal photons which could be produced at plausible Jovian interior temperatures. Thus we may neglect any radiative contribution to the thermal conductivity of the metallic portion of Jupiter (this result obviously also applies to the metallic cores of other planets, such as the earth). Electrons are the main carriers of energy and electrical currents in the deep Jovian interior because of their light masses and high velocities. To order of magnitude, the thermal conductivity of metallic hydrogen is given by[52]

$$K \simeq \hbar^3 n_e^{4/3} k/(m_e e)^2, \qquad (8\text{-}41)$$

where $\hbar$ is Planck's constant divided by 2π, and k is Boltzmann's constant. This formula gives the thermal conductivity due to electron transport; the energy transport by the protons is much smaller. The expression is independent of temperature because to a first approximation, the heat capacity of the electrons and the cross-section for scattering the electrons are both proportional to the temperature, and the expression for K depends on the quotient of these quantities. Thus, to this approximation, the result can be used for the solid or liquid phase of metallic hydrogen. Numerically, one finds $K \sim 10^8$ ergs/cm/s/°K for $\rho \sim 1$ g/cm³. This number is within an order of magnitude of the thermal conductivity of lithium at room temperature ($\sim 10^7$ ergs/cm/s/°K), which is a plausible result since metallic hydrogen is an alkali metal.

We will now combine the estimate of the thermal conductivity of metallic hydrogen with the observed intrinsic Jovian heat flow presented in Table 5-1. Assuming that the heat sources (which we will discuss presently) are distributed approximately uniformly within the metallic region of Jupiter, the thermal gradient within this region will be given to order of magnitude by

$$\Delta T/\Delta r \sim 5000/10^8 = 5 \times 10^{-5} \text{ °K/cm.} \qquad (8\text{-}42)$$

This is a very large temperature gradient. If it extended over a significant fraction of the planet, it would imply central temperatures on the order of several $\times 10^5$ °K, a result which we have already ruled out on other grounds.

There is a further problem with the hypothesis of conductive heat transport in Jupiter. Figure 3-3 shows the Grüneisen parameter of metallic hydrogen as a function of pressure. For typical Jovian interior pressures, we may adopt $\gamma = (\partial \ln T/\partial \ln \rho)_S \simeq 0.64$. Using a polytropic model of Jupiter, we find that the density gradient about midway between the center and the surface is $(d\rho/dr) \simeq 8.5 \times 10^{-10}$ g/cm⁴. Taking $\rho \sim 1$ g/cm³, the adiabatic temperature gradient within Jupiter is of order

$$(dT/dr)_{\text{adiabatic}} \sim T \cdot (5 \times 10^{-10}) \text{ °K/cm,} \qquad (8\text{-}43)$$

where T is the local temperature. Thus, unless T exceeds $\sim 10^5$ °K, the conductive thermal gradient will exceed the adiabatic gradient, and convection must ensue (Chapter 5) if the material has negligible viscosity. Figure 3-4 shows approximate melting temperatures for hydrogen at high pressures. Note that hydrogen has an extremely low melting temperature (~ 1000 °K in the Jovian pressure range), and therefore it is essentially certain that the metallic hydrogen in Jupiter is liquid and thus convects readily (the viscosity of liquid metallic hydrogen is calculated to be about equal to that of liquid water). We will now consider the energy balance of the Jovian heat flow, which will confirm that temperatures within the planet must be substantially above the melting temperature of hydrogen.

Energetics of Jovian Heat Flow. According to Table 5-1, the specific intrinsic luminosity of Jupiter is 1.7×10^{-6} erg/g/s. This number exceeds the earth's specific luminosity by a factor of about 30, and thus rules out any possibility that the Jovian heat flow could be produced by radioactivity. Even if Jupiter were composed entirely of silicates instead of hydrogen, a cosmic abundance of radioactive nuclei in the silicates could only produce $\frac{1}{30}$th of the required heating rate. Jovian interior temperatures are well below the values required for thermonuclear reactions. Of the mechanisms for intrinsic planetary luminosity listed in Chapter 5, only gravitational energy release or cooling appear plausible. We will first assess the energetics of a simple cooling model, assuming that there is no release of gravitational energy through rearrangement of interior chemical constituents (*e.g.,* core formation).

According to liquid-state calculations for metallic mixtures of hydrogen and helium,[53,54] the internal energy per gram of a solar mixture of liquid hydrogen and helium can be approximately expressed in the form

$$E = E_0 + AT \qquad (8\text{-}44)$$

(*cf* Eq. 5-19), where E_0 is the energy of the mixture at some reference temperature ($\sim 0°K$), and A is a constant which turns out to be approximately $1.66 \, k/m_p$ for the Jovian temperature range which we will shortly establish. If the Jovian luminosity is entirely derived from loss of internal energy due to a temperature decline, we may write (see Chapter 5)

$$L_i \simeq -1.66 \, (k/m_p) \int_M dm \, dT/dt, \qquad (8\text{-}45)$$

where the integral is taken over the entire mass M of the planet. Now define an average internal temperature $\langle T \rangle$ by

$$\int_M dm\, T = M \langle T \rangle, \tag{8-46}$$

and write

$$-dT/dt \sim T/\tau, \tag{8-47}$$

where τ is a characteristic e-folding time for the planet's cooling. Using $L_i/M = 1.7 \times 10^{-6}$ erg/g/s, we then find

$$\tau = 1.5 \times 10^6 \langle T \rangle \text{ yr}, \tag{8-48}$$

where $\langle T \rangle$ is in °K. A minimum temperature for the interior of Jupiter is established by requiring that τ be substantially greater than the age of the solar system, $\approx 4.6 \times 10^9$ years, since otherwise the Jovian luminosity would be unable to persist to the present. This then implies $\langle T \rangle \gg 3000$ °K, which in turn implies that the metallic hydrogen in Jupiter is molten.

Jupiter: A Failed Star? Thus far we have demonstrated that Jupiter has a composition very similar to the primordial sun, that its interior is composed primarily of metallic hydrogen, that the metallic part of the interior must be molten, at temperatures 10^4 °K, and convective if the chemical composition is uniform. The temperature distribution therefore closely approximates an adiabat in this portion of the planet. We have also seen that the temperature distribution in the deep Jovian atmosphere is also adiabatic. Does the adiabatic temperature distribution which starts in the Jovian atmosphere extend without interruption to the metallic hydrogen region? As a related issue, are there further phase transitions in the deep atmosphere and interior of the planet, analogous to the condensation layers for water and ammonia discussed above, which could affect the assumed chemical homogeneity in the interior? The answers to these questions are not well known because they require a better understanding of the physics of the deep molecular hydrogen envelope than is presently available. The principal advantage of the assumption of a continuous adiabat is that a quantitative calculation of the thermal evolution of Jupiter is then possible. The calculated evolution is analogous to the evolution of extremely low-mass stars.

As discussed in Chapter 5, if efficient convection dominates the interior heat transport of a planet, then the escape of heat from the interior is entirely regulated by the radiative properties of the atmosphere in the vicinity of the tropopause. Assume that Jupiter can be described by such a model, so that the troposphere extends from a pressure of ~ 1 bar to great depths within the planet; possibly to the vicinity of the center of the planet. The interior temperature distribution can then be calculated from a knowledge of the heat flux passing through the planet's atmosphere, together with the surface gravity g. If the diffusion approximation for photon energy transport is valid, then in the portion above the atmosphere where radiative heat transport takes place, one has

$$H = -K \, dT/dr, \tag{8-49}$$

where H is the total heat flux through the atmosphere, K is the effective thermal conductivity due to infrared photon transport, and dT/dr is the temperature gradient. Dividing Eq. 8-49 by the equation of hydrostatic equilibrium, $-g\rho = dP/dr$, we have

$$d \ln T/d \ln P = R_g H/(Kg\mu), \tag{8-50}$$

where R_g is the gas constant and μ is the mean molecular weight. The photon thermal conductivity K is obviously inversely proportional to $\kappa\rho$, the photon absorption probability per unit path length, which becomes very small in the tenuous upper atmosphere. These quantities are related by[55]

$$K = 16\sigma T^3/(3\kappa\rho). \tag{8-51}$$

Thus the atmosphere tends to become isothermal in the region where the optical depth is small, and therefore $d \ln T/d \ln P$ decreases with altitude. The optically thin part of the atmosphere is always stable to convection. But as the altitude decreases and the opacity increases, $d \ln T/d \ln P$ can increase. When it exceeds the value of $(\partial \ln T/\partial P)_S$, the atmosphere becomes unstable to convection (Fig.

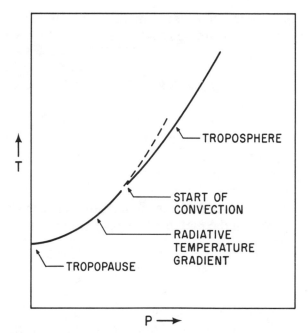

Fig. 8-24. Schematic diagram of the structure of a Jovian planet atmosphere. Dashed line shows the radiative gradient, which would continue if efficient convection did not occur.

8-24). In Jupiter, the convection zone starts between the tropopause ($P \sim 0.1$ bar) and a pressure level of about 1 bar. Thus we may characterize the Jovian adiabat by the temperature at a pressure of 1 bar. The value of this temperature depends in turn on the total heat flux H (or equivalently, on the effective temperature T_e), the surface gravity g, and on the opacity law for the atmosphere.

The major source of infrared photon opacity in a hydrogen-rich atmosphere turns out to be dipole transitions in H_2 molecules.[56] Although the H_2 molecule has no permanent dipole moment, a pressure-induced moment appears when collisions are sufficiently frequent. It is this phenomenon which is apparently responsible for the appearance of a deep Jovian troposphere. At high pressures (but still below the molecular-metallic transition pressure), opacity due to absorptions by trace molecules such as H_2O may also be important for maintaining convective instability. This matter requires further study.

When detailed model atmospheres for Jupiter are calculated,[58] making allowance for the wavelength-dependence of the opacity and for long photon mean free paths in the optically thin part of the atmosphere, the results can be represented by the following analytic fit:

$$T_1^3 = 3.8 \ T_e^{3.73}/g^{1/2}, \tag{8-52}$$

where T_1 is the temperature in °K at a pressure of 1 bar, T_e is the effective temperature in °K, and g is the surface gravity. This relation says that as the starting temperature for the interior adiabat decreases, i.e. as the planet becomes colder, the effective temperature of the atmosphere also declines. Thus the planetary luminosity declines as the planet cools. A calculation of the relation between the adiabatic temperature profile in the metallic hydrogen interior and T_1 gives the following result[57]:

$$T = 50 \ T_1 \rho^{0.64}, \tag{8-53}$$

where T is in °K and ρ is in g/cm³. Thus, using the observed value $T_1 = 165$ °K, we find that the temperature of a deep interior layer with $\rho = 4$ g/cm³ is about 20000 °K for a fully adiabatic temperature distribution.

Equations 8-52 and 8-53 can be combined with Eqs. 5-6 and 5-10 to derive a differential equation for the thermal evolution of Jupiter. Assuming that the radius of the planet remains constant throughout its most recent cooling history, we have for an increment in time

$$dt = -\alpha' T_e^{-3.757} \ [1 - (T_s/T_e)^4]^{-1} \ dT_e, \tag{8-54}$$

where α' is a constant which is equal to about 3.2×10^{23} cgs units for Jupiter. The bracketed quantity represents a correction due to the fact that a portion of the energy being radiated by Jupiter is derived not from loss of heat from the deep interior but from the conversion of sunlight to infrared radiation.

According to Eq. 8-54, the planet initially cools rapidly, when T_e and the luminosity are very high. Cooling then becomes very slow, and a very gradual drop in temperature proceeds over the billions of years leading to the present. Integration of Eq. 8-54, taking into account corrections due to changing radius during the initial, high-temperature epoch of the planet's existence,[58] leads to the result that it takes about 5×10^9 years for Jupiter to cool from an arbitrarily high initial value of T_e to the present observed value. The precise starting temperature of the planet is unimportant, provided only that it is much greater than the present value, since the initial high-temperature evolution proceeds very rapidly and contributes little to the planet's lifetime.

Consider the evolution of a proto-Jupiter which initially has a very high luminosity and interior temperature distribution. If the interior temperature is sufficiently high, the material acts as an ideal gas, and the planet will evolve in the manner described in Chapter 5. That is, the interior temperature *rises* as the planet radiates energy and contracts. This process terminates when the planetary radius begins to approach $a_{T=0}$, which is the zero-temperature radius appropriate to the object's mass and chemical composition. At this point, the planet begins to be supported by the zero-temperature (nonideal) pressure rather than by ideal gas pressure. As discussed in Chapter 5, the interior temperature then *declines* as the planet radiates energy and contracts. Consequently there exists a maximum in the central temperature which can be achieved by any contracting object (in the absence of other energy sources). Figure 8-25 schematically shows the evolution of the central temperature and radius of a contracting proto-Jupiter.

The maximum central temperature of a contracting object can be estimated by setting the central thermal pressure approximately equal to the central pressure of the zero-temperature object. We have already carried out this exercise for Jupiter, finding a maximum central temperature $\sim 10^5$ °K. Now since the thermal pressure scales as MT/a^3 and the central pressure scales as M^2/a^4, it follows that the maximum central temperature scales as $M/a_{T=0}$. Recall that the zero-temperature radius is essentially independent of mass for objects of solar composition and with masses $\sim$ the mass of Jupiter. This remains true up to a mass of about 100 Jupiter masses. Thus the maximum central temperature which can be achieved by an object of mass M is $\sim (M/M_J) \times 10^5$ °K, valid for $1 \lesssim M/M_J \lesssim 100$ ($M_J =$ Jovian mass). A central temperature $\sim 5 \times 10^6$ °K will initiate proton-proton reactions at a rate high enough to stabilize the object, preventing its further contraction until the hydrogen is exhausted at the center. The object then lies on the main sequence as defined in Section II, and can be considered a true star. Thus objects with

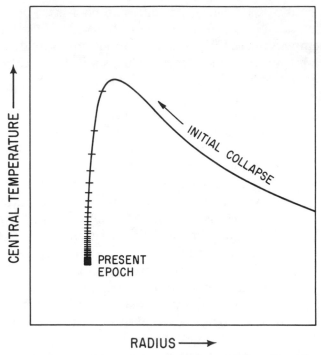

Fig. 8-25. Schematic diagram of the evolution of a Jovian-type planet. Equal time intervals as determined by Eq. 8-54 are shown.

masses $\lesssim 50\ M_J$ are not stars. Their central temperatures never become high enough to fuse hydrogen. After reaching their maximum central temperature, they then continue to contract and cool without interruption, and would resemble Jupiter at present. There is a class of objects in the mass range $10 \lesssim M/M_J \lesssim 50$ which we might call *substars*. Although they never reach the main sequence, they resemble stars in the sense that their central temperatures become high enough to burn the small amount of deuterium present in primordial cosmic matter.[58] Thus their contraction is briefly delayed (for about 10^7 years) while they derive the energy available from converting D^2 to He^3. Jupiter was never a member of the "deuterium main sequence" because of its low mass. However, its present intrinsic luminosity is consistent with an evolutionary history resembling that of a low-mass protostar. Of all the planets, Jupiter may have a bulk composition which most closely resembles that of the primordial sun.

Differentiation of Jupiter. The model for the evolution of Jupiter which we have just discussed is based upon analogy with the evolution of low-mass stars. Interior temperatures are assumed to be everywhere high enough that differentiation of the interior into zones with different chemical composition does not occur. At the same time, convection in the metallic hydrogen zone is sufficiently rapid that diffusive separation of chemical components cannot proceed (the mixing-length theory of convection of Chapter 5, combined with the observed value of L_i, gives convective velocities $\sim 10^0$– 10^1 cm/s in the liquid metallic layer). Thus the simplest model is chemically homogeneous, of primordial solar composition, and derives all of its internal power by cooling from an initially high temperature established by processes of planetary formation.

As we will see below, the simplest model does not provide an adequate fit to observational constraints on the Jovian interior. As we will discuss, Jupiter is (a) not of bulk solar composition, and (b) not chemically homogeneous. There are two independent mechanisms which could lead to chemical layering in the planet. First, it is possible that the accumulation of Jupiter took place in stages. The initial stage could have started with formation of a solid protoplanetary nucleus composed of materials which were present as condensates in the primordial solar nebula. Such materials would include rock and iron (the components of the terrestrial planets), as well as more volatile components such as H_2O and possibly NH_3 and CH_4 at the low temperatures prevailing in the outer solar system. A sufficiently large solid nucleus, once accumulated, then might trigger an instability in the surrounding gas of hydrogen and helium, causing it to collapse onto the solid core. In this way a giant planet would be formed, but since the solar ratio of hydrogen-helium/rock-ice would be unlikely to be preserved in such a two-stage formation process, one would produce a planet having a nonsolar value of this ratio, and possibly a nonsolar ratio of ice/rock as well. Furthermore, because accumulation takes place in two or more stages, the less volatile (and denser) material would then form a planetary core, while the envelope would be composed of noncondensible materials such as H and He, together with any substances from the core which might redistribute themselves into the envelope. If this process is important, atmospheric abundances might give clues as to which components are trapped in the core.

An alternative model for the formation of Jupiter assumes that it originated in a manner similar to a star, via a large-scale gravitational instability in a primordial nebula containing solar proportions of all components (gas and condensate). Thus the planet would contain a solar proportion of all components. However, even in this case, which can be ruled out by model studies, chemical layering would in general be expected. Recall that in a solar composition, convective atmosphere which is in precise thermodynamic equilibrium at every level, the chemical composition must still vary with altitude because of condensation processes. Although we discussed condensation in the accessible portion of the atmosphere, in principle there could be analogous processes at much deeper levels, which could lead to the depletion of certain species at higher levels. The study of such high-pressure phase transformations is still at an early stage, and most of the information comes from theory. Chapter 3 presented some results from an extremely simplified model of the high-pressure behavior of mixtures of hydrogen and other elements. These results showed that in the high-pressure limit, such mixtures would be expected to display phase-separation behavior, with the critical temperature increasing as a function of Z. However, electronic structure is still important even at tens of megabars pressure, and correlations of behavior with position in the periodic table may be more important than correlations with Z alone.

H-He is the most important binary system in Jupiter. Calculations for this system, assuming full pressure ionization, indicate that the critical temperature is on the order of 5000 – 10,000 °K at a pressure of ~ 10 megabars[53] (Fig. 3-8). Corrections for electron screening based on perturbation theory do not change this result very much. Thus, if perturbation theory can be trusted for H-He in this pressure range, one expects hydrogen and helium to undergo a fluid-fluid phase transition at temperatures on the same order as the interior temperature of Jupiter. This would lead to the following scenario for the evolution of the Jovian interior. The starting point is the same as the one for chemically homogeneous evolution: we start with a high-temperature, convective, chemically uniform mixture of fluid hydrogen and helium. This object evolves following Eq. 8-54 until a point in the interior cools to a temperature on the phase transition curve. Since perturbation-theory calculations indicate that critical temperatures *decrease* with increasing pressure, the first onset of the phase transition may appear at the upper boundary of the metallic hydrogen region. At this point, droplets of helium-rich condensate will appear in the fluid. Being denser than the surrounding hydrogen, they form a rain which falls to deeper layers in Jupiter. The rainfall terminates not on a solid surface, but at a deeper, hotter layer where thermodynamic conditions cause the raindrops to go back into solution (desert dwellers on the earth are familiar with this phenomenon).

After the process of planetary cooling and helium differentiation has proceeded over a period of time, the following situation is observed. The outermost layers of the planet are depleted in helium, while below the current region of phase transition there is an enrichment of helium. The gravitational binding energy of the planet has increased (in absolute value) because the planet is more centrally condensed. The increment in gravitational plus internal energy has gone into heating the planet; frictional heating of falling raindrops provides the mechanism for energy transfer. Thus the helium differentiation mechanism provides a stable, long-term source of gravitational energy to the planet.[59,60]

It is instructive to compare the helium differentiation mechanism in Jupiter with the process of formation of an iron core in a terrestrial planet. Suppose a terrestrial planet initially forms as a uniform mixture of iron and magnesium silicates. Initial heating is supplied by accretion and/or radioactivity. When the temperature becomes high enough to produce partial melting, iron droplets appear and migrate toward the center (they are immiscible with the molten silicates). This liberates gravitational energy, producing more heating and more melting, etc. Obviously, the process goes rapidly to completion because of its unstable nature (Fig. 8-26a). A terrestrial planet which forms a distinct core will normally do so very early in its history, and the energy of core formation can be simply added to the initial complement of thermal energy in the body. In contrast, helium differentiation in a Jovian-type planet proceeds as a very stable self-regulated process (Fig. 8-26b). Deposition of gravitational energy results in an increase of the temperature of the planetary interior, which in turn causes some of the helium to go back into solution. The process acts as a thermostat, tending to hold the interior planetary temperature within the critical temperature range until a substantial amount of unmixing has occurred. Unlike core formation in a

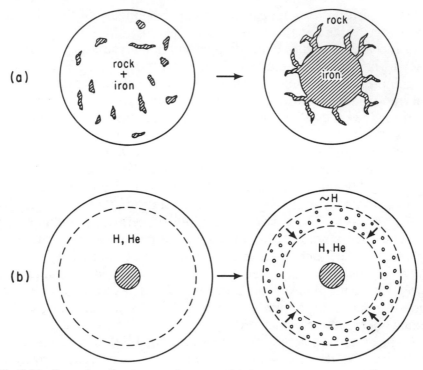

Fig. 8-26a. Formation of an iron core in a terrestrial planet.

Fig. 8-26b. Hypothetical formation of a helium-enriched layer by helium rain-out in Jupiter. The dashed curve (left) shows the approximate locus of the molecular-metallic hydrogen interface, while the shaded region shows the dense core (both to scale).

terrestrial planet, helium "core" formation can proceed over a very extended time interval. In the event that simple cooling models provide an inadequate source energy source for the Jovian luminosity, the helium unmixing model is a potentially attractive alternative.

How adequate an energy source does helium unmixing provide? We can estimate this quantity to order of magnitude as follows. The gravitational binding energy per gram for Jupiter is $\Omega_0 \sim -GM/a \sim -10^{13}$ erg/g. Let F be the fractional change in Ω_0 produced by an increasing degree of central concentration resulting from unmixing of helium. In order for F to have a reasonably large value, there must be a substantial change in the degree of central concentration, which means that the substance which unmixes must be relatively abundant and have a sizable density contrast with the primary constituent. Helium satisfies these conditions; assuming that all of the helium unmixes and settles to the central regions of the planet, one finds $F \sim 0.03$. Thus the gravitational energy which is potentially available is $\sim -F\Omega_0 \sim 3 \times 10^{11}$ erg/g. Dividing this number by the age of the solar system, $\simeq 1.5 \times 10^{17}$ s, we obtain a corresponding specific luminosity equal to $\sim 2 \times 10^{-6}$ erg/g/s. Note from Table 5-1 that this number is approximately equal to the present specific luminosity of Jupiter. Thus, if helium unmixing does come into play, it is capable of extending the Jovian cooling time by approximately a factor of two.

Available evidence suggests that helium unmixing has not yet begun in Jupiter. First, the atmospheric helium abundance is very similar to the solar helium abundance. Second, computed cooling times for chemically homogeneous models come out very close to the presumed age of Jupiter, so that an additional energy source is not really required. Finally, the computed temperatures near the top of the metallic hydrogen zone in adiabatic Jovian models are typically somewhat higher than the maximum temperatures for the phase transition, and the differences depend on which theory is used to describe the phase transition. As we have mentioned previously, theoretical results for the phase transition temperatures depend substantially on the method adopted for treating the electron states.[61]

Jovian Interior Abundances. The chemical composition of Jupiter's interior is constrained by computing models which provide a match to the observed mass, radius, and zonal gravity harmonics via the planet's response to the rotational perturbation. Because of Jupiter's liquid interior, the external gravitational potential conforms closely to the symmetry expected for strict hydrostatic equilibrium. Tesseral and odd zonal harmonics are absent or are present only at undetectably low levels. The orders of magnitude of the even zonal harmonics obey the scaling law of Eq. 4-37. The terrestrial planets Mars and earth are in hydrostatic equilibrium at most only to a level corresponding to the linear response to the rotational perturbation, so that only the zonal harmonic J_2 can be used as a constraint on interior structure. But in Jupiter, it is also possible to use the first nonlinear response, J_4, as a constraint. Efforts are being made to determine the term which describes the next higher nonlinear response, J_6, with enough precision to be useful. The fundamental constraints on Jupiter interior structure which are available at present are the mass M, equatorial radius at the 1-bar pressure level a_1, and the zonal harmonics J_2 and J_4. An accurate rotation period of the planet is also needed since this is the quantity which governs the amplitude of the nonspherical part of the potential field.

It is difficult to obtain an intuitive grasp of the effect of these constraints on possible Jupiter models because the nonlinear nature of the J_4 constraint. However, it is possible to deduce one general result. Because the expected interior pressure-density relation is expected to resemble that of hydrogen, which in turn resembles the $n = 1$ polytrope, and because the observed value of J_2 is close to that which an $n = 1$ polytrope would have, we conclude that if Jupiter has a dense core, the mass of this core is quite small compared with the total planetary mass. This conclusion is confirmed by more detailed studies which produce computer-generated pressure-density relations for the hydrogen-rich envelope of Jupiter.[62] The relation is required to agree exactly with theoretical and experimental results for hydrogen and helium mixtures in the limits of high and low pressures, as discussed in Chapter 3, and is also constrained, along with an appropriately-chosen core mass, to yield an exact match to the gravitational constraints. The typical "best-fit" interior model which emerges from this type of study is definitely not of precise solar composition. Table 8-2 lists some of the important parameters of the model.

Model studies of the Jovian interior typically find a central core mass of 10–30 earth masses. The density (and therefore composition) of the core is not well constrained by observation; we can only specify that the core have a density much higher than the corre-

Table 8-2. Properties of a Jupiter Interior Model

Mass of central rock-iron-ice core = 15 earth masses
Mass of H_2O, CH_4, and NH_3 in envelope = 15 earth masses
Mean radius of central rock-iron-ice core = 15,000 km
Pressure at center of planet = 100 Mbar
Pressure at surface of central core = 40 Mbar
Density at surface of central core = 3.6 g/cm³
Fractional radius at 3-Mbar pressure (H_2 − metallic H boundary) = 0.78
Approximate central temperature = 20000 °K
Approximate temperature at metallic-molecular interface = 10000 °K

sponding central density of a solar composition model (about 4.5 g/cm^3). It is altogether reasonable to assume that high-Z materials such as silicates and iron would remain insoluble in metallic hydrogen, and would naturally segregate to form a discrete core, even if they were not already present as a core at the time of formation of the planet. In a solar-composition planet with a mass equal to Jupiter's, how large would such a core be? The question is easily answered by totaling up the contributions of the various rock-forming elements listed in Tables 2-1 – 2-2, and then multiplying by the total mass of Jupiter. Although there are some chemical uncertainties (for example, do we include sulfur in the core, or assume that it is mostly present in the hydrogen envelope in the form of H_2S?), the result is always about 1.0 – 1.5 earth masses. This is considered suggestive by some investigators, who note that if the missing solar complement of material is restored to the larger terrestrial planets, the total resulting mass is on the order of a Jupiter or Saturn. However, modeling studies of Jupiter's interior generally show that the planet has a dense core at least an order of magnitude more massive than the earth. Thus Jupiter, like the earth and the other terrestrial planets, has also lost a considerable amount of the hydrogen and helium gas which was present at the time that the planet formed. In the case of the earth, essentially all of this component was lost. Most of the H-He component was also lost in the case of Jupiter, but a significant fraction ($\sim$ 10 – 50%) was retained as an envelope.

Modeling studies also indicate that Jupiter's H-He envelope is slightly denser than would be expected for pure hydrogen and helium in solar proportions. If volatile components such as H_2O, CH_4, and NH_3 are present as well, and in solar proportions, then one would expect to find these in the envelope in a total amount equal to about 3 – 5 earth masses. But studies indicate that the non-H-He material in the envelope has a total mass of about 15 – 30 earth masses. This result is at least qualitatively consistent with atmospheric abundance measurements which show that methane is enhanced by about a factor of two relative to solar abundance.

In summary, Jupiter has enhanced amounts of heavy material (i.e., material other than hydrogen and helium) in both its core and its envelope. This enhanced material probably includes both rock and iron (in the core) and some or all of the "ices," H_2O, CH_4, and NH_3. Some of the latter may be present in both the core (perhaps as a primordial component of the protoplanetary nucleus) and in the envelope. The planetary mass ratio of H/He is considered to be solar in all models, because neither gas condenses under plausible nebular conditions, and neither can escape from Jupiter's powerful gravitational attraction once the planet is formed.

Other Constraints on Interior Structure. Several observed phenomena provide indirect evidence for the picture of a liquid, convective Jovian interior. These are (a) the existence of a substantial intrinsic magnetic moment, (b) the constancy of the planetary effective temperature from equator to pole, and (c) the existence of close-in, massive satellites (the Galilean satellites of Jupiter).

The first of these phenomena can be interpreted using the hydromagnetic dynamo theory mentioned in Chapter 6, and has implications for the dynamo parameters of the deep Jovian interior. As discussed in Chapter 6, the existence of an intrinsic dynamo depends upon the presence of a liquid, convecting, electrically conducting dynamo within the planetary interior. There is little doubt that such a dynamo can be sustained in Jupiter. The electrical conductivity of molten metallic hydrogen in Jupiter is similar to that of an ordinary alkali metal near its melting point. For the Jovian interior, $(T \sim 10^4 \, ^\circ K, \rho \sim 1 \text{ g/cm}^3)$ we have[63,64] $\sigma_c \sim 10^4 - 10^5 \text{ ohm}^{-1}\text{cm}^{-1}$; for comparison, the electrical conductivity of lithium at one atmosphere pressure and at a temperature just above its melting point is $4 \times 10^4 \text{ ohm}^{-1} \cdot \text{cm}^{-1}$. The presence of impurities may lower the calculated value somewhat, but this will not change the conclusions which are presented in the following paragraph.

The magnetic diffusion coefficient for the metallic hydrogen interior of Jupiter is $\eta \sim 10^3 - 10^4 \text{ cm}^2/\text{s}$ (Eq. 6-4). Taking a typical Jovian length scale to be $10^9 - 10^{10}$ cm, the decay time for an initial field is found to be $\sim 10^{14} - 10^{17}$ s. Thus it is unlikely that the Jovian field which we now observe, the most powerful among the planets, is a remnant of a primordial field (the age of the solar system is about 1.5×10^{17} s). Now the magnetic Reynolds number is essentially the ratio of the field decay time to the convective circulation time in Jupiter. The latter is estimated to be on the order of decades to centuries on the basis of generalizations of the mixing length theory discussed in Chapter 5, and so the magnetic Reynolds number in Jupiter is many orders of magnitude greater than unity. There is little doubt that a substantial dynamo can be sustained by such highly turbulent magnetohydrodynamic flow. If this picture is correct, one might expect to observe some changes in the basic Jovian magnetic field geometry over the period of a convective circulation time. Observations of the Jovian magnetic field with the requisite precision do not yet extend over a sufficiently large time base for such changes to be detected.

Observations of the latitude dependence of the effective temperature provide an interesting test of the model of convective transport in the Jovian interior. We showed earlier that the interior temperature distribution in a fully convective planet depends only on the values of T_e and g in the planetary atmosphere. But recall that T_e is a parameter which gives the *total* (i.e., intrinsic plus thermalized solar) radiation passing through the planetary atmosphere. The fact that part of the radiation which passes through the atmosphere does not originate in the interior is irrelevant as long as the intrinsic heat flow is large enough to maintain convection throughout the interior, and as long as the sunlight is thermalized at a deeper layer than the effective planetary photosphere located at ~ 0.5 bar pressure. If a component of the atmospheric heat flow comes from thermalized sunlight, this component substitutes for the heat flow from the interior which would otherwise be supplied as appropriate to the given boundary conditions T_e and g. Thus we may think of thermalized sunlight as having the effect of partially choking off escape of interior heat.

Jupiter has an obliquity of only 3°, and so the diurnally averaged solar energy incident on a parcel of atmosphere is essentially proportional to sin θ, where θ is the colatitude. The contribution of the solar energy flux to the total heat flow varies from 0% at the pole to about 60% at the equator. If the total heat flow was a sum of two entirely independent components, the flow from the deep interior and the flow resulting from thermalized sunlight, then we should see a substantial pole-to-equator variation in the total heat flow, by something like a factor of two. But in fact, the total heat flow is observed to be essentially constant from pole to equator, varying by at most a few percent.[65] This is a remarkable result, totally unlike the situation in terrestrial planets such as the earth. It implies that a sort of conspiracy exists between the deep interior and the atmospheric layers responsible for supplying infrared radiation by converting solar visual photons. Where there is a deficit of the latter (the poles), the internal heat source makes it up. Where a large fraction of the heat flow can be supplied by thermalized sunlight (the equator), the contribution from the deep interior is reduced (Fig. 8-27).

These results can be understood in terms of the properties of convective energy transport. Recall that if efficient convective heat transport takes place in a liquid, low-viscosity object, then the temperature distribution becomes very close to isentropic. If, for example, incident solar energy causes an equatorial region to increase in temperature and develop a high specific entropy compared with the rest of the planet, convection quickly reduces the discrepancy.[65] We have seen that the specific entropy of the interior depends only on T_e and g. Thus it follows that if g is constant, T_e must be constant. Convection is the agent which reroutes part of the internal heat flow from the equator to the pole.

In reality, g is not strictly constant, because of the Jovian oblateness. Because g is higher at the poles, the atmosphere must maintain a slightly higher total heat flux at the poles in order to keep the specific entropy constant. Thus T_e may be slightly *higher* at the poles (by a degree or two) than at the equator. This effect has not been observed, and it may be obscured by meteorological effects in the Jovian atmosphere.

Consideration of the tidal evolution of the Galilean satellites provides a third constraint on the properties of the Jovian interior. These

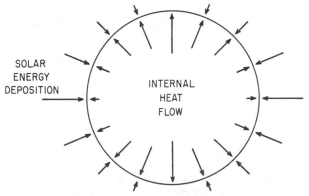

Fig. 8-27. In a fully convective body, internal heat flow tends to be regulated so that the sum of internal heat flow plus solar energy deposition tends to be constant everywhere over the surface.

four satellites (Io, Europa, Ganymede, and Callisto; Table 1-1) are massive and comparatively close to Jupiter. Let us focus on the closest of these, Io, since it should have undergone the most significant tidal evolution. From Eq. 4-73, the tidal torque per unit mass acting on Io is

$$T_t = 9\Lambda_{2,0}\, GM_{Io}\, (a^5/R^6)\, \delta, \qquad (8\text{-}55)$$

where M_{Io} is the mass of Io, a is the radius of Jupiter, R is the separation between the centers of mass of Jupiter and Io, $\Lambda_{2,0}$ is the dimensionless Jovian tidal response coefficient, and δ is the phase shift caused by dissipation in Jupiter of the tides produced by Io. Since Io is outside the orbital distance corresponding to a period synchronous with Jupiter's rotation, tidal dissipation in Jupiter causes angular momentum to be transferred from Jupiter's rotation to Io's orbital angular momentum, causing R to gradually increase. It is a straightforward matter to show that the rate of increase of Io's specific angular momentum is given by

$$T_t = \frac{1}{2}\, (GM_J)^{1/2}\, R^{-1/2}\, \frac{dR}{dt}, \qquad (8\text{-}56)$$

where M_J is the mass of Jupiter. Equating relations 8-55 and 8-56, and then integrating, we find that the time required for Io to evolve from $R \simeq 2.4\, a$ (the Roche limit) to its present orbital distance is given by

$$t = 2.3 \times 10^{11}\, \delta^{-1}\ \text{s}, \qquad (8\text{-}57)$$

where δ is assumed to be a constant and we have taken $\Lambda_{2,0} = 0.17$, the value appropriate to the $n = 1$ polytrope. In order for Io to be able to exist in its present orbit, then, we require that the value of the Jovian Q, averaged over the evolution of Io's orbit, obey the constraint[66]

$$Q \gtrsim 3 \times 10^5. \qquad (8\text{-}58)$$

Because of an orbital resonance between the periods of the inner three Galilean satellites, one cannot accurately treat the evolution of

Io separately from that of Europa and Ganymede. The angular momentum which must be transferred to change Io's orbit must also include a portion for Europa and Ganymede. When this is taken into account, the lower limit for Q is reduced by a factor ~ 5, but the principal conclusion remains unchanged: Jupiter's tidal Q must be $\sim 10^5$ or greater. Since this number is much larger than the value of Q which is customarily associated with solid materials (~ 100), we conclude that the existence of Io in its present orbit confirms that most of the interior of Jupiter cannot be solid. In a following section which deals with the interiors of the Galilean satellites, we will discuss a corresponding process of tidal dissipation in Io, which also leads to evolution of Io's orbit. Consideration of this process leads to a relationship between the Q's of Jupiter and Io. If the latter is fixed by a measurement of Io's heat flow, one finds Q (Jupiter) $\sim 4 \times 10^4$, a somewhat embarrassing result, since it is too small by about a factor of two to be compatible with Io's present orbit. The reason for the discrepancy is not understood.

Although Jupiter's tidal Q is in any case much larger than the Q of terrestrial-type planets, it is not easy to find a mechanism which gives a present value of Q in Jupiter as *small* as 10^6. Ordinary molecular viscosity in a liquid planet can produce only a very small amount of dissipation in tidal flows, and the resulting planet-averaged Q is many orders of magnitude greater than 10^6. Turbulent viscosity is likewise inadequate, although it could have been significantly larger in the early epoch of Jupiter when convective currents were more rapid.[67,68] Dissipation in a solid core has been suggested as a mechanism, but would require a core with substantially larger volume and different characteristics than is indicated by interior models.[69] Finally, dissipation related to a lag in the completion of an interior phase transition (e.g., hydrogen-helium immiscibility) has been proposed as a source of Jupiter's anomalously low tidal Q.[70]

Saturn

Atmosphere.   There are many similarities between the atmospheres of Jupiter and Saturn. Saturn's atmosphere is also predominantly composed of H_2, and exhibits a cloud structure which is caused by the condensation of minor constituents at pressure levels on the order of one bar and deeper. As Fig. 8-21 shows, if the composition of Saturn's atmosphere is solar and the temperature distribution nearly adiabatic, then the atmosphere should have condensation layers which correspond to those in Jupiter's atmosphere, but at somewhat deeper levels.

As in the case of Jupiter, there appears to be vigorous convective stirring of the gases in the Saturnian atmosphere. Atmospheric abundances thus may be representative of the deeper interior with the proviso that phase transitions could lead to alterations in the distributions of some elements.

The He/H ratio in Saturn's atmosphere is a key diagnostic of interior processes in Saturn. We expect the total He/H ratio in the planet to be essentially solar, because neither of these gases could have been in a condensed phase at the time that Saturn was forming, and there is probably no other efficient process for fractionating them at that time. Thus, if the present atmospheric ratio is solar, this would tend to indicate that no interior separation of H from He has occurred. We have already noted that the He/H number ratio in Jupiter's atmosphere, 0.063 ± 0.02, is close to the solar value. A preliminary analysis of infrared data from the *Voyager* spacecraft gives He/H $\simeq 0.03$ (no error bar yet) for Saturn's atmosphere,[71] while an earlier analysis of cruder data from the *Pioneer 11* spacecraft's Saturn flyby[72] gave He/H $= 0.055 \pm 0.02$. Analysis of the spacecraft data is continuing at present. There appears to be some indication that helium is depleted in Saturn's atmosphere relative to Jupiter and to the sun. We will discuss below what this result may imply about Saturn's interior.

Since methane does not condense in Saturn's atmosphere, observations of its abundance need not take this complication into account, but derivation of the Saturnian atmospheric C/H ratio from remote spectral data is still a complicated and model-dependent process. The results give C/H $\sim 10^{-3}$, or about twice the solar ratio.[73,74] It appears that methane is enhanced in both Jupiter and Saturn by a similar factor.

Because it condenses in the upper Saturnian atmosphere, ammonia is difficult to observe directly. The ammonia abundance is determined from its effect on the thermal microwave emission spectrum of the planet.[75] Values inferred in this manner for the lower troposphere give the result N/H $= 0.7 \times 10^{-4}$ to 3.9×10^{-4}, corresponding to slightly below the solar ratio to about four times the solar ratio. As in the case of Jupiter, we can only say that nitrogen is present in roughly solar abundance, and may be somewhat enhanced.

Water vapor has not so far been detected in Saturn's atmosphere, and according to the chemical equilibrium models discussed earlier, it may be present only at deeper levels. As in the case of Jupiter, phosphine (PH_3) is observed in Saturn's atmosphere in nonequilibrium concentrations, and it is similarly interpreted as a quenched product of equilibrium chemistry in deeper and hotter planetary layers. The abundance of the phosphine is similar to that in Jupiter, i.e., similar to solar. Finally, the D/H ratio in Saturn's atmosphere[76,77] is also very similar to Jupiter's, and is approximately equal to the presumed primordial solar value of 2×10^{-5}.

We may summarize the observed abundances in Saturn's atmosphere as follows. Overall, the abundances are very similar to those in Jupiter's atmosphere, taking into account the well-understood effects of condensation processes in Saturn's slightly colder atmosphere.

Saturn Interior Structure. As the mass-radius diagram (Fig. 8-20) shows, Saturn is predominantly composed of hydrogen. However, unlike Jupiter, Saturn's interior structure cannot be approximated by a polytrope of index one. This is most readily demonstrated by computing the ratio $J_2/q \simeq \Lambda_{2,0}$ for Saturn. The rotation rate of Saturn's deeper layers is presumably identical to the planet's magnetospheric rotation period of $10^h 39^m 24^s$. The equatorial radius of Saturn at one bar pressure is determined to be $60,200 \pm 200$ km (recent Voyager measurements have adopted $a = 60,330$ km) (see Table 1-1). However, for historical reasons the normalizing radius for the gravitational harmonics is taken to be $a = 60,000$ km. For consistency we will use the latter number in the following discussion, although for accurate work the distinction must be observed. We then have $q = 0.153$ for Saturn, leading to $\Lambda_{2,0} \simeq 0.11$, substantially smaller than the response coefficient for the $n = 1$ polytrope, $\Lambda_{2,0} = 0.173$, or the value for Jupiter, $\Lambda_{2,0} = 0.166$. This result indicates that Saturn is considerably more centrally condensed than Jupiter. Since the equation of state of a solar-composition mixture is approximately represented by the $n = 1$ polytrope, this difference between the two bodies is to be interpreted as follows. Both Jupiter and Saturn have extensive hydrogenic envelopes, but Saturn has a massive dense core which comprises a much larger fraction of the total planetary mass than Jupiter's core. Saturn's response to the rotational perturbation is principally determined by the hydrogenic envelope, but the presence of the massive central core, which does not respond appreciably, reduces the total induced mass quadrupole moment.

The total mass of Saturn's nonhydrogenic central core is not well constrained because of uncertainties in the hydrogen equation of state, but it almost certainly lies in the range of 10–30 earth masses. Thus Saturn's core is probably very similar in mass to Jupiter's, but as a fraction of the total mass of the planet, it is much larger. For overall solar composition, Saturn's central core would at most comprise only about one earth mass, and so Saturn is clearly enriched in elements heavier than hydrogen and helium. Saturn deviates even more from solar composition than does Jupiter. This result may at first seem somewhat surprising, since Saturn's mean density (0.7 g/cm^3) is smaller than Jupiter's (1.3 g/cm^3), and is in fact the lowest known mean density in the solar system. The explanation is that hydrogen is a highly compressible substance, and in massive bodies such as Jupiter and Saturn, where the hydrogen is strongly compressed, the body's mean density is not strongly correlated with its mean chemical composition.

Saturn interior models are thus analogous to Jupiter interior models, and are calculated in the following manner. We begin by assuming the presence of a hydrogen-helium envelope of roughly solar composition, and with a pressure-density relation similar to that of Jupiter's envelope, modified for lower temperatures. The temperature distribution is assumed to be adiabatic, using the same arguments as were used for Jupiter. This envelope is placed around a core whose precise structure is not constrained by the gravitational field, but whose total mass is. Some models assume that this core includes rock and iron as well as lower-temperature condensates such as H_2O, while other models assume that these condensates are dissolved in the mantle. Figure 8-28 shows the temperature-pressure relation in a typical Saturn interior model, while Table 8-3 summarizes its important parameters. The model is reminiscent of Jupiter models, but shows an effect with which we have become familiar in the context of the terrestrial planets. If we decrease the total mass of a body of specified chemical composition, the locations of the phase bounda-

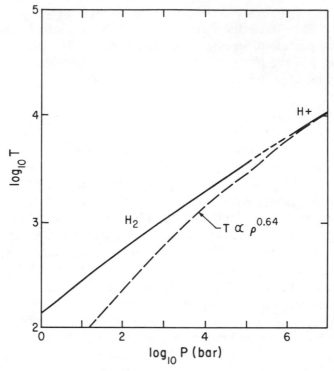

Fig. 8-28. Temperature versus pressure for an adiabatic Saturn model. Dashed curve shows Eq. 8-59.

ries move deeper, and the original structure of the outermost layers is "stretched" over the entire planet.

The metallic hydrogen region of Saturn is thus a much smaller fraction of the planet than in Jupiter, and is located much deeper. The strongly-interacting molecular hydrogen zone, which extends

Table 8-3. Properties of a Saturn Interior Model

Mass of central rock-iron-ice core = 19 earth masses
Mass of H_2O, CH_4, and NH_3 in envelope: uncertain
Mean radius of central rock-iron-ice core = 16,000 km
Pressure at center of planet = 50 Mbar
Pressure at surface of central core = 7 Mbar
Density at surface of central core = 1.7 g/cm³
Fractional radius at 3-Mbar pressure (H_2 − metallic H boundary) = 0.50
Approximate central temperature = 10000 – 15000 °K
Approximate temperature at metallic-molecular interface = 8000 °K

from a radius of about 30,000 km out to about 50,000 km, comprises a major fraction of Saturn's interior. It is also the most poorly understood region of the planet from the point of view of molecular physics; shock compression experiments only span the pressure range from the surface ($r = 60,000$ km) to $r \simeq 40,000$ km, and small errors in the pressure-density relation can have a major effect on the deduced chemical composition. Furthermore, such experiments have not so far given information about the radiative conductivity of these layers, which is needed to establish whether they are in fact in convective equilibrium.

Energetics of Saturn's Heat Flow. Assume that Saturn, like Jupiter, has an adiabatic internal temperature distribution starting with the convective layer in the atmosphere. In this case the temperature distribution deep within the planet, in the metallic hydrogen layer, is given approximately by the relation

$$T = 7000 \; °K \cdot \rho^{0.64}, \qquad (8\text{-}59)$$

where the specific entropy is constrained to be equal to that of the atmospheric layer which is at a pressure of 1 bar and a temperature of about 135 °K, according to spacecraft measurements.[78] This relation provides a good fit to the temperature in the metallic zone, but underestimates the temperature in the molecular zone by up to about 50% (see Fig. 8-28). We now go through a similar exercise to that carried out for Jupiter. The conductive temperature gradient in the metallic hydrogen zone of Saturn will be of order $2000/10^8 = 2 \times 10^{-5}$ °K/cm, assuming that heat sources are distributed roughly uniformly throughout the planet. The density gradient in the same region is about 3.7×10^{-10} g/cm^4, so that the adiabatic temperature gradient is calculated to be

$$(dT/dr)_{\text{adiabatic}} \sim T \cdot (2.35 \times 10^{-10}) \; °K/cm, \qquad (8\text{-}60)$$

where T is the local temperature in °K. Again, unless the local temperature is about 10^5 °K or greater, the conductive gradient will exceed the adiabatic gradient and convection will ensue.

This argument cannot be applied to Saturn's extensive molecular hydrogen zone since we do not know whether radiative heat transport is likely to be important in this layer; certainly ordinary molecular heat conduction is not. However, a reasonably self-consistent picture can be constructed if we assume that the photon opacity is high enough in this region for it to be convective also. Consideration of the opacity produced by pure molecular hydrogen along with solar amounts of H_2O, CH_4, and NH_3 indicates that this is probably correct.[79]

We will now calculate a thermal evolution model for Saturn using the same approach as that used earlier for Jupiter, namely, by assuming that the intrinsic luminosity is derived from the cooling/contraction of a chemically invariant planet. The planet's evolution is then governed by the same differential equation (Eq. 8-54), except that different values are used for T_s and α'. It is straightforward to show that α' is given by

$$\alpha' = (4\pi a^2 \sigma)^{-1}\ 1.243\ T_e^{-1.243} \int dm\ C_V T, \qquad (8\text{-}61)$$

using the analytic approximations given previously; here all quantities are expressed in c.g.s. units. Note also that α' is independent of T_e and so can be calculated using the temperature distribution of any epoch, such as the present one. A conservative estimate of this quantity is obtained by evaluating the integral in Eq. 8-61 using Eq. 8-59 for the temperature distribution in the hydrogen-rich envelope, and assuming that the temperature is constant within the dense core. The actual temperature will be higher in both the core and the molecular hydrogen region, so this procedure gives a lower limit to α'. The result is $\alpha' = 7.3 \times 10^{22}$ c.g.s. units, or about a factor of four smaller than the value for Jupiter. The lower value for Saturn is not surprising: Saturn cools more rapidly than Jupiter because it has a lower mass but a similar surface area.

Substituting observed values for Saturn's present T_e and T_s, and using theoretical estimates of the heat capacity of dense molecular and metallic hydrogen, one obtains a time interval of 4×10^9 years for Saturn to cool from a very high initial effective temperature to the present value of $T_e = 95$ °K. This time interval is less than the age of the solar system, but not significantly so, and thus simple cooling appears to be a marginally viable mechanism for explaining Saturn's intrinsic heat flow (radioactivity can be ruled out as in the case of Jupiter). However, there are obvious uncertainties in the calculation, and it is possible that the model is inadequate, as has been suggested by other theoretical calculations.[80]

If the time scale for cooling is too short, it can be extended by invoking gravitational differentiation of a major component, such as helium. Calculation of the size of this effect is carried out in a manner analogous to the corresponding calculation for Jupiter. First, we estimate the gravitational binding energy per gram released by unmixing *all* of the helium in the planet to be $-F\Omega_0 \sim 1 \times 10^{11}$ erg/g. Dividing by the approximate age of the solar system, one obtains an average specific luminosity of $\sim 0.7 \times 10^{-6}$ erg/g/s, comparable to the observed value for Saturn. Not all of this energy is actually available since the phase separation process is assumed to occur only in the metallic hydrogen phase. In any case, as in Jupiter, this mechanism is a *potential* source of energy for the planet, which under certain circumstances could extend its thermal age substantially.[79]

Saturn's Magnetic Field.   As we have discussed in Chapter 6, Saturn possesses an intrinsic magnetic field which is somewhat weaker than the Jovian field. It is probable that the field is produced by a similar mechanism of dynamo action in a convective layer of molten metallic hydrogen. It is easy to show that an initial, primordial field would diffuse out of the metallic zone in about 10^8 years.

Saturn's magnetic field is remarkably symmetric compared with other planetary magnetic fields. The relative weakness of high-order components in the multipole expansion of the magnetic field may be related to the small size of Saturn's metallic hydrogen region compared with the planetary radius. However, the very small tilt of the equivalent dipole to the spin axis requires another explanation. It has been noted[81] that the nonaxisymmetric components of the field could be filtered out by a conducting layer which is convectively stable (and hence does not participate in the dynamo action) but which rotates differentially with respect to the field-producing layers. The mechanism works as follows. Suppose that Saturn's magnetic field is produced by dynamo action within its convective liquid-metallic hydrogen core. Further suppose that the field which is produced by this dynamo is intrinsically similar to the fields produced by dynamos in other planets which we have discussed: earth, Mercury, and Jupiter. A well-known theorem[82] (Cowling's theorem) states that an axisymmetric magnetic field cannot be maintained or generated by fluid motions. Thus an axisymmetric external field can appear only as a random, transient configuration. Its "normal" or typical state will be nonaxisymmetric. And indeed, the known large-scale intrinsic planetary fields mentioned above have a small but measurable tilt, $\sim 10°$, to the rotation axis (Table 6-1). The intrinsic planetary field in Saturn may also have this property (dashed magnetic field lines in Fig. 8-29). However, the field lines which are actually observed (solid field lines) could have a much greater degree of axisymmetry, provided that they pass through a stable, conducting region which differentially rotates.

Recall (Chapter 7) the electromagnetic skin depth $\delta = c/(2\pi\sigma_c\omega)^{1/2} = (2\eta/\omega)^{1/2}$, where c is the speed of light, σ_c is the electrical conductivity of a medium, ω is the frequency of a time-varying electromagnetic signal which impinges on the medium, and η is the magnetic diffusion coefficient defined earlier. Induced currents will screen out the signal over a distance scale on the order of δ. The

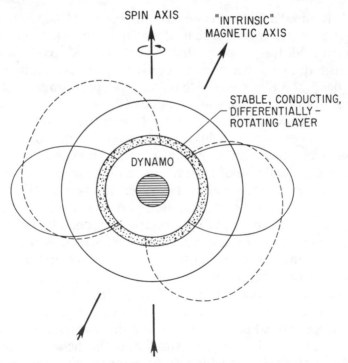

Fig. 8-29. Hypothetical mechanism for producing an axisymmetric Saturn magnetic field. Stippled region shows the region just below the molecular-metallic hydrogen interface, where helium rain-out is assumed to occur. Dynamo action occurs in the metallic hydrogen region. Shaded region is the dense core (all regions are shown to scale). Dashed curves are original field lines produced by dynamo, while solid curves are field lines "processed" by the differentially-rotating layer.

skin depth δ is just the electromagnetic analog of the thermal skin depth l defined in Eq. 5-39. Now suppose that the dynamo region in Saturn is enclosed by a conducting shell (Fig. 8-29) which rotates at a different rate. In the frame in which the conducting shell is stationary, the axisymmetric components of the field produced by the dynamo do not oscillate, but the nonaxisymmetric components do, and are therefore subject to attenuation by the skin effect. The skin depth can be readily calculated for an assumed value of σ_c and the differential rotation frequency ω. Let us take $\eta \sim 10^4$ cm^2/s, or approximately the Jovian value for the metallic hydrogen layer. The differential rotation rate is harder to quantify. The value which gives a skin depth of 1000 km turns out to correspond to a differential rotation period of millions of years. Thus, with more plausible shorter periods, the skin depth could be much smaller than reasonable values for the shell thickness, and so all nonaxisymmetric components of the intrinsic magnetic field produced by the dynamo would be strongly attenuated before they reach the surface.

Although the rotating-layer model seems to provide a plausible explanation of the axisymmetry of Saturn's field, it does not explain the origin of the layer itself. A separate assumption must be made in this regard. One plausible assumption[81] is that the layer is produced by helium-hydrogen differentiation, and is stable to convection because of a chemical gradient. On the other hand, the layer may also be produced by purely meteorological effects. Of course, the somewhat unlikely possibility that Saturn's field geometry is nothing but a random occurrence cannot be completely discarded.

Finally, note that if Saturn's field were precisely axisymmetric, it would be impossible to determine the planet's rotation period from observations of its magnetic field, since the field would then be completely spin-invariant. This frustrating state of affairs obtained following the first Saturn flyby by Pioneer 11. However, subsequent observations by the Voyager spacecraft of radio emission from plasma trapped in field lines showed that there was a periodic behavior of the emission, related to sectorial structure of the magnetic field. It is not yet clear how the radio "hot spot" is related to the nonaxisymmetric topology of the magnetic field.[83]

Saturn's Differential Rotation. Cloud patterns in the atmospheres of both Saturn and Jupiter exhibit considerable differential rotation.

The differential rotation takes the form of a mean zonal flow pattern, which is particularly pronounced in the case of Saturn. If we choose the magnetospheric rotation rate of Saturn as a reference, then Saturn's equatorial zone rotates eastward more rapidly, than the bulk of the planet, with a velocity of 0.5 km/s at the equator. Considering that the "background" rotation rate corresponds to an equatorial velocity of 10 km/s, this is a substantial perturbation.

Saturn's zonal flow pattern shows a considerable amount of north-south symmetry. The equatorial jet drops to zero relative velocity at about ±40° latitude, but other less pronounced jets are seen at higher latitudes.[84,85]

There are two interpretations of Saturn's striking pattern of pronounced differential rotation. First, it may be a meteorological phenomenon confined entirely to the planet's gaseous atmosphere. But if the geostrophic approximation is valid, that is, if the wind currents flow without friction and are not otherwise accelerated, then a large equator-to-pole temperature difference is required to maintain equilibrium.[84,85,86] Alternatively, the currents could extend to very deep within the planet, and would then be maintained in equilibrium by the very large mass of differentially rotating fluid. The latter interpretation is demanded by a model of Saturn in which the specific entropy is everywhere constant. In this case, there exists a unique barotropic relation between the mass density and pressure, and it can be shown[87] that in the absence of friction, this implies that the total (centrifugal plus gravitational) force on an element of fluid must be derivable from a potential. A further consequence is that the differential rotation must take place on coaxial cylinders.

The observed north-south symmetry of Saturn's differential rotation is certainly consistent with the hypothesis that the planet is rotating on cylinders (Jupiter's differential rotation is considerably less pronounced, and has less north-south symmetry). But since the symmetry persists from the equator to at least ±60° latitude, this implies that differential rotation on cylinders persists from the equator inward to a radius of about 30,000 km. According to interior models, at about this radius, the molecular hydrogen envelope gives way to a metallic hydrogen layer. It is possible that pronounced differential rotation on cylinders cannot persist in the electrically conducting layers because of strong hydromagnetic stresses. It has also been suggested[81] that the deep, differentially rotating, electrically conducting extension of the atmospheric rotation pattern may be the layer responsible for transforming the planetary magnetic field into an axisymmetric external configuration, as discussed above.

If Saturn is rotating on cylinders, then a correction must be applied to the coefficients of the external gravity field, J_2, J_4, J_6, . . . , to allow for the fact that the planet is not rotating as a solid body. It has been found that for Saturn, J_2 should be reduced by 0.5%, J_4 by 2.5%, and J_6 by about 10% in order to obtain the coefficients appropriate for a planet rotating as a solid body with the period equal to the magnetospheric rotation period. These corrections are largely independent of the assumed interior pressure-density relation.[88]

Tides in Saturn. As in the case of Jupiter, the existence of satellites close to Saturn can be used to place a constraint on the rate of tidal friction in Saturn averaged over the planet's lifetime. However, except for Titan, which is too distant to impose a strong constraint, Saturn has no Galilean-size satellites, and so the constraint is weaker than in the case of Jupiter.

Saturn actually has several satellites within or very close to the Roche limit, but these are too small to induce a significant tidal bulge in the planet. The most massive close-in satellites are Mimas (mass = 4.5×10^{22} g, distance = 3.075 Saturn radii) and Enceladus (mass = 7.6×10^{22} g, distance = 3.946 Saturn radii) (see Table 1-1). Because the tidal evolution time scale varies as the $13/2$ power of the radial distance (assuming time-independent Q), Mimas is expected to evolve faster than the more massive Enceladus, and thus imposes the strongest constraint on Saturn's Q. Again requiring that Mimas tidally evolve from the Roche limit to its present orbital distance in a time greater than or equal to the age of the solar system, we find

$$Q \gtrsim 3 \times 10^4 \qquad (8\text{-}62)$$

for Saturn, where we have taken $\Lambda_{2,0} = 0.11$. This lower limit for Q is about an order of magnitude smaller than the limit for Jupiter, but still rules out solid-body-type dissipation in Saturn. Although there are indications of (possibly intermittent) tidal dissipation in Enceladus, the origin of this dissipation is not well-understood, as we shall discuss below, and thus it cannot be used to establish an independent constraint on Saturn's Q as was done in the case of Jupiter. Saturn's Q could be much larger than the limit given in inequality 8-62.

Uranus and Neptune

It is convenient to discuss Uranus and Neptune interior models simultaneously because the available constraints are generally very similar for both planets. This state of affairs mainly reflects our state of ignorance about the interior structure of both bodies, and almost certainly will not persist after more information becomes available following spacecraft flybys and further groundbased investigation. Hints that Uranus and Neptune could have radically different interior structures already exist.

The most general quantities which pertain to the interior structure of Uranus and Neptune are their mass and radius, which are now well-known as a result of measurement of satellite orbits and observations of stellar occultations by these planets. As Fig. 8-20 clearly shows, Uranus and Neptune differ fundamentally from Jupiter and Saturn because they cannot contain a large component of hydrogen. But apart from this one firm conclusion, there is considerable non-uniqueness in possible compositions. For example, planets composed of pure helium would have a mass-radius curve which passes close to the values for Uranus and Neptune. Alternatively, one could contrive models with iron cores and hydrogen envelopes which would also reproduce the observed masses and radii. Either type of model would be highly implausible, however. There is no way that helium could be preferentially condensed from a protoplanetary cloud and incorporated alone into a planet. Similarly, it would be difficult to condense iron and then enclose the resulting core in hydrogen without incorporating materials of intermediate volatility.

We must therefore attempt to infer the interior compositions of Uranus and Neptune by a process which should now be familiar. Once again we refer to the table of cosmic abundances of the elements and to the relative volatilities of the compounds which can be formed from them. It is reasonable to suppose that Uranus and Neptune formed at substantially lower temperatures than the terrestrial planets, and should thus contain a full solar complement of rock and iron. The "ices" CH_4, NH_3, and H_2O could in principle be somewhat depleted relative to the rock and iron, if temperatures were never low enough for all of them to be fully condensed and then incorporated in the planet in a solar ratio to rock and iron. However, observations of the composition of the atmospheres of Uranus and Neptune, to which we now turn, suggest that all three of these volatile components may have been incorporated.

Atmospheres of Uranus and Neptune. Viewed through a telescope, Uranus and Neptune have a greenish appearance, in contrast to Jupiter and Saturn, which appear rather yellow. This difference is probably mainly attributable to a difference in atmospheric temperature and secondarily to a difference in chemical composition.

Spectroscopic observations have shown that H_2 is the principal molecular constituent in the atmospheres of all four Jovian planets.[89] If Uranus and Neptune's atmospheric composition were precisely solar, then the atmospheric temperature-pressure relation and condensation loci would be as shown in Fig. 8-21. Uranus and Neptune's atmospheric temperatures are colder than Jupiter and Saturn's because of lower solar influx and because of smaller internal heat sources (discussed below). As a result, the principal cloud layers in a solar composition model atmosphere for Uranus or Neptune arc predicted to occur at pressures approximately one order of magnitude higher than in Jupiter and Saturn. According to this picture, visible-wavelength light from the sun passes through the Uranian/ Neptunian atmosphere until it is reflected by cloud layers. Since such layers are much deeper in Uranus and Neptune than in Jupiter or Saturn, more absorption of visible sunlight takes place in the clear atmosphere. Molecular hydrogen is responsible for rather little of such absoprtion, but methane has strong absorption bands in the red portion of the visible spectrum and in the near infared. It is these bands which are responsible for the blue-green appearance of Uranus and Neptune.

Since visible sunlight passes through a larger amount of clear atmosphere in Uranus and Neptune than in Jupiter and Saturn, it is evident that methane absorption features will be stronger in the former even if the methane/hydrogen ratio is identical in all four planets. Although hydrogen absorptions are also enhanced, correcting for the effect of deeper cloud layers is an uncertain process, and there is considerable scatter in reported values of C/H in Uranus and

Neptune. Results range from C/H ~ 2 × solar to ~ 100 × solar.[90,91] The consensus is that Uranus' value of C/H is several times larger than the solar ratio, and is probably substantially larger than the ratio in Jupiter or Saturn as well. The same result is obtained for Neptune, but with less certainty.

H_2O vapor is not observed in Uranus and Neptune, and is not expected to be observable because of the depth of the cloud layers. Although NH_3 likewise condenses out deep in the atmosphere, it could in principle be detected through its effect on the microwave opacity as in the case of Jupiter and Saturn. The warm, deep layers in an adiabatically stratified Uranus model atmosphere should be observable at microwave wavelengths (~ 10 cm), with their "visibility" regulated by the opacity of ammonia at these wavelengths. But curiously, the microwave disc temperatures of Uranus at these wavelengths are substantially higher than expected.[92] If we were looking into an adiabatically stratified hydrogen-rich atmosphere with a solar mixing ratio of NH_3, the disc temperature would be about 150 °K at a wavelength of 10 cm, but it is observed to be about 75 °K higher. A possible interpretation of this measurement is that the atmosphere is more transparent than assumed, and that we are observing to deeper (and therefore hotter) layers. This interpretation then implies that NH_3 is substantially *depleted* in Uranus relative to the solar mixing ratio with hydrogen.

Deuterium has been detected in Uranus' atmosphere: the D/H ratios deduced from these measurements are approximately the same as the values in Jupiter and Saturn. HD measurements[77,93] give D/H = $(4.8 \pm 1.5) \times 10^{-5}$ and $(3.0 \pm 1.2) \times 10^{-5}$.

Composition of Uranus. The atmospheric composition measurements are used as a starting point for assumptions about interior composition. Since the atmosphere is hydrogen-rich, one class of models assumes that the atmosphere comprises a veneer of solar-composition material, perhaps modified in composition by the mixing of additional material from deeper layers. The high abundance of methane in the atmosphere is taken as evidence of such a process, for there is no other mechanism for increasing its concentration in a solar composition mixture. Furthermore, since methane is enhanced, it follows that the other two major "ices," H_2O and NH_3, are also enhanced, because they condense before methane.

The simplest model of Uranus' interior which seems consistent with atmospheric abundances and with the planet's mean density is then as follows. We start with a solar composition mixture and deplete only the most volatile components in it. These components are hydrogen and helium (and neon), which remain gases under all plausible conditions of planetary origin. We then assemble Uranus out of the remaining material, which is crudely subdivided into "rock", iron, and "ice," as defined in Chapter 3. The proportions of these components are not arbitrary, for we assume that they are given by the solar values. Finally, a veneer of hydrogen and helium (in solar proportions to each other) is added to the planetary core. The mass of hydrogen which is added is a free parameter. Because hydrogen has far lower density than the other components, the effect of adding this final layer is to increase the planet's radius without causing much change in its mass. A final embellishment, which has a modest effect on the deduced properties of the model, is to enhance the density of the atmosphere to account for the fact that it contains nonsolar amounts of CH_4 and possibly other components, which have dissolved into it from deeper layers.

There is some uncertainty about the role of temperature in the interior of Uranus. Although a finite temperature does not affect the pressure-density relation as significantly as in the hydrogen-rich planets Jupiter and Saturn, the value of the temperature has a bearing on another issue: the degree of internal differentiation in the planet. So little is known about Uranus' thermal state that a simple approximation is used. The intrinsic heat flow (so far undetected; see Table 5-1) is assumed to be large enough to maintain convection within the planet, leading to an adiabatic temperature profile. The model is self-consistent to the extent that an adiabatic temperature distribution leads to temperatures which are undoubtedly well above the melting temperatures of the "ices" everywhere in the planet. Using experimental values for the Grüneisen parameter in shocked liquid water, methane, and ammonia at high pressures, this model leads to temperatures of ~ 5000 °K in the deep interior of Uranus.[94] This result does not seem implausible, for adiabatic compression of icy planetesimals to pressures on the order of a megabar would lead to a similar result.

If this assumption is correct, the rock and iron component in Uranus may have separated from the ices, leading to a planetary model which has three principal layers. At the center we have a core of approximately earthlike composition. The iron is probably solid, but the remainder of the core could be liquid or solid, or both, for it may not be far from the solidus of dense "rock." Thus the inner rock/iron core could in principle be differentiated as well, like the interior of a terrestrial planet. Above this region we would find an immense, hot, dense, liquid "ocean" of water, methane, and ammonia in roughly cosmic proportions. The outermost layer is then the hydrogen-rich envelope, whose temperature also follows an adiabatic lapse rate, declining to very low values in the observable atmosphere.

A three-layer model of Uranus, which has the structure outlined above (rock-iron core, ice envelope, deep solar-composition atmosphere), has the following properties. The rock-iron core has a radius of about 8000 km and comprises 24% of the planetary mass. The ice layer has a radius of 18,000 km and comprises 65% of the planetary mass as required by the hypothesis of a solar ratio of ice to rock-iron. Since the 1-bar radius of the planet is 25,650 km, the solar-composition "atmosphere" has a depth of about 8000 km (nearly one-third of the planet's radius!), but comprises only about 11% of Uranus' mass. The pressure at the base of this atmosphere is about 220 kilobar, and the adiabatic temperature would be about 2500 °K. Thus there can be no metallic hydrogen in Uranus. The pressure at the base of the ice layer is computed to be about 5.8 megabar. This is only about a factor of two higher than the highest pressure achieved in H_2O shock compression experiments, so the equation of state of water, probably the principal constituent of Uranus, can be considered to be well-known in this planet.

Although the above model for the interior of Uranus seems to be a very reasonable one, it fails. It does not yield a value of $\Lambda_{2,0}$ which is in agreement with observation. The value of J_2 for Uranus which is given in Table 4-1 is derived from observations of the precession of Uranian rings under the influence of Uranus' oblate gravitational field. For some time, this result was useless for constraining interior models because the rotation period of Uranus was unknown. There is still some uncertainty in the rotation period of Uranus, with recently published values ranging from about 13 hours to about 24 hours (Table 1-1). There is probably something wrong with the 24-hour period, for it leads to rather implausible interior models, which have a very low degree of central condensation. Two careful investigations obtain very similar results for the value of the rotation period: 16.3 $\pm$ 0.3 hours[95] and 15.5 $\pm$ 1.3 hours.[96] We can estimate the value of $\Lambda_{2,0}$ by dividing Uranus' J_2 by the value of q (the rotation parameter defined earlier) obtained using a period of 15.5 hours. For consistency, we will use a normalizing radius of 26,200 km, which is the radius used to normalize J_2 and J_4 (and which corresponds to the pressure level observed in stellar occultations, on the order of a microbar). This yields $q = 0.039$, and we then estimate $\Lambda_{2,0} \simeq 0.085$. This result is substantially smaller than the corresponding values for Jupiter and Saturn ($\Lambda_{2,0} = 0.17$ and 0.11 respectively), but it is still too large to be compatible with the three-layer model described above, which has $\Lambda_{2,0} = 0.067$.

It is not clear how a Uranus interior model should be modified to be compatible with the observed $\Lambda_{2,0}$. We need to find a way to make the planet somewhat less centrally-condensed. The three-layer model is substantially differentiated, but is not as fully differentiated as might be imagined, since the rock and iron are assumed to be mixed in the central core. One could make the planet less centrally-condensed by supposing that the rock-iron component and ice component are undifferentiated, but this seems implausible unless central temperatures are remarkably low. The best answer to the problem at the moment appears to be a relaxation of the assumption that the hydrogen-rich "atmosphere" is essentially solar in composition. As discussed above, there is substantial evidence that Uranus' hydrogen-rich atmosphere has a nonsolar composition. If the density of this atmosphere is increased by adding substantial amounts of CH_4 and, in deeper, unobservable layers, H_2O, then the model value of $\Lambda_{2,0}$ can be brought into agreement with observation. For this purpose, the amount of methane in the atmosphere needs to be close to the upper limit permitted by observations, or about 100 times solar. Figure 8-30 shows a proposed model for the interior of Uranus (or Neptune) which may be compatible with available constraints. According to this model, which is quite tentative at present, Uranus is about 50% H_2O by mass.

What sort of process could have produced the Uranian atmosphere, which appears to be so heavily enriched in methane? There are two possibilities. First, a thin solar-composition atmosphere was captured onto a rock-ice core, and was then enriched in methane (and possibly water and ammonia) as these constituents dissolved in it. The second possibility is that the atmosphere is of secondary origin, analogous to the atmospheres of the terrestrial planets. In this case, hydrogen would be produced by decomposition of methane at the high pressures and temperatures obtaining at the base of the "ice"-composition envelope. Assuming that the envelope contains $\simeq 3.5$ earth masses of methane, decomposition of all of this could yield on the order of an earth mass of hydrogen. Not all of the methane could have been destroyed, but enough hydrogen to yield the observed atmosphere could conceivably be produced. It is not clear that this process actually occurs in Uranus. The hypothesis could be tested by measuring the helium abundance in the atmosphere. If the atmosphere is of secondary origin, helium may well be absent.

Composition of Neptune. Neptune models are basically similar to Uranus models. Neptune is slightly more massive than Uranus and has a slightly smaller radius, but the three-layer Uranus models described above can be readily adapted to Neptune's parameters. Figure 8-30 would also serve to describe this class of models for Neptune. In this model, Neptune also has a rock-iron core with a radius of about 8000 km, comprising 25% of the planet's mass. The H_2O-CH_4-NH_3 layer extends to a radius of 19,000 km and comprises 68% of the planet's mass. Thus the iron-rock-ice portion of Neptune is slightly more massive than the corresponding portion of Uranus. Since we are on the rising part of the radius vs. mass curve for masses in this range, the lower radius of Neptune is attributed to a smaller hydrogen abundance. A fit is obtained by assuming that the third layer, which has solar composition, comprises only about 6% of Neptune's mass (compared with 11% for Uranus). Temperatures and pressures within Neptune (assuming an adiabatic model) are very similar to the values in Uranus.

Although such three-layer models are a useful starting point for the discussion of Neptune's interior abundances, as in the case of Uranus they appear to be ruled out by a consideration of the planet's gravity field. We found for Uranus that the three-layer model gave a value of $\Lambda_{2,0}$ which was only about 25% smaller than the observed value. Thus there is a possibility of obtaining agreement by making small modifications in this model, e.g. by allowing for increased amounts of methane in the atmosphere. The disagreement appears to be far more

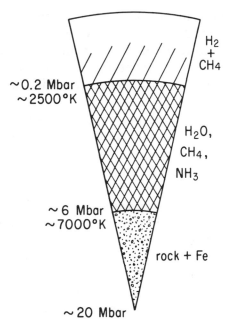

Fig. 8-30. Schematic interior model for the interior of Uranus or Neptune.

serious in the case of Neptune. It is not clear whether there are problems with the data or whether Neptune's interior structure differs in a very drastic way from Uranus' and from our preconceptions.

The value of Neptune's J_2 is derived from observations of its large, retrograde satellite Triton (see Table 1-1). The result,[97] normalized to $a = 24550$ km, is $J_2 = 0.0041 \pm 0.0004$. This result could conceivably be in error by a larger amount due to uncertainty about the position of Neptune's pole.[98] The rotation period of Neptune has been estimated from observations of variations in the reflectivity of the planet, presumed to be caused by rotating cloud layers, with typical results of about 18 hours.[99] This period gives $q = 0.020$, and so we find $\Lambda_{2,0} = 0.20$! If this result is taken seriously, it implies that Neptune's interior density profile is far more uniform than that of any other Jovian planet. Is it possible that Neptune has much lower interior temperatures than Uranus and is therefore almost completely undifferentiated? Heat flow measurements (discussed in the following subsection) indicate the contary if anything. It seems far

Neptune's is that Uranus, being nearer to the sun, receives a substantially larger heat flow component from it. Convective heat transport to the atmosphere is inhibited by the extra solar heating, as in the case of Jupiter. This model clearly hinges on the assumption that a residual heat flow exists in both Neptune and Uranus, and that it is adequate to maintain convective heat transport throughout both planets. There may be other explanations of the heat flow difference between Uranus and Neptune, but none has yet been advanced.

The Deuterium Problem. We mentioned above that the D/H ratio in Uranus' atmosphere is approximately equal to the value in Jupiter and Saturn. The ratio in Neptune's atmosphere is unknown. What values do we expect to observe in the Jovian planets? In the primordial solar composition nebula, the D/H ratio was about 2×10^{-5}, and this ratio presumably obtained in the gaseous phase at sufficiently high temperatures. As the temperature decreased into the range where hydrogen-bearing solid compounds could form (principally condensates of H_2O and CH_4 at sufficiently low temperatures), the D/H ratio in the gaseous and solid phases changed. Subject to the condition that the total number of deuterons in the gaseous and solid phases be conserved, thermodynamic equilibrium requires that the D/H ratio increase in the condensed phases and decrease in the gaseous phase. When a deuteron is substitued for a proton in species such as H_2, H_2O, CH_4, and NH_3, the extra mass causes a decrease in frequencies of normal modes of vibration of the molecule. Since the binding energy of the molecule includes its zero-point vibrational energy, the deuterated species are more tightly bound than the undeuterated species. It turns out that this effect is slightly greater in deuterated forms of H_2O, CH_4, and NH_3 as compared with H_2, and so the concentration of deuterium tends to be enhanced in the former. The energy differences are rather slight, equivalent to 10^2 °K in temperature, but the concentration effect can evidently become quite significant at the low temperatures at which solid "ice" phases start to form.

Figure 8-32 shows results of calculations of the deuterium concentration in a solar composition mixture of gases as a function of temperature. At temperatures below 200 °K, the D/H ratio becomes very considerably enhanced in H_2O, CH_4, and NH_3. The D/H ratio initially changes very little in H_2 because of its great abundance.

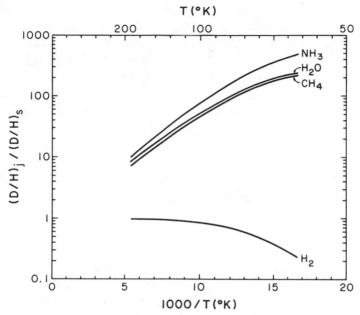

Fig. 8-32. Equilibrium concentration of deuterium in the *j*th molecular species, relative to solar concentration, in a solar composition mixture of gases. [From Hubbard, W. B., and MacFarlane, J. J. Theoretical predictions of deuterium abundances in the Jovian planets. *Icarus* **44**: 676–682 (1980). Copyright, Academic Press.]

According to this picture, if isotopic equilibrium can be established at the time that icy planetesimals are forming in the outer solar system (at temperatures below ~200 °K), the D/H ratio in these bodies would be at least one order of magnitude greater than the primordial D/H ratio. In fact, just this mechanism has been suggested as the origin of the ten-fold enhancement of D/H in the earth's oceans. Taking the number ten as a reasonable enhancement factor for D/H in the "ice" component, we then predict an overall enhancement in the planetary D/H in all of the Jovian planets. This enhancement is negligible in Jupiter and Saturn, because the amount of H in these planets is very large compared with the amount of "ice." On the other hand, H is a minor component in Uranus and Neptune compared with "ice," and the enhancement should be substantial. Assuming that the deuterium reequilibrates at the high internal temperatures in these bodies, we find that the atmospheric D/H ratio in both planets should be $\geq 10^{-4}$, or at least a factor of five greater than the observed value in Uranus.

Thus, the observation of "normal" D/H in the atmosphere of Uranus does not necessarily reflect a "normal" situation. On the contrary, it suggests that there is a problem with the current model of the origin and internal structure of Uranus and Neptune. There are two reasonable explanations for the discrepancy. First, at the low temperatures where equilibrium deuterium fractionation becomes large, the equilibration reactions may not be able to proceed rapidly enough to effect a significant concentration in the "ices." Second, Uranus may be extremely inactive, owing to the absence of an internal heat source, and the atmosphere may not have ever interacted chemically with the deep interior. If the latter explanation is correct, we would expect to see enhanced deuterium in Neptune's atmosphere.

Magnetic Fields?   To date, no definite evidence about the existence of an intrinsic magnetic field on Uranus or Neptune has been obtained. Our knowledge about the interior structure of these planets is too uncertain to permit a secure theoretical prediction. However, consideration of available data on the behavior of "ices" at high pressures and temperatures suggests the possibility of dynamo action.

As we have seen in Chapter 3 (Fig. 3-13), shock compression of water and ammonia to pressures and temperatures comparable to those within the interiors of Uranus and Neptune produces a marked enhancement in electrical conductivity, presumably due to formation of ions. In the case of water, a saturation in the conductivity profile is observed at the highest pressures, with the conductivity leveling off at about 10 (ohm · cm)$^{-1}$ ($\sim 10^{13}$ s^{-1} in gaussian units). Adopting this value for the interior of Uranus or Neptune, we obtain a magnetic diffusion coefficient $\eta \sim 7 \times 10^6$ cm^2/s. For a length scale $\sim 10^9$ cm, this gives a magnetic diffusion time scale $\sim 10^{11}$ s. A dynamo could conceivably exist if this time scale is longer than a typical convective time scale, which is given by the ratio of the length scale to a convective velocity. The latter is unknown, but dimensional considerations suggest that it may be on the order of 1 cm/s in Neptune. This yields a convective scale $\sim 10^9$ s, and thus the condition may be satisfied, at least in Neptune. However, results are very sensitive to assumed interior parameters. Calculations for Uranus[100] indicate that this planet may not have an intrinsic magnetic field produced by conducting H_2O.

Tidal Evolution. Consideration of limits on the tidal evolution of close satellites of Uranus and Neptune leads to limits on the value of the planetary Q, as in the case of Jupiter and Saturn. The limits are consistent with the hypothesized liquid state of Uranus and Neptune. Again we obtain a lower limit on the planetary Q by employing Eqs. 8-55 and 8-56 to calculate the radial distance of a satellite as a function of time, starting with the condition that the satellite be at or outside of the Roche limit at $t = 0$ and that it be at its present location at $t =$ present.

Similar limits on Uranus' Q are obtained by considering its two closest satellites, Miranda ($M = 0.3 \times 10^{-6} M_{Uranus}$, $R = 5.13\ a$), and Ariel ($M = 6 \times 10^{-6} M_{Uranus}$, $R = 7.54\ a$) (see Table 1-1). The former gives

$$Q_{Uranus} \gtrsim 3000, \qquad (8\text{-}63)$$

while the latter gives the slightly more stringent limit

$$Q_{Uranus} \gtrsim 5000. \qquad (8\text{-}64)$$

We have used the observed value of $\Lambda_{2,0}$ in this calculation.

It is not possible to perform an equivalent calculation for Neptune, since Neptune's massive satellite Triton (see Table 1-1) is in a retrograde orbit (assuming prograde rotation of Neptune), and is thus evolving toward the Roche limit rather than away from it. However, a recent occultation observation[101] has disclosed a probable Neptune satellite with dimensions similar to Nereid (Neptune's other known satellite) but probably in a much closer orbit. The data are consistent with an object of about 100 km radius moving in an equatorial orbit at a distance of $R \simeq 3a$. Assuming that this orbit is circular and prograde, and that the satellite has a mass of about 6×10^{21} g, the existence of the satellite implies

$$Q_{Neptune} \gtrsim 36,000, \qquad (8\text{-}65)$$

where we have used Uranus' value for $\Lambda_{2,0}$.

Jovian Planet Satellites

The earth's moon is the only significant satellite among the terrestrial planets. Mars' two natural satellites have dimensions of only tens of kilometers. In contrast, the known number of natural satellites in the outer solar system is very large, much larger than the total number of planets in the solar system, and there is little doubt that the inventory is still incomplete. Many satellites in the outer solar system are so large that they would be considered planets in their own right were they in independent orbits about the sun (Table 1-1). However, we are at a very early stage in the study of the interiors of the outer planet satellites, and relatively little is known about most of them. Recent spacecraft encounters have expanded the available data base for a few of these objects sufficiently to permit a meaningful discussion of their interior structure. Some of the phenomena which have been discovered have analogs in the inner solar system, but are nevertheless qualitatively distinct because of a different chemical, thermal, and gravitational environment.

The following discussion will be primarily limited to the major satellites of Jupiter and Saturn. Not enough is now known about the satellites of Uranus and Neptune to provide many constraints on their interior structures, and the same is true of the double planet Pluto/Charon.

We have already noted that the composition and structure of the Jovian planets seem to reflect their origin from an initial agglomeration of low-temperature condensates, rich in volatile materials such as water and other ices. If this picture is correct, one expects the compositions of the satellites to be similar to the presumed initial composition of Jovian planet cores. However, in some cases the compositions may have been modified by the presence of a massive primary body. The proximity of the primordial earth may have been a factor in causing the moon to have an unusual chemical composition, depleted in iron and volatiles. Similarly, the bulk compositions of Jovian planet satellite systems appear to reflect the influence of a massive and initially luminous neighbor.

Neglecting this effect for the moment, we may assume as a zeroth approximation that the satellites of a Jovian planet were made up of the same planetesimals that formed the initial core of the planet itself. Because of the presumed low temperatures in the outer solar system at the time of planetary formation, the composition of these planetesimals may correspond roughly to a chemical equilibrium mixture of condensates which are stable at temperatures on the order of 150 °K and lower, at pressures on the order of millibars and lower. According to this model, then, such satellites are composed of a mixture of rock, iron, and ice. It is not altogether clear what the composition of the ice component should be.

Recall that the sequence in which the major "ices" appear in condensates, in order of decreasing temperature, is H_2O, NH_3, and CH_4. Comets are thought to be agglomerations of primitive material, possibly similar in composition to the primordial planetesimals which accreted to form the Jovian planet cores and satellites. They are rich in volatiles (the "ices"), but also contain dust grains ("rock") mixed with the ice. The composition of the ice component cannot be directly determined at present, because spectral observations of comets only reveal the presence of neutral and ionized molecular fragments in the gassy envelope, not the parent molecules in the cometary nucleus. The chemistry of the nucleus is probably not simple, and may not correspond to the abundances predicted for thermodynamic equilibrium low-temperature condensation from a solar composition nebula. For example, the carbon may be present in species such as CO and CO_2 rather than CH_4. Obviously, these circumstances will have repercussions on the predicted mass ratio of ice to rock plus iron in the planetesimals. If oxygen is partially depleted in the initial gas via formation of CO, less H_2O can be formed.[102] Furthermore, as we shall discuss further, the accretion temperature (or environmental temperature) may rise sufficiently to drive off some of the more volatile ices at the time of formation of the satellite or planetary core.

Depending on the extent to which "ice" is fully condensed, one obtains the following percentages for the mass of iron plus rock relative to the total amount of condensed matter. If H_2O alone has condensed, but in its full solar abundance, Pagel's abundances (Table 2-1) give about 35% for the fraction of rock plus iron, and the other 65% is H_2O. If NH_3 has condensed as well, the rock/iron fraction falls to 33%. If C is present as CH_4, the mass of ice increases substantially to 76%, and the rock/iron fraction falls to only 24%. Small (i.e. satellite-sized) objects with any of these compositions will have very low mean densities, on the order of 1.5 g/cm³, and should thus be readily distinguishable from objects which have rock/iron composition and densities on the order of $3-4$ g/cm^3.

The Galilean Satellites of Jupiter. *Mean Densities and Average Composition.* The four great satellites of Jupiter were discovered by Galileo, and represent one of the most remarkable groups of objects in the solar system. Table 1-1 indicates that Io is very similar in size and mass to the earth's moon, but, as we shall see, its interior structure is altogether different. The other Galilean satellites are in the same general range of size and mass. Prior to the initial spacecraft observations of the Jupiter system, there was a vague indication of a systematic trend in density among the Galilean satellites. The masses and radii of these objects have now been measured quite precisely, and the trend is clearly evident. Io, with a mean density of 3.55 g/cm^3, could be composed entirely of rock and iron. Europa's density of 3.04 g/cm^3 suggests the presence of a lighter component (ice), while Ganymede (mean density $=$ 1.93 g/cm^3) and Callisto (mean density $=$ 1.83 g/cm^3) evidently must contain substantial amounts of ice.

Information about the surface compositions of the Galilean satellites can be obtained from measurements of surface reflectivity as a function of wavelength. The inferred surface compositions are consistent with the observed trend in the mean densities. No ice features are found in Io's spectrum. Instead, one sees sulfur compounds— SO_2 in particular (radically different from the surface chemistry of the earth's moon!). On the other hand, the infared absorption features of water ice are clearly evident in the spectra of Europa, Ganymede, and Callisto. The features are strongest in the spectra of Europa and Ganymede, and correspond to a surface composition of greater than 90% water ice by mass for Europa, and about 90% by mass for Ganymede. Callisto is much darker than the other Galilean satellites, but its surface is also predominantly composed of H_2O ice, with a mass abundance of about $30-90\%$.[103,104] These observations imply that Callisto is the most primitive of the Galilean satellites, and that its surface is composed of "dirty" primordial icy material, which is well-mixed with dust grains ("rock"), while Europa and Ganymede have possibly undergone some degree of differentiation which has led to a purer ice surface. No evidence is seen in the spectra of any of the Galilean satellites for the presence of carbon- or nitrogen-bearing ices.

Geological observations of the surfaces of the three outer Galilean satellites confirm the picture outlined above. Europa's surface shows no impact craters, but only a complex system of fissures in an icy surface devoid of significant relief. Ganymede's surface is somewhat cratered, but large regions have evidently been resurfaced since the initial formation of the satellite. Callisto's geological appearance nicely confirms the hypothesis that its surface is primordial — a saturated density of impact craters is seen everywhere.[105,106]

The bulk composition of the Galilean satellites is deduced from static models which are constrained by the evidence mentioned above. Thus, interior models of Io contain no ice, since no ice is seen on the surface, and the mean density is in any case consistent with a pure rock/iron composition. We shall say more about the interior of this remarkable satellite in a moment.

Models of the interior of Europa assume that the object is composed of a mixture of H_2O ice and rock/iron. These two components are assumed to be differentiated, since the surface characteristics of Europa suggest that this has occurred. The uncompressed density of a solar-composition rock/iron mixture is[107] about 3.66 g/cm^3. This material is taken to comprise the core of Europa, and then a relatively thin layer of H_2O is placed above it. The observed mass and radius of Europa can be reproduced if the thin layer of water ice has a depth of about 140 km (which is still a rather deep "ocean" by terrestrial standards!). The phase of this layer depends upon the thermal model for Europa's envelope, which we discuss below. The pressure at the base of the H_2O layer is only about a kilobar, and so the solid H_2O layer remains everywhere within the phase boundaries of Ice I (refer to the phase diagram of H_2O, Fig. 3-11), unless the temperature at the lower boundary is quite low, $\lesssim 150\ °K$. This model then implies that Europa is about 92% rock/iron by mass.

A model of Ganymede is constructed in a similar manner, again assuming that the object is composed of fully differentiated layers of H_2O and rock/iron. The principal difference with Europa is that the H_2O layer is far more extensive. The pressure at the base of this layer approaches 20 kilobar, and so account must be taken of the different phases of ice. Referring to Fig. 3-11, we see that ice in Ganymede passes through the stability fields of ice I, II, V, and VI, and approaches the field of ice VII. The density of H_2O at the base of the layer is nearly 1.4 g/cm^3, which is produced partly by compression of ice VI but primarily as a result of discontinuous density increases across the phase boundaries. When allowance is made for these transitions, one finds that Ganymede's H_2O layer is about 1000 km deep, and that the satellite is about 45% H_2O and 55% rock-iron.[108]

Models of Callisto are quite similar to those of Ganymede. If Callisto is assumed to be completely differentiated, the ice layer is also about 1000 km deep, and a similar series of ice phase transitions is encountered. According to this model, the satellite is about 50% H_2O and 50% rock/iron, and thus approaches the solar ratio for these quantities. As we have discussed, there is evidence that Callisto has not differentiated fully. A more appropriate model might therefore be one with a uniform mixture of ice and silicate particles (in some respects similar to a comet). The deduced proportions of H_2O and rock/iron in such a model would differ slightly from the values given because the central pressures would be lower.

Figure 8-33 summarizes results for the bulk compositions of the Galilean satellites. The figure is analogous to Fig. 8-19, only here iron/rock plays the role of iron, and H_2O plays the role of rock. The clear trend in bulk composition as a function of Jovicentric distance suggests that the Jovian system may be to some extent analogous to a miniature solar system, since the composition of the major satellites becomes more volatile-rich with distance from the primary. The "starlike" early evolution of Jupiter discussed earlier could provide enough central heat to vaporize water in the inner part of the Jovian system, thus explaining the absence of water in Io and its severe depletion in Europa. Calculations indicate that proto-Jupiter could have maintained temperatures at this level during the first 10^5 to 10^6 years of its existence (the reader can reproduce these results using the cooling theory presented earlier), depending on the opacity of the early proto-Jovian nebula. Thus accumulation of the satellites should have been completed within this interval.[109]

Thermal Structure of the Galilean Satellites. *Energetics of Io.* It is immediately clear from Table 5-1 that the energetics and thermal structure of Io must be very unusual. This small body is about the size of the moon, but has an observed average intrinsic surface heat flux about 100 times larger than the moon's, and except for the sun, its specific luminosity is the largest in the solar system, exceeding Jupiter's by a factor of four. This heat flow is accompanied by extensive

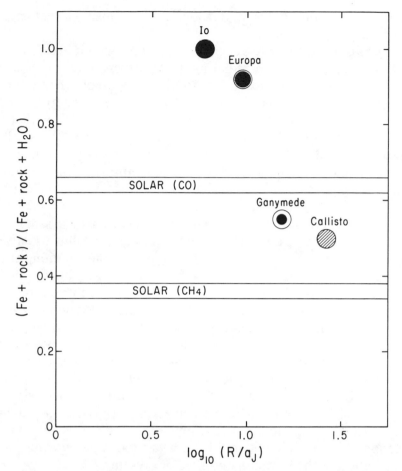

Fig. 8-33. Iron-rock content of Jovian satellites (as derived from models) as a function of distance from Jupiter (in units of the Jupiter radius). Solid shading within each symbol shows proportional size of the rock-iron core. Shading for the Callisto symbol indicates that the model is assumed to be undifferentiated (cf Fig. 8-19). The solar ratio of iron-rock to iron-rock-H_2O is shown, based on assumptions that C is present at the time of condensation in the form of either CO (upper line) or CH_4 (lower line).

active volcanism, which produces ejecta enriched with sulfur compounds. Io is more volcanically active than any other solar system body, including the earth.

We have already noted that radioactive heating cannot account for Jupiter's intrinsic luminosity, and the same argument is even more conclusive for Io. Jupiter's heat flow is almost certainly derived from slow loss of primordial heat, but such a mechanism cannot explain Io's heat flow.

The requirements on Io's energy sources are seen more clearly if we assume that Io has been radiating at its present rate for its entire lifetime of $\simeq 4.6 \times 10^9$ years. The total energy involved amounts to $\sim 10^{38}$ ergs, which exceeds Io's gravitational binding energy by one to two orders of magnitude. Clearly, there is no plausible way that this much energy could have been primordially stored in the satellite. Actually, the energy is continually transferred to Io from Jupiter, and the way that this occurs is a fascinating example of the magnified importance of tidal dissipation in a close satellite of a massive primary.

We discussed the mechanism of tidal heating of a satellite in Chapter 4. There we noted that tidal torques on a satellite usually act quickly to bring a close-in satellite into synchronous rotation with its primary. Nevertheless, the tidal potential experienced by the satellite will still be time-variable if the satellite is in an eccentric orbit. The accompanying dissipation in the satellite is described by Eq. 4-86. If the satellite is isolated and not being acted upon by any other bodies other than the primary, this dissipation is accompanied by a gradual reduction of the satellite's orbital eccentricity. The final state is one in which the satellite rotates synchronously *and* has a perfectly circular orbit, so that the tidal stresses are constant in time and no dissipation can occur. If an isolated Io were prepared in an initial orbit with high eccentricity in order to produce a large rate of tidal dissipation in its interior, this initial state would die away too rapidly to afford an explanation of the heat flux observed at present.

The dilemma is resolved by noting that the next Galilean satellite, Europa, is in a near-resonance with Io. That is, the mean motion of Io (n_1) is related to the mean motion of Europa (n_2) by[110]

$$n_1 - 2n_2 = 0.7395 \text{ °/day} \simeq 0. \qquad (8\text{-}66)$$

Were the right-hand side of this equation exactly equal to zero, Io and Europa would always be in conjunction at the same orbital longitude. Io and Europa conjunctions always occur at Io perijove and Europa apojove.

It is now important to note that the more closely $n_1 - 2n_2$ approaches zero, the greater the eccentricity of Io's orbit becomes. This result is intuitively reasonable, because if Io and Europa always encounter at the same orbital longitude, their mutual perturbations must add in a resonant fashion.[110] The eccentricity which is forced by repeated near-resonant encounters with Europa, called the *forced eccentricity,* is substantially greater than the eccentricity which would remain in Io's orbit after all perturbers were slowly removed (the latter is called the *free eccentricity*). Thus, according to Eq. 4-86, tidal dissipation in Io is enhanced by the near-resonance with Europa. Furthermore, tidal dissipation keeps the resonance from becoming exact, since it opposes an indefinite increase in the eccentricity. This suggests that we may be observing some type of quasi-equilibrium, which cannot under any circumstances be a complete steady-state equilibrium, because of the existence of an irreversible transformation of mechanical energy into heat.

In order to understand how such a quasi-equilibrium state might come about, it is important to consider the following features of tidal interactions. As we have already mentioned, tidal dissipation in a satellite tends to reduce the satellite's orbital eccentricity. Tidal torques on a synchronously rotating satellite are strongest at periapse, where they tend to increase the rotation of the satellite. The transfer of kinetic energy and angular momentum from orbital motion to spin, which thus takes place primarily at periapse, must reduce the orbital eccentricity.

Tides raised on the primary by the satellite have just the opposite effect. Again, such tidal torques are strongest at periapse. But the planet rotates more rapidly than the satellite's orbital motion, and so the effect of tidal dissipation in the primary is to transfer kinetic energy and angular momentum from the primary's rotation to the satellite's orbital motion. Since this transfer occurs mainly at periapse, the satellite's eccentricity is increased.

Before discussing possible scenarios for the origin of Io's forced eccentricity, we should note that the near-resonance between Io and Europa is clearly not accidental. Most remarkably, precisely the same near-resonance is observed between Europa and Ganymede. That is,

$$n_2 - 2n_3 = 0.7395 \text{ °/day,} \tag{8-67}$$

where n_3 is the mean motion of Ganymede. It follows that

$$n_1 - 3n_2 + 2n_3 = 0, \tag{8-68}$$

the famous Laplace resonance. As far as anyone can determine, this relationship among the inner three Galilean satellites is exactly satisfied. As a consequence of the nearness of Europa and Ganymede to a 2 : 1 orbital resonance, Europa also has a substantial forced eccentricity. The forced eccentricity of Io is 0.0041, while that of Europa is 0.0101.

First assume that there is no tidal dissipation in Jupiter, so that Q_J, the Jovian tidal Q, is infinite. In this case we must assume that the resonance and near-resonances among the inner three Galilean satellites are primordial, since dissipation in the satellites reduces the eccentricities and therefore moves the system further from resonance, and there is no countervailing effect from Jupiter to restore the eccentricities. That is, in this case we must assume that the system started out in a highly resonant state, with *much higher* forced eccentricities than are observed today, and that the system is therefore in the process of decaying through its present state to a final state with low forced eccentricity, far from resonance. This model can be tested by requiring that it produce an Io heat flux consistent with current active volcanism and the present value of the heat flow. The test of the model is in fact reminiscent of the tests applied earlier to models of the intrinsic heat flow of Jupiter and Saturn.

The second possibility is that the inner three Galilean satellites are in a state of quasi-equilibrium, such that tidal dissipation in Jupiter keeps the system in a state of partial balance, supplying eccentricity to Io and Europa's orbits at the same rate that tidal dissipation in these satellites removes it. According to this model, the initial state of the Galilean satellite system was quite random, with no resonances or near-resonances. Since Io was affected the most by tides in Jupiter, Io initially moved outward under their effect until it approached resonance with Europa and began to suffer a substantial increase in its forced orbital eccentricity. Thereafter, Io and Europa moved outward together until they approached resonance with Ganymede. The system is now observed in a state of quasi-equilibrium where Jovian tides are simultaneously moving all three satellites outward. The flow of energy and angular momentum is as follows. Jupiter's rotation is the ultimate energy source. It supplies eccentricity to the orbits of the inner three Galilean satellites. This portion of the Jovian rotational energy is dissipated as heat in their interiors as the eccentricities are damped. The system comes into a state of balance so that the forced eccentricities and nearness to resonance remain constant. The angular momentum derived from slowing down Jupiter's rotation is transferred into a gradual expansion of the satellite system, and this expansion also uses up the remainder of the energy.

The condition of quasi-equilibrium for the satellite eccentricities can be used to set constraints on dissipation in the satellites and in Jupiter. The quasi-equilibrium value of Io's forced eccentricity once the three-body resonance has been established is given by[111]

$$e_{lf}^2 = (\tfrac{1}{13})(k/k_J)(a_J/a)^5(M/M_J)^2(Q/Q_J), \qquad (8\text{-}69)$$

where the subscripted quantities refer to Jupiter and the unsubscripted quantities refer to Io. Let us now substitute some numbers in this formula in order to determine the ratio of Q/Q_J required for quasi-equilibrium. There is some uncertainty in the correct value of Io's Love number. For a start, we adopt a uniform, solid, moon-like model with shear modulus $\mu = 6.5 \times 10^{11}$ dyne/cm^2 (equal to the value of the shear modulus in the moon's outer layer). Formula 4-64 then gives $k \simeq 0.027$ for Io, and we will take $k_J \simeq 0.38$ for Jupiter.[112] Equation 8-69 then implies

$$Q_J/Q \simeq 10^4. \qquad (8\text{-}70)$$

Thus if Io's $Q \simeq 100$, the "canonical" value for solid bodies, then quasi-equilibrium requires $Q_J \simeq 10^6$. So far nothing is unreasonable, although there are no quantitative theories which predict a Jovian Q in this range. At least such a Jovian Q is consistent with the constraint discussed earlier which requires that the tidal evolution of the inner Galilean satellites not be so rapid that Io would have to originate within the Roche distance from Jupiter.

A problem arises when we examine the observed value of Io's heat flow (Table 5-1). Rewrite Eq. 4-86 in the form

$$Q \sim n^5 a^5 e^2 \rho^2 / (4\pi\mu H_i), \qquad (8\text{-}71)$$

where H_i is Io's observed intrinsic heat flux as determined by remote infrared observations. Substituting the other observed parameters for Io and continuing to assume that $\mu \simeq 6.5 \times 10^{11}$ dyne/cm^2, we find

$$Q \sim 600/H_i, \qquad (8\text{-}72)$$

for H_i in c.g.s. units. In order to explain Io's observed heat flux by the tidal dissipation mechanism, we must assume that Io's Q is of order unity! This result has other disconcerting implications. If Io's Q is of order unity, then the condition for quasi-equilibrium, Eq. 8-70, implies that the Jovian Q is so low that inequality 8-58 is violated. Both Jupiter and Io are then so dissipative that Io's orbit evolves outward far too quickly to be consistent with Io's age. This dilemma is reminiscent of a much milder problem which we encountered in connection with the moon's tidal evolution.

One can explore various possibilities for resolving the problem. First, the inner Galilean satellites may not be in a state of quasi-equilibrium, owing to a lack of dissipation in Jupiter's interior. Second, we may need to use a more elaborate model for heat dissipation in Io's interior. Third, there may be other important heating mechanisms beside tidal dissipation.

The first possibility has difficulties because it requires that the inner three Galilean satellites start out much closer to resonance than they are today, with correspondingly larger forced orbital eccentricities. They must then decay to the present state over a period of 4.6×10^9 years. This in turn requires that Io's Q be $\sim 10^3$, since otherwise the decay proceeds too rapidly. But if this is the case, there is not enough tidal dissipation in Io to account for the satellite's enormously enhanced heat flow.

The second possibility is explored by noting that the tidal dissipation rate in Io can be enhanced significantly if the satellite has a molten interior. For a uniform, solid Io with the adopted value of μ, the Love number $k \simeq 0.027$. If the satellite is entirely molten, the Love number approaches 1.5, giving a much higher tidal amplitude, but the heat dissipation goes to zero because the material cannot support stress. But if the satellite has a thin solid crust above a molten interior, the tidal amplitude in the crust is about the same as the amplitude for a liquid body, and therefore tidal dissipation occurs at a high rate in the solid crust. For crusts having a thickness on the order of 5% of the total radius, the enhancement in tidal amplitude compensates the most for the diminished volume of solid material, and the total tidal dissipation rate in the satellite is enhanced by about a factor of ten relative to the rate in a completely solid planet.[108] Such a model obviously ameliorates the heat flow problem, but does not solve it.

Is Io heated by some exotic mechanism other than tidal dissipation? Electrical heating has been considered as a possibility because Io moves in the strong Jovian magnetic field, and induced currents flow along the Jovian field lines which thread Io, completing the circuit in the Jovian ionosphere. However, if these currents are to produce ohmic heating in Io, the maximum power that they could supply would be about 10^{19} erg/s[108], while the observed luminosity of Io is 6×10^{19} erg/s. Moreover, ohmic heating of Io's interior can proceed only if the currents are routed through the body of the satellite. It is more likely that the induced currents are short-circuited by Io's conducting ionosphere.

Energetics of Europa. Europa is substantially smaller than Io and is farther from Jupiter, so tidal dissipation is not expected to be as important in Europa, even though Europa's equilibrium forced eccentricity is somewhat larger than Io's. Once more using Eq. 8-86, but with the parameters for Europa, we find

$$L_i/M \simeq 0.3 \times 10^{-8} \text{ erg/g/s} \tag{8-73}$$

for Europa's intrinsic luminosity per unit mass produced by tidal dissipation. We have used $Q = 100$ and $\mu = 6.5 \times 10^{11}$ dyne/cm^2, as appropriate for a largely rocky body. The present radioactive heat production rate for a solar-composition mixture of rock and iron would be about an order of magnitude higher (see Table 5-1), and therefore would govern Europa's energetics. However, tidal dissipation does not fall far short of being important, and it is possible to imagine models of Europa where it could be the dominant energy-producing mechanism. For example, if the initial heating of Io during accumulation resulted in an object with a differentiated water envelope, and if this envelope were liquid at the bottom, then the outer icy crust would be tidally decoupled from the interior, and could undergo substantial tidal "working" with concomitant heating. It is also conceivable that Europa's orbit was even more eccentric in the past, leading to increased tidal dissipation. The craterless, cracked ice on Europa's surface suggests that substantial thermal evolution of the outer layers has taken place.

Energy transport in Ganymede and Callisto. As we have mentioned, Ganymede's surface shows some evidence of internal activity, but

Callisto's surface does not. What does this imply about these satellites' interior state?

As a simple starting point, consider an interior model which is a homogeneous mixture of H_2O ice and rock/iron. We wish to determine whether radioactive heating in such a mixture can eventually cause the ice to melt, leading a two-layer model with a rock/iron core and an ice/water mantle. The calculation is entirely analogous to one performed earlier in connection with the problem of the formation of Mars' iron core. For an approximately equal mixture of rock/iron and ice by mass, we adopt a heat capacity of about 1.5×10^7 erg/g/°K. The radioactive heating rate ϵ_N in such a mixture will be variable over the age of the solar system (see Fig. 5-1), but a typical value for a chondritic mixture of rock/iron would be about 10^{-7} erg/g/s. Allowing for a dilution factor of ½, we then adopt $\epsilon_N \simeq 0.5 \times 10^{-7}$ erg/g/s, and so the rate of temperature increase in the homogeneous body is initially

$$dT/dt \sim \epsilon_N/C_P \sim 200 \ °\mathrm{K}/(10^9 \ \text{years}). \qquad (8\text{-}74)$$

This stage of the thermal evolution occurs when $t' \ll 1$, where t' is the dimensionless time variable defined in terms of Eq. 8-25. Because the thermal diffusivity χ of ice is only about 10^{-2} cm²/s, we are guaranteed that this inequality holds reasonably well in objects the size of the Galilean satellites and larger, over the age of the solar system. Referring to the phase diagram of H_2O ice shown in Fig. 3-11, we see that a temperature increase of $\sim 200 - 400$ °K is sufficient to completely melt the ice in the Galilean satellites.

Although heating due to accretional energy has been entirely neglected in this calculation, Eq. 8-74 seems to suggest that the Galilean satellites, and any other icy bodies of comparable size, should have completely differentiated over the age of the solar system, and may still have extensive liquid water envelopes. Only an outer ice crust, exposed to the cold of space, would remain solid. Once melting occurs, the liquid water layers would tend to have adiabatic temperature profiles maintained by convection. But this model does not agree well with the observations. According to this model, both Callisto and Ganymede should have differentiated and formed new surfaces in the process. If a silicate-laden primordial ice layer found itself above a pure layer of lower-density water, instability would rapidly ensue and lead to destruction of the initial crust. All of the dark, primordial ice would be devoured.

The solution to the dilemma appears to be a relaxation of the requirement that convection occur only in the liquid water phase. As we have discussed in Chapter 5, convection provides an efficient mechanism for removing internal heat. If convection occurs in solid icy satellites, it could in principle remove the radiogenic heat which would otherwise produce liquid water layers. The process of solid state convection is, as we have discussed, suspected to be of some importance in the rocky bodies of the inner solar system. Convincing evidence of the occurrence of solid state convection in icy satellites also would provide a significant confirmation of the theoretical basis for this process.

Unfortunately, the viscosity of solids plays a key role in evaluating possibilites for solid state convection, and this parameter varies over an enormous range as a function of temperature. According to Eq. 5-46, convection occurs in a viscous material when the Rayleigh number Ra exceeds a certain critical value $\sim 10^3$. Let us substitute the numbers for the appropriate parameters in an icy, outer-solar system satellite. Let the temperature difference $T_1 - T_2 = 300$ °K, as appropriate to a satellite which is on the verge of melting. We take $\alpha = 3 \times 10^{-5}$ (°K)$^{-1}$, $g = 180$ cm/s^2, $d = 1000$ km, and $\chi = 10^{-2}$ cm^2/s.[113] Then $Ra \sim 10^{26}$ (cm^2/s)$/v$, where v is the kinematic viscosity. According to one proposed model for creep in ice I,[114]

$$v = 10^{14} \ (\text{cm}^2/\text{s}) \ \exp \ [25 \ (T_m/T - 1)], \qquad (8\text{-}75)$$

where T_m is the ice melting temperature. Substituting this expression in the definition of the Rayleigh number, one finds that Ra exceeds the critical value for the onset of convection at temperatures well below T_m. Even after making allowances for substantial uncertainty in the parameters in expression 8-75, it turns out that an icy mantle can convect if its temperatures exceed about one-half the value of the melting temperature. Calculations of the energy transported by convection under these circumstances[113] indicate that such convection can transport the heat generated by radioactivity for $T/T_m \simeq 0.75$.

Thus, a resolution of the dilemma posed by the primitive appearance of Callisto's surface is provided by the mechanism of solid state convection. A primitive Galilean satellite composed of a uniform, equal mixture of ice and silicates will not melt as a result of radioactive heating. Long before the material approaches melting temperatures, convection carries the radioactive heat to the surface. Like terrestrial planets, ice bodies should have a surface layer where temperatures are sufficiently low that the material is unable to flow. A conductive temperature profile will still be found in this layer, which acts as a rigid material. Primordial craters should be preserved in such an icy "lithosphere."

Although the mechanism of solid state convection seems to provide a reasonable explanation for the observed properties of the surfaces of Ganymede and Callisto, a problem still remains. The surface of Ganymede shows some evidence of alteration due to internal dynamics. Although the slow process of radioactive heating probably cannot account for this difference with Callisto, perhaps the rapid process of initial accumulation is responsible. We do not know how rapid this process was, but it is conceivable that enough accretional energy could have been deposited in the outer layers of the icy satellite to partially melt them. This would result in a degree of segregation of the silicates in these layers, leading to production of a purified ice layer at the surface. However, the masses and sizes of Ganymede and Callisto are not very different, and so it is not easy to argue that this mechanism would have been much more efficient in the former than in the latter.

The Satellites of Saturn. *Mean Densities and Average Composition.* Important parameters of Saturn satellites are given in Table 1-1 (where known). Only one Saturn satellite, Titan, is in the same size class as the Galilean satellites. All of the rest are much smaller. Perhaps this is related to the fact that the neat correlation between volatile content and distance from the primary which is noted in the Galilean satellites disappears totally in the Saturn satellite system. Figure 8-34 shows a plot of the inferred rock-iron content for several of the satellites as a function of mean distance from Saturn. No point is plotted for satellites whose mean density is very poorly determined, such as Enceladus. It should be noted that the rock-iron content is inferred from an interior model which contains only water ice and rock-iron.

In any case, it seems clear that the clear trend of increasing volatile content with increasing density from the primary which is observed

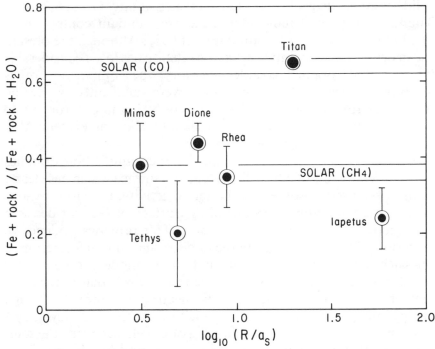

Fig. 8-34. Iron-rock content of Saturn satellites (as derived from models) as a function of distance from Saturn (in units of the Saturn radius). Solid shading within each symbol shows proportional size of the rock-iron core (*cf* Figs. 8-19 and 8-33). The solar ratio of iron-rock to iron-rock-H_2O is shown, based on assumptions that C is present at the time of condensation in the form of either CO (upper line) or CH_4 (lower line):

in the Galilean satellites is not reproduced in the Saturn system. *All* of the small Saturn satellites are rich in volatiles, and the error bars on mean densities are such that a clear trend is not likely to appear even after the mean densities have been better determined. It would appear that proto-Saturn had less of a thermal effect on the chemistry of aggregating satellites than did proto-Jupiter. Qualitatively, this result seems reasonable because the lower mass of Saturn would result in a lower maximum central temperature during the phase of initial contraction of the entire system.

Study of the interior structures of the Saturn satellites is still in an early phase because several important pieces of information are lacking. For example, the surface of Titan is entirely shrouded in clouds, and nothing is known about whether the surface is liquid or solid, or about its geological characteristics, if any. Observations of Titan's atmosphere[78] indicate that the predominant component is N_2, but with a substantial admixture of CH_4. Although the present mean density of Titan implies that the satellite has lost a considerable fraction of its original volatile inventory, the composition of its atmosphere suggests that temperatures were sufficiently low during initial accretion for substantial amounts of NH_3 and CH_4 (or CO) to be present. In contrast, the Galilean satellites show no evidence that such species were ever present.

The Problem of Enceladus. The satellite Enceladus, whose orbit lies between Mimas and Tethys, displays an icy surface which has clearly been reprocessed by some sort of internal activity. Extensive regions of the surface resemble portions of the surface of Ganymede, with long grooves and an absence of craters. This provides evidence of extrusion of new material from the interior since initial formation of the satellite.[115] However, it is difficult to find a model for the interior which can account for this phenomenon. As we shall now verify, neither radioactive heating nor tidal heating can provide enough energy input to produce significant melting in this small, icy body.

We first consider radioactive heating of Enceladus, using the solution for the temperature distribution in a uniform body with a uniform distribution of radioactive sources. Assuming that Enceladus is about ⅓ rock-iron and ⅔ water ice, the present average radioactive heating rate would be about 0.3×10^{-7} erg/g/s. With a radius $a = 251$ km, thermal diffusivity $\chi \simeq 10^{-2}$ cm²/s, heat capacity $C_P \simeq 1.5 \times 10^7$ erg/g/°K, and mean density $\rho \simeq 1.2$ g/cm³, we compute the unit of temperature (see Eqs. 8-24 and 8-25 and Fig. 8-17) and time:

$$T_u \simeq 20 \ °K; \tag{8-76}$$

$$t_u \simeq 2 \times 10^9 \text{ years.} \tag{8-77}$$

Thus such a small object can increase its central temperature by only about 20°K over the age of the solar system. Although more rapid radioactive heating actually takes place early in the history of the solar system, allowance for this effect does not change the conclusion that radioactivity alone probably cannot melt Enceladus, or any other icy satellite of similar composition and size.

The problem of melting Enceladus is reduced somewhat if we postulate that the water ice in the satellite also contains ammonia ice in roughly solar proportions. The presence of ammonia can reduce the melting temperature of pure H_2O by as much as 100 °K.[115] However, a significant problem still remains, and there is as yet no direct evidence for the presence of NH_3 in Enceladus. Moreover, the conductive solution discussed above does not take into account the possibility of solid-state convection, which could substantially reduce the maximum central temperature.

Principally because of Enceladus' small dimensions, tidal heating is not very effective in this object either. Using similar parameters to those used for the Galilean satellites, one finds that the tidal heating rate is about equal to the radioactive heating for a $Q \simeq 20$. However, in principle, tidal heating could be a significant factor for Enceladus, for its orbital period is close to a 1 : 2 resonance with the orbital period of Dione, and thus Enceladus has a forced eccentricity of 0.0044. This forced eccentricity would need to be about a factor of twenty larger in order for tidal heating to be responsible for the resurfacing of Enceladus.

Mimas and Tethys are similarly close to a 2 : 1 orbital resonance, although again they are apparently too small for tidal heating to be a major factor in their internal energetics. Nevertheless, it seems clear that tides, which effectively transfer kinetic energy from the massive primary to the satellite, are likely to be a major factor in the interior structure of the satellites of the major planets. Unraveling the previous history of the interiors of these bodies may well require a detailed understanding of their orbital histories as well.

REFERENCES

1. Ringwood, A. E. and Kesson, S. E. Composition and origin of the moon. *Proc. 8th Lunar Sci. Conf.* **1**: 371–398 (1977).
2. Sonett, C. P. Electromagnetic induction in the moon. *Rev. Geophys. and Spa. Phys.* **20**: 411–455 (1982).
3. Sonett, and Duba, A. Lunar temperature and global heat flux from laboratory electrical conductivity and lunar magnetometer data. *Nature* **258**: 118–121 (1975).
4. Ferrari, A. J., Sinclair, W. S., Sjogren, W. L., Williams, J. G., and Yoder, C. F. Geophysical parameters of the earth-moon system. *J. Geophys. Res.* **85**: 3939–3951 (1980).

5. Bills, B. G., and Ferrari, A. J. A harmonic analysis of lunar gravity. *J. Geophys. Res.* **85**: 1013–1025 (1980).
6. Bills, B. G. and Ferrari, A. J. A harmonic analysis of lunar topography. *Icarus* **31**: 244–259 (1977).
7. Nakamura, Y., Latham, G., Lammlein, D., Ewing, M., Duennebier, F., and Dorman, J. Deep lunar interior inferred from recent seismic data. *Geophys. Res. Lett.* **1**: 137–140 (1974).
8. Nakamura, Y., Latham, G. V., and Dorman, H. J. Apollo lunar seismic experiment—final summary. *J. Geophys. Res. (Suppl.)* **87**: A117–A123 (1982).
9. Keihm, S. J. and Langseth, M. G. Lunar thermal regime to 300 km. *Proc. 8th Lunar Sci. Conf.* **1**: 371–398 (1977).
10. Strom, R. G., Trask, N. J., and Guest, J. E. Tectonism and volcanism on Mercury. *J. Geophys. Res.* **80**: 2478–2507 (1975).
11. Melosh, H. J. Global tectonics of a despun planet. *Icarus* **31**: 221–243 (1977).
12. Melosh, H. J. and Dzurisin, D. Mercurian global tectonics: a consequence of tidal despinning? *Icarus* **35**: 227–236 (1978).
13. Pechmann, J. B. and Melosh, H. J. Global fracture patterns of a despun planet: application to Mercury. *Icarus* **38**: 243–250 (1979).
14. Esposito, P. B., Anderson, J. D., and Ng, A. T. Y. Experimental determination of Mercury's mass and oblateness. *Space Res.* **17**: 639–644 (1977).
15. Siegfried, R. W., II and Solomon, S. C. Mercury: internal structure and thermal evolution. *Icarus* **23**: 192–205 (1974).
16. Solomon, S. C. Some aspects of core formation in Mercury. *Icarus* **28**: 509–521 (1976).
17. Cassen, P., Young, R. E., Schubert, G., and Reynolds, R. T. Implications of an internal dynamo for the thermal history of Mercury. *Icarus* **28**: 501–508 (1976).
18. Goettel, K. A., Shields, J. A., and Decker, D. A. Density constraints on the composition of Venus. *Proc. Lunar Sci. Conf.* **12B**: 1507–1516 (1981).
19. Ringwood, A. E. and Anderson, D. L. Earth and Venus: a comparative study. *Icarus* **30**: 243–253 (1977).
20. Lewis, J. S. Metal/silicate fractionation in the solar system. *Earth and Planet. Sci. Lett.* **15**: 286–290 (1972).
21. Sill, G. T. Sulphuric acid in the clouds of Venus. *Commun. Lunar and Planet. Lab.* No. **171**: 191–198 (1972).
22. Surkov, Yu. A. Natural radioactivity of the moon and planets. *Proc. Lunar and Planet. Sci.* **12B**: 1377–1386 (1981).
23. Toksoz, M. N., Hsui, A. T., and Johnston, D. H. Thermal evolutions of the terrestrial planets. *The Moon and Planets* **18**: 281–320 (1978).
24. Seiff, A., Kirk, D. B., Young, R. E., Blanchard, R. C., Findlay, J. T., Kelly, G. M., and Sommer, S. C. Measurements of thermal structure and thermal contrasts in the atmosphere of Venus and related dynamical observations: results from the four Pioneer Venus probes. *J. Geophys. Res.* **85**: 7903–7933 (1980).
25. Pollack, J. B., Toon, O. B., and Boese, R. Greenhouse models of Venus' high surface temperature, as constrained by Pioneer Venus measurements. *J. Geophys. Res.* **85**: 8223–8231 (1980).
26. Anderson, D. L. Plate tectonics on Venus. *Geophys. Res. Lett.* **8**: 309–311 (1981).
27. Ananda, M. P., Sjogren, W. L., Phillips, R. J., Wimberly, R. N., and Bills, B. G.

A low-order gravity field of Venus and dynamical implications. *J. Geophys. Res.* **85:** 8303–8318 (1980).

28. Sjogren, W. L., Bills, B. G., Birkeland, P. W., Esposito, P. B., Konopliv, A. R., Mottinger, N. A., Ritke, S. J., and Phillips, R. J. Venus gravity anomalies and their correlations with topography. *J. Geophys. Res.* **85:** 8303–8318 (1983).

29. Anderson, D. L., Miller, W. F., Latham, G. V., Nakamura, Y., Toksoz, M. N., Dainty, A. M., Duennebier, F. K., Lazarewicz, A. R., Kovach, R. L., and Knight, T. C. D. Seismology on Mars. *J. Geophys. Res.* **82:** 4524–4546 (1977).

30. Toulmin, P., III, Baird, A. K., Clark, B. C., Keil, K., Rose, H. J., Jr., Christian, R. P., Evans, P. H., and Kelliher, W. C. Geochemical and mineralogical interpretation of the Viking inorganic chemical results. *J. Geophys. Res.* **82:** 4625–4634 (1977).

31. McGetchin, T. R. and Smyth, J. R. The mantle of Mars: some possible geological implications of its high density. *Icarus* **34:** 512–536 (1978).

32. Christensen, E. J. and Balmino, G. Development and analysis of a twelfth degree and order gravity model for Mars. *J. Geophys. Res.* **84:** 7934–7953 (1979).

33. Stacey, F. D. *Physics of the earth.* New York: Wiley & Sons, 1969.

34. Reasenberg, R. D. The moment of inertia and isostasy of Mars. *J. Geophys. Res.* **82:** 369–375 (1977).

35. Kaula, W. M. The moment of inertia of Mars. *Geophys. Res. Lett.* **6:** 194–196 (1979).

36. Johnston, D. H., McGetchin, T. R., and Toksoz, M. N. The thermal state and internal structure of Mars. *J. Geophys. Res.* **79:** 3959–3971 (1974).

37. Morgan, J. W. and Anders, E. Chemical composition of Mars. **43:** 1601–1610 (1979).

38. Toksoz, M. N. and Hsui, A. T. Thermal history and evolution of Mars. *Icarus* **34:** 537–547 (1978).

39. Zapolsky, H. S. and Salpeter, E. E. The mass-radius relationship for cold bodies of low mass. *Astrophys. J.* **158:** 809–813 (1969).

40. Weidenschilling, S. J. and Lewis, J. S. Atmospheric and cloud structures of the Jovian planets. *Icarus* **20:** 465–476 (1973).

41. Lindal, G. F., *et al.* The atmosphere of Jupiter: an analysis of the Voyager radio occultation measurements. *J. Geophys. Res.* **86:** 8721–8727 (1981).

42. Ridgway, S. T., Larson, H. P., and Fink, U. The infrared spectrum of Jupiter. In *Jupiter* (T. Gehrels, ed.). University of Arizona Press: Tucson, 1976.

43. Gautier, D., Conrath, B., Flasar, M., Hanel, R., Kunde, V., Chedin, A., and Scott, N. The helium abundance of Jupiter from Voyager. *J. Geophys. Res.* **86:** 8713–8720 (1981).

44. Knacke, R. F., Kim, S. J., Ridgway, S. T., and Tokunaga, A. T. The abundances of CH_4, CH_3D, NH_3, and PH_3 in the troposphere of Jupiter derived from high-resolution 1100–1200 cm^{-1} spectra. *Astrophys. J.* **262:** 388–395 (1982).

45. Kunde, V., Hanel, R., Maguire, W., Gautier, D., Baluteau, J. P., Marten, A., Chedin, A. Husson, N., and Scott, N. The tropospheric gas composition of Jupiter's north equatorial belt (NH_3, PH_3, CH_3D, GeH_4, H_2O) and the Jovian D/H isotopic ratio. *Astrophys. J.* **263:** 443–467 (1982).

46. Barshay, S. and Lewis, J. S. Chemical structure of the deep atmosphere of Jupiter. *Icarus* **33:** 593–611 (1978).

47. Prinn, R. G. and Lewis, J. S. Phosphine on Jupiter and implications for the great red spot. *Science* **190:** 274–276 (1975).

48. Prinn, R. G. and Barshay, S. S. Carbon monoxide on Jupiter and implications for atmospheric convection. *Science* **198**: 1031–1034 (1977).
49. Prinn, R. G. and Owen, T. Chemistry and spectroscopy of the Jovian atmosphere. In *Jupiter* (T. Gehrels, ed.). University of Arizona Press: Tucson, 1976.
50. Berge, G. L. and Gulkis, S. Earth-based observations of Jupiter: millimeter to meter wavelengths. In *Jupiter: Studies of the interior, atmosphere, magnetosphere, and satellites* (T. Gehrels, ed.). University of Arizona Press: Tucson, 1976.
51. Hubbard, W. B. Gravitational field of a rotating planet with a polytropic index of unity. *Soviet Astron.—AJ* **18(5)**: 621–624 (1975).
52. Hubbard, W. B. and Lampe, M. Thermal conduction by electrons in stellar matter. *Astrophys. J. (Suppl.)* **18**: 297–346 (1969).
53. Stevenson, D. J. Thermodynamics and phase separation of dense fully-ionized hydrogen-helium fluid mixtures. *Phys. Rev. B* **12**: 3999–4007 (1975).
54. DeWitt, H. E. and Hubbard, W. B. Statistical mechanics of light elements at high pressure. IV. A model free energy for the metallic phase. *Astrophys. J.* **205**: 295–301 (1976).
55. Cox, J. P. and Giuli, R. T. *Stellar Structure: Physical Principles*. Gordon & Breach: New York, 1968, p. 385.
56. Trafton, L. M. On the He-H_2 thermal opacity in planetary atmospheres. *Astrophys. J.* **179**: 971–976 (1973).
57. Hubbard, W. B. The Jovian surface condition and cooling rate. *Icarus* **30**: 305–310 (1977).
58. Graboske, H. C., Pollack, J. B., Grossman, A. S., and Olness, R. J. The structure and evolution of Jupiter: the fluid contraction phase. *Astrophys. J.* **199**: 265–281 (1975).
59. Salpeter, E. E. On convection and gravitational layering in Jupiter and in stars of low mass. *Astrophys. J.* **181**: L83–L86 (1973).
60. Stevenson, D. J. and Salpeter, E. E. Interior models of Jupiter. In *Jupiter* (T. Gehrels, ed.). University of Arizona Press: Tucson, 1976.
61. MacFarlane, J. J. and Hubbard, W. B. Statistical mechanics of light elements at high pressure. V. Three-dimensional Thomas-Fermi-Dirac theory. *Astrophys. J.*, **272**: 301–310 (1983).
62. Hubbard, W. B. and Horedt, G. P. Computation of Jupiter interior models from gravitational inversion theory. *Icarus* **54**: 456–465 (1983).
63. Hubbard, W. B. Thermal Structure of Jupiter. *Astrophys. J.* **152**: 745–754 (1968).
64. Stevenson, D. J. and Ashcroft, N. W. Conduction in fully ionized liquid metals. *Phys. Rev. A* **9**: 782–789 (1974).
65. Ingersoll, A. P. Pioneer 10 and 11 observations and the dynamics of Jupiter's atmosphere. *Icarus* **29**: 245–253 (1976).
66. Goldreich, P. and Soter, S. Q in the solar system. *Icarus* **5**: 375–389 (1966).
67. Hubbard, W. B. Tides in the giant planets. *Icarus* **23**: 42–50 (1974).
68. Goldreich, P. and Nicholson, P. Turbulent viscosity and Jupiter's tidal Q. *Icarus* **30**: 301–304 (1977).
69. Dermott, S. F. Tidal dissipation in the solid cores of the major planets. *Icarus* **37**: 310–321 (1979).
70. Stevenson, D. J. Anomalous bulk viscosity of two-phase fluids and implications for planetary interiors. *J. Geophys. Res.* **88**: 2445–2455 (1983).
71. Hanel, R., Conrath, B., Flasar, F. M., Kunde, V., Maguire, W., Pearl, J., Pirraglia, J., Samulson, R., Herath, L., Allison, M., Cruikshank, D., Gautier,

D., Gierasch, P., Horn, L., Koppany, R., and Ponnamperuma, C. Infrared observations of the Saturnian system from Voyager 1. *Science* **212**: 192–200 (1981).

72. Orton, G. S. and Ingersoll, A. P. Saturn's atmospheric temperature structure and heat budget. *J. Geophys. Res.* **85**: 5871–5881 (1980).

73. Buriez, J. C., and de Bergh, C. A study of the atmosphere of Saturn based on methane line profile near 1.1 microns. *Astron. Astrophys.* **94**: 382–390 (1981).

74. Encrenaz, T. and Combes, M. On the C/H and D/H ratios in the atmospheres of Jupiter and Saturn. *Icarus* **52**: 54–61 (1982).

75. Marten, A., Courtin, R., Gautier, D., and Lacombe, A. Ammonia vertical density profiles in Jupiter and Saturn from their radioelectric and infrared emissivities. *Icarus* **41**: 410–422 (1980).

76. Larson, H. P., Fink, U., Smith, H. A., and Davis, D. S. The middle-infrared spectrum of Saturn: evidence for phosphine and upper limits to other trace constituents. *Astrophys. J.* **240**: 327–337 (1980).

77. Macy, W., Jr., and Smith, W. H. Detection of HD on Saturn and Uranus and the D/H ratio. *Astrophys. J.* **222**: L137–L140 (1978).

78. Tyler, G. L., Eshleman, V. R., Anderson, J. D., Levy, G. S., Lindal, G. F., Wood, G. E., and Croft, T. A. Radio science investigations of the Saturn system with Voyager 1: preliminary results. *Science* **212**: 201–206 (1981).

79. Stevenson, D. J., and Salpeter, E. E. The dynamics and helium distribution in hydrogen-helium fluid planets. *Astrophys. J. Suppl.* **35**: 239–261 (1977).

80. Grossman, A. S., Pollack, J. B., Reynolds, R. T., Summers, A. L., and Graboske, H. C., Jr. The effect of dense cores on the structure and evolution of Jupiter and Saturn. *Icarus* **42**: 358–379 (1980).

81. Stevenson, D. J. Saturn's luminosity and magnetism. *Science* **208**: 746–748 (1980).

82. Parker, E. N., *Cosmical Magnetic Fields—Their Origin and Activity*. Clarendon Press: Oxford, 1979, pp. 538–541.

83. Desch, M. D. and Kaiser, M. L. Voyager measurement of the rotation period of Saturn's magnetic field. *Geophys. Res. Let.* **8**: 253–256 (1981).

84. Smith, B. A., *et al.* Encounter with Saturn: Voyager 1 imaging science results. *Science* **212**: 163–191 (1981).

85. Smith, B. A., *et al.* Voyager 2 encounter with the Saturnian system. *Science* **215**: 499–537 (1983).

86. Allison, M. and Stone, P. H. Saturn meteorology: a diagnostic assessment of thin-layer configurations for the zonal flow. *Icarus* **54**: 296–308 (1983).

87. Tassoul, J.-L. *Theory of Rotating Stars*. Princeton: Princeton University Press, 1978.

88. Hubbard, W. B. Effects of differential rotation on the gravitational figures of Jupiter and Saturn. *Icarus* **52**: 509–515 (1982).

89. Hunt, G. E. The atmospheres of the outer planets. *Ann. Rev. Earth & Plant. Sci.* **11**: 415–459 (1983).

90. Owen, T., and Cess, R. D. Methane absorption in the visible spectra of the outer planets and Titan. *Astrophys. J.* **197**: L37–L40 (1975).

91. Wallace, L. The structure of the Uranus atmosphere. *Icarus* **43**: 231–259 (1980).

92. Gulkis, S., Janssen, M. A., and Olsen, E. T. Evidence for depletion of ammonia in the Uranus atmosphere. *Icarus* **34**: 10–19 (1978).

93. Trafton, L. and Ramsey, D. A. The D/H ratio in the atmosphere of Uranus: detection of the $R_5(1)$ line of HD. *Icarus* **41**: 423–429 (1980).

94. Hubbard, W. B. and MacFarlane, J. J. Structure and evolution of Uranus and Neptune. *J. Geophys. Res.* **85:** 225–234 (1980).
95. Goody, R. M. The rotation of Uranus. In *Uranus and the Outer Planets,* (G. E. Hunt, ed.). Cambridge University Press: Cambridge, 1982.
96. Elliot, J. L. Rings of Uranus: a review of occultation results. In *Uranus and the Outer Planets,* (G. E. Hunt, ed.). Cambridge University Press: Cambridge, 1982.
97. Peale, S. J. The gravitational fields of the major planets. *Space Sci. Rev.* **14:** 412–423 (1973).
98. Harris, A. W. Where is Neptune's Pole? *Bull. Amer. Astron. Soc.* **12:** 705 (1980).
99. Slavsky, D. B. and Smith, H. J. The rotation period of Neptune. *Astrophys. J.* **226:** L49–L52 (1978).
100. Torbett, M. and Smoluchowski, R. The structure and magnetic field of Uranus. *Geophys. Res. Lett.* **6:** 675–676 (1979).
101. Reitsema, H. J., Hubbard, W. B., Lebofsky, L. A., and Tholen, D. J. Occultation by a possible third satellite of Neptune. *Science* **215:** 218–291 (1982).
102. Prinn, R. G. and Lewis, J. S. Kinetic inhibition of CO and N_2 reduction in the solar nebula. *Astrophys. J.* **238:** 357–364 (1980).
103. Morrison, D. Introduction to the satellites of Jupiter. In *Satellites of Jupiter* (D. Morrison, ed.). Tucson: University of Arizona Press, 1982, pp. 3–43.
104. Sill, G. T. and Clark, R. N. Composition of the surfaces of the Galilean satellites. In *Satellites of Jupiter* (D. Morrison, ed.). Tucson: University of Arizona Press, 1982, pp. 174–212.
105. Woronow, A. Strom, R. G., and Gurnis, M. Interpreting the cratering record: Mercury to Ganymede and Callisto. In *Satellites of Jupiter* (D. Morrison, ed.). Tucson: University of Arizona Press, 1982, pp. 237–276.
106. Lucchitta, B. K. and Soderblom, L. A. The geology of Europa. In *Satellites of Jupiter* (D. Morrison, ed.). Tucson: University of Arizona Press, 1982, pp. 521–555.
107. Lupo, M. J. and Lewis, J. S., Mass-radius relationships in icy satellites. *Icarus* **40:** 157–170 (1979).
108. Cassen, P., Peale, S., and Reynolds, R. Structure and thermal evolution of the Galilean satellites. In *Satellites of Jupiter* (D. Morrison, ed.), Tucson: University of Arizona Press, 1982.
109. Cameron, A. G. W. and Pollack, J. B. Origin of the solar system and Jupiter. In *Jupiter,* (T. Gehrels, ed.). Tucson: University of Arizona Press, 1976, pp. 61–84.
110. Greenberg, R. Orbital evolution of the Galilean satellites. In *Satellites of Jupiter* (D. Morrison, ed.), Tucson: University of Arizona Press, 1982, pp. 65–92.
111. Yoder, C. F. and Peale, S. J. The tides of Io. *Icarus* **47:** 1–35 (1981).
112. Gavrilov, S. V. and Zharkov, V. N. Love numbers of the giant planets. *Icarus* **32:** 443–449 (1977).
113. Schubert, G., Stevenson, D. J., and Ellsworth, K. Internal structures of the Galilean satellites. *Icarus* **47:** 46–59 (1981).
114. Hughes, T. J. The theory of thermal convection in polar ice sheets. *J. Glaciol.* **16:** 41–71 (1976).
115. Squyres, S. W., Reynolds, R. T., Cassen, P. M., and Peale, S. J. The evolution of Enceladus. *Icarus* **53:** 319–331 (1983).

General Bibliography

Allen, C. W. *Astrophysical Quantities,* London: Athlone Press, 1973.

Allison, M. and Stone, P. H. Saturn meteorology: a diagnostic assessment of thin-layer configurations for the zonal flow. *Icarus* **54**: 296–308 (1983).

Al'tshuler, L. V., Bakanova, A. A., and Trunin, R. F. Shock adiabats and zero isotherms of seven metals at high pressures. *Sov. Phys. JETP* **15**: 65–74 (1962).

Al'tshuler, L. V. and Sharipjanov, I. I. Additive equations of state of silicates at high pressures. *Bull. Acad. Sci. USSR Earth Phys.* **3**: 11–28 (1971).

Ananda, M. P., Sjogren, W. L., Phillips, R. J., Wimberly, R. N., and Bills, B. G. A low-order gravity field of Venus and dynamical implications. *J. Geophys. Res.* **85**: 8303–8318 (1980).

Anders, E. and Ebihara, M. Solar-system abundances of the elements. *Geochim. Cosmochim. Acta* **46**: 2363–2380 (1982).

Anderson, D. L. Plate tectonics on Venus. *Geophys. Res. Lett.* **8**: 309–311 (1981).

Anderson, D. L., Miller, W. F., Latham, G. V., Nakamura, Y., Toksoz, M. N., Dainty, A. M., Duennebier, F. K., Lazarewicz, A. R., Kovach, R. L., and Knight, T. C. D. Seismology on Mars. *J. Geophys. Res.* **82**: 4524–4546 (1977).

Barshay, S. and Lewis, J. S. Chemical structure of the deep atmosphere of Jupiter. *Icarus* **33**: 593–611 (1978).

Berge, G. L. and Gulkis, S. Earth-based observations of Jupiter: millimeter to meter wavelengths. In *Jupiter: Studies of the interior, atmosphere, magnetosphere, and satellites* (T. Gehrels, ed.). University of Arizona Press: Tucson, 1976.

Bills, B. G. and Ferrari, A. J. A harmonic analysis of lunar topography. *Icarus* **31**: 244–259 (1977).

Bills, B. G. and Ferrari, A. J. A harmonic analysis of lunar gravity. *J. Geophys. Res.* **85**: 1013–1025 (1980).

Brown, J. M. and McQueen, R. G. Melting of iron under core conditions. *Geophys. Res. Lett.* **7**: 533–536 (1980).

Brown, L. W. Possible radio emission from Uranus at 0.5 MHz. *Astrophys. J.* **207**: L209–L212 (1976).

Bruston, P., *et al.* Physical and chemical fractionation of deuterium in the interstellar medium. *Astrophys. J.* **243**: 161–169 (1981).

Buriez, J. C. and de Bergh, C. A study of the atmosphere of Saturn based on methane line profile near 1.1 microns. *Astron. Astrophys.* **94**: 382–390 (1981).

Busse, F. H. Generation of planetary magnetism by convection. *Phys. Earth and Plan. Interiors* **12**: 350–358 (1976).

Cameron, A. G. W. Abundances of elements in the solar system. *Space Sci. Rev* **15:** 121–146 (1973).

Cameron, A. G. W. and Pollack, J. B. Origin of the solar system and Jupiter. In *Jupiter,* (T. Gehrels, ed.). Tucson: University of Arizona Press, 1976, pp. 61–84.

Cassen, P., Young, R. E., Schubert, G., and Reynolds, R. T. Implications of an internal dynamo for the thermal history of Mercury. *Icarus* **28:** 501–508 (1976).

Cassen, P., Peale, S., and Reynolds, R. Structure and thermal evolution of the Galilean satellites. In *Satellites of Jupiter* (D. Morrison, ed.). Tucson: University of Arizona Press, 1982.

Christensen, E. J. and Balmino, G. Development and analysis of a twelfth degree and order gravity model for Mars. *J. Geophys. Res.* **84:** 7943–7953 (1979).

Christensen-Dalsgaard, J. and Gough, D. On the interpretation of 5-minute oscillations in the solar spectrum. *Mon. Not. R. A. S.* **198:** 141–171 (1982).

Cox, J. P. and Giuli, R. T. *Stellar Structure: Physical Principles.* Gordon & Breach: New York, 1968, p. 385.

Cruikshank, D. The satellites of Uranus, in *Uranus and the Outer Planets* (G. E. Hunt, ed.). Cambridge: Cambridge University Press, 1982.

Dermott, S. F. Tidal dissipation in the solid cores of the major planets. *Icarus* **37:** 310–321 (1979).

Desch, M. D. and Kaiser, M. L. Voyager measurement of the rotation period of Saturn's magnetic field. *Geophys. Res. Let.* **8:** 253–256 (1981).

DeWitt, H. E. and Hubbard, W. B. Statistical mechanics of light elements at high pressure. IV. A model free energy for the metallic phase. *Astrophys. J.* **205:** 295–301 (1976).

Donivan, F. F. and Carr, T. D. Jupiter's decametric rotation period. *Astrophys. J.* **157:** L65–L68 (1969).

Dyal, P., Parkin, C. W., and Daily, W. D. Magnetism and the interior of the moon. *Rev. Geophys. and Spa. Phys.* **12:** 23–70 (1974).

Dyal, P., Parkin, C. W., and Daily, W. D. Structure of the lunar interior from magnetic field measurements. *Proc. Lunar Sci. Conf.* **7:** 3077–3095 (1976).

Dziewonski, A. M. and Anderson, D. L. Preliminary reference earth model. *Phys. Earth and Plan. Int.* **25:** 297–356 (1981).

Elliot, J. L. Rings of Uranus: a review of occultation results, in *Uranus and the Outer Planets* (G. E. Hunt, ed.). Cambridge: Cambridge University Press, 1982.

Encrenaz, T. and Combes, M. On the C/H and D/H ratios in the atmospheres of Jupiter and Saturn. *Icarus* **52:** 54–61 (1982).

Esposito, P. B., Anderson, J. D., and Ng, A. T. Y. Experimental determination of Mercury's mass and oblateness. *Space Res.* **17:** 639–644 (1977).

Fazio, C. G., Traub, W. O., Wright, E. L., Low, F. J., and Trafton, L. The effective temperature of Uranus. *Astrophys. J.* **209:** 633–637 (1976).

Ferrari, A. J., Sinclair, W. S., Sjogren, W. L., Williams, J. G., and Yoder, C. F. Geophysical parameters of the earth-moon system. *J. Geophys. Res.* **85:** 3939–3951 (1980).

Franks, F. *Water: A Comprehensive Treatise.* New York: Plenum Press, 1972.

Freeman, K. C. and Lynga, G. Data for Neptune from occultation observations. *Astrophys. J.* **160:** 767–780 (1970).

Gaposchkin, E. M. Global gravity field to degree and order 30 from Geos 3 satellite altimetry and other data. *J. Geophys. Res.* **85:** 7221–7234 (1980).

Garland, G. D. *Introduction to Geophysics.* Philadelphia: Saunders, 1971.

Gautier, D. Conrath, B., Flasar, M., Hanel, R., Kunde, V., Chedin, A., and Scott, N. The helium abundance of Jupiter from Voyager. *J. Geophys. Res.* **86:** 8713–8720 (1981).

Gavrilov, S. V. and Zharkov, V. N. Love numbers of the giant planets. *Icarus* **32:** 443–449 (1977).

Goettel, K. A., Shields, J. A., and Decker, D. A. Density constraints on the composition of Venus. *Proc. Lunar Sci. Conf.* **12B:** 1507–1516 (1981).

Goldreich, P. and Nicholson, P. Turbulent viscosity and Jupiter's tidal Q. *Icarus* **30:** 301–304 (1977).

Goldreich, P. and Peale, S. Spin-orbit coupling in the solar system. *Astron. J.* **71:** 425–438 (1966).

Goldreich, P. and Soter, S. Q in the solar system. *Icarus* **5:** 375–389 (1966).

Goody, R. M. The rotation of Uranus. In *Uranus and the Outer Planets,* (G. E. Hunt, ed.). Cambridge University Press: Cambridge, 1982.

Graboske, H. C., Jr. and Grossman, A. S. Evolution of low-mass stars. IV. Effects of multilevel atomic partition functions for the ideal-gas region. *Astrophys. J.* **170:** 363–370 (1971).

Graboske, H. C., Pollack, J. B., Grossman, A. S., and Olness, R. J. The structure and evolution of Jupiter: the fluid contraction phase. *Astrophys. J.* **199:** 265–281 (1975).

Greenberg, R. Orbital evolution of the Galilean satellites. In *Satellites of Jupiter* (D. Morrison, ed.). Tucson: University of Arizona Press, 1982, pp. 65–92.

Grossman, A. S. and Graboske, H. C., Jr. Evolution of low-mass stars. V. Minimum mass for the deuterium main sequence. *Astrophys. J.* **180:** 195–198 (1973).

Grossman, A. S., Pollack, J. B., Reynolds, R. T., Summers, A. L., and Graboske, H. C., Jr. The effect of dense cores on the structure and evolution of Jupiter and Saturn. *Icarus* **42:** 358–379 (1980).

Grossman, L. Condensation in the primitive solar nebula. *Geochim. Cosmochim. Acta* **36:** 597–619 (1972).

Gulkis, S., Janssen, M. A., and Olsen, E. T. Evidence for depletion of ammonia in the Uranus atmosphere. *Icarus* **34;** 10–19 (1978).

Hager, B. H. and O'Connell, R. J. Lithospheric thickening and subduction, plate motions and mantle convection. In *Physics of the Earth's Interior—Proc. Int. School of Physics "Enrico Fermi"* (A. M. Dziewonski, ed.), Course LXXVIII. North-Holland: Amsterdam, 1980.

Hanel, R., *et al.* Infrared observations of the Saturnian system from Voyager 1. *Science* **212:** 192–200 (1981).

Hanel, R. A., Conrath, B. J., Herath, L. W., Kunde, V. G., and Pirraglia, J. A. Albedo, internal heat, and energy balance of Jupiter: preliminary results from the Voyager infrared investigation. *J. Geophys. Res.* **86:** 8705–8712 (1981).

Hanel, R. A., Conrath, B. J., Kunde, V. G., Pearl, J. C., and Pirraglia, J. A. Albedo, internal heat flux, and energy balance of Saturn. *Icarus* **53:** 262–285 (1983).

Harris, A. W. Where is Neptune's Pole? *Bull. Amer. Astron. Soc.* **12:** 705 (1980).

Harrington, R. S. and Christy, J. W. The satellite of Pluto. III. *Astron. J.* **86:** 442–443 (1981).

Hayashi, C. Evolution of protostars. *Ann. Rev. Astron. and Astrophys.* **4:** 171–192 (1966).

Hazen, R. M., Mao, H. K., Finger, L. W., and Bell, P. M. Crystal structures and compression of Ar, Ne, and CH_4 at 20 °C to 90 kbar. *Carnegie Inst. of Wash. — Yearbook* **79:** 348–351 (1980).

Hood, L. L., Herbert, F., and Sonett, C. P. The deep lunar electrical conductivity profile: structural and thermal inferences. *J. Geophys. Res.* **87:** 5311–5326 (1982).

Howard, T., *et al.* Mercury: results on mass, radius, ionosphere, and atmosphere from Mariner 10 dual frequency radio signal. *Science* **185:** 179–180 (1974).

Hubbard, W. B. Thermal Structure of Jupiter. *Astrophys. J.* **152:** 745–754 (1968).

Hubbard, W. B. Gravitational field of a rotating planet with a polytropic index of unity. *Soviet Astron.—AJ* **18(5):** 621–624 (1975).

Hubbard, W. B. Tides in the giant planets. *Icarus* **23:** 42–50 (1974).

Hubbard, W. B. The Jovian surface condition and cooling rate. *Icarus* **30:** 305–310 (1977).

Hubbard, W. B. Effects of differential rotation on the gravitational figures of Jupiter and Saturn. *Icarus* **52:** 509–515 (1982).

Hubbard, W. B. and Horedt, G. P. Computation of Jupiter interior models from gravitational inversion theory. *Icarus* **54:** 456–465 (1983).

Hubbard, W. B. and Lampe, M. Thermal conduction by electrons in stellar matter. *Astrophys. J. (Suppl.)* **18:** 297–346 (1969).

Hubbard, W. B. and MacFarlane, J. J. Structure and evolution of Uranus and Neptune. *J. Geophys. Res.* **85:** 225–234 (1980).

Hubbard, W. B. and MacFarlane, J. J. Theoretical predictions of deuterium abundances in the Jovian planets. *Icarus* **44:** 676–682 (1980).

Hubbard, W. B., MacFarlane, J. J., Anderson, J. D., Null, G. W., and Biller, E. D. Interior structure of Saturn from Pioneer 11 gravity data. *J. Geophys. Res.* **85:** 5909–5916 (1980).

Hubbard, W. B. and Slattery, W. L. Statistical mechanics of light elements at high pressures. I. Theory and results for metallic hydrogen with simple screening. *Astrophys. J.* **168:** 131–139 (1971).

Hughes, T. J. The theory of thermal convection in polar ice sheets. *J. Glaciol.* **16:** 41–71 (1976).

Hunt, G. E. The atmospheres of the outer planets. *Ann. Rev. Earth and Plant. Sci.* **11:** 415–459 (1983).

Ingersoll, A. P. Pioneer 10 and 11 observations and the dynamics of Jupiter's atmosphere. *Icarus* **29:** 245–253 (1976).

Johnston, D. H., McGetchin, T. R., and Toksoz, M. N. The thermal state and internal structure of Mars. *J. Geophys. Res.* **79:** 3959–3971 (1974).

Kaula, W. M. *Theory of Satellite Geodesy.* Blaisdell: Waltham, Mass., 1966.

Kaula, W. M. The moment of inertia of Mars. *Geophys. Res. Lett.* **6:** 194–196 (1979).

Keihm, S. J. and Langseth, M. G. Lunar thermal regime to 300 km. *Proc. 8th Lunar Sci. Conf.* **1:** 499–514 (1977).

Kirzhnits, D. A. *Field Theoretical Methods in Many-Body Systems*. Oxford: Pergamon, 1967.

Klaasen, K. P. Mercury's rotation axis and period, *Icarus* **28**: 469–478 (1976).

Knacke, R. F., Kim, S. J., Ridgway, S. T., and Tokunaga, A. T. The abundances of CH_4, CH_3D, NH_3, and PH_3 in the troposphere of Jupiter derived from high-resolution 1100–1200 cm^{-1} spectra. *Astrophys. J.* **262**: 388–395 (1982).

Kunde, V., *et al.* The tropospheric gas composition of Jupiter's north equatorial belt (NH_3, PH_3, CH_3D, GeH_4, H_2O) and the Jovian D/H isotopic ratio. *Astrophys. J.* **263**: 443–467 (1982).

Landau, L. D. and Lifshitz, E. M. *Electrodynamics of Continuous Media*. Reading, Mass.: Addison-Welsey, 1960, pp. 193–194.

Landau, L. D. and Lifshitz, E. M. *Statistical Physics*. London: Pergamon, 1969.

Larson, H. P., Fink, U., Smith, H. A., and Davis, D. S. The middle-infrared spectrum of Saturn: evidence for phosphine and upper limits to other trace constituents. *Astrophys. J.* **240**: 327–337 (1980).

Levy, E. H. Generation of planetary magnetic fields. *Ann. Rev. Earth and Plan. Sci.* **4**: 159–185 (1976).

Lewis, J. S. Low temperature condensation from the solar nebula. *Icarus* **16**: 241–252 (1972).

Lewis, J. S. Metal/silicate fractionation in the solar system. *Earth and Plant. Sci. Lett.* **15**: 286–290 (1972).

Lindal, G. F., *et al.* The atmosphere of Jupiter: an analysis of the Voyager radio occultation measurements. *J. Geophys. Res.* **86**: 8721–8727 (1981).

Liu, L. G. Calculations of high-pressure phase transitions in the system $MgO\text{-}SiO_2$ and implications for mantle discontinuities. *Phys. Earth and Plan. Int.* **19**: 319–330 (1979).

Liu, L. G. and Bassett, W. A. The melting of iron up to 200 Kbar. *J. Geophys. Res.* **80**: 3777–3782 (1975).

Lowenstein, R. F., Harper, D. A., and Moseley, H. The effective temperature of Neptune. *Astrophys. J.* **218**: L145–L146 (1977).

Lucchitta, B. K. and Soderblom, L. A. The geology of Europa. In *Satellites of Jupiter* (D. Morrison, ed.). Tucson: University of Arizona Press, 1982, pp. 521–555.

Lupo, M. J. and Lewis, J. S. Mass-radius relationships in icy satellites. *Icarus* **40**: 157–170 (1979).

MacFarlane, J. J. and Hubbard, W. B. Statistical mechanics of light elements at high pressure. V. Three-dimensional Thomas-Fermi-Dirac theory. *Astrophys. J.* **272**: 301–310 (1983).

Macy, W., Jr. and Smith, W. H. Detection of HD on Saturn and Uranus and the D/H ratio. *Astrophys. J.* **222**: L137–L140 (1978).

Mansoori, G. A. and Canfield, F. B. Variational approach to the equilibrium thermodynamic properties of simple liquids. I. *J. Chem. Phys.* **51**: 4958–4967 (1969).

Marten, A., Courtin, R., Gautier, D., and Lacombe, A. Ammonia vertical density profiles in Jupiter and Saturn from their radioelectric and infrared emissivities. *Icarus* **41**: 410–422 (1980).

McGetchin, T. R. and Smyth, J. R. The mantle of Mars: some possible geological implications of its high density. *Icarus* **34**: 512–536 (1978).

McQueen, R. G., Marsh, S. P., Taylor, J. W., Fritz, J. N., and Carter, W. J. The equation of state of solids from shock wave studies. In *High Velocity Impact Phenomena,* pp. 293–568. New York: Academic Press, 1970.

Melosh, H. J. Global tectonics of a despun planet. *Icarus* **31:** 221–243 (1977).

Melosh, H. J. and Dzurisin, D. Mercurian global tectonics: a consequence of tidal despinning? *Icarus* **35:** 227–236 (1978).

Mitchell, A. C. and Nellis, W. J. Equation of state and electrical conductivity of water and ammonia shocked to the 100 GPa (1 Mbar) pressure range. *J. Chem. Phys.* **76:** 6273–6281 (1982).

Morgan, J. W. and Anders, E. Chemical composition of Mars. **43:** 1601–1610 (1979).

Morrison, D. Introduction to the satellites of Jupiter. In *Satellites of Jupiter* (D. Morrison, ed.). Tucson: University of Arizona Press, 1982, pp. 3–43.

Nakamura, Y. Seismic velocity structure of the lunar mantle. *J. Geophys. Res.* **88:** 677–686 (1983).

Nakamura, Y., Latham, G. V., and Dorman, H. J. Apollo lunar seismic experiment —final summary. *J. Geophys. Res. (Suppl.)* **87:** A117–A123 (1982).

Nakamura, Y., Latham, G., Lammlein, D., Ewing, M., Duennebier, F., and Dorman, J. Deep lunar interior inferred from recent seismic data. *Geophys. Res. Lett.* **1:** 137–140 (1974).

Nellis, W. J., Ree, F. H., van Thiel, M., and Mitchell, A. C. Shock compression of liquid carbon monoxide and methane to 90 GPa (900 kbar). *J. Chem. Phys.* **75:** 3055–3063 (1981).

Nellis, W. J., Ross, M., Mitchell, A. C., van Thiel, M., Young, D. A., Ree, F. H., and Trainor, R. J. Equation of state of molecular hydrogen and deuterium from shock-wave experiments to 760 kbar. *Phys. Rev. A* **27:** 608–611 (1983).

Null, G. W. Gravity field of Jupiter and its satellites from Pioneer 10 and Pioneer 11 tracking data. *Astron. J.* **81:** 1153–1161 (1976).

Null, G. W., Lau, E. L., Biller, E. D., and Anderson, J. D. Saturn gravity results obtained from Pioneer 11 tracking data and Earth-based Saturn satellite data. *Astron. J.* **86:** 456–468 (1981).

Orton, G. S. and Ingersoll, A. P. Saturn's atmospheric temperature structure and heat budget. *J. Geophys. Res.* **85:** 5871–5881 (1980).

Owen, T. and Cess, R. D. Methane absorption in the visible spectra of the outer planets and Titen. *Astrophys. J.* **197:** L37–L40 (1975).

Pagel, B. E. J. Solar abundances: a new table (October 1976). In *Physics and Chemistry of the Earth* **11:** 79–80 (L. H. Ahrens, ed.). Oxford: Pergamon, 1977.

Parker, E. N. *Cosmical Magnetic Fields—Their Origin and Activity.* Clarendon Press: Oxford, 1979, pp. 538–541.

Peale, S. J. The gravitational fields of the major planets. *Spa. Sci. Rev.* **14:** 412–423 (1973).

Peale, S. J., Cassen, P., and Reynolds, R. T. Melting of Io by tidal dissipation. *Science* **203:** 892–894 (1979).

Pearl, J. C. and Sinton, W. M. Hot spots of Io. In *Satellites of Jupiter* (D. Morrison, ed.). Tucson: University of Arizona Press, 1982, pp. 724–755.

Pechmann, J. B. and Melosh, H. J. Global fracture patterns of a despun planet: application to Mercury. *Icarus* **38**: 243–250 (1979).

Pettengill, G., *et al.* Pioneer Venus radar results: altimetry and surface properties. *JGR* **85**: 8261–8270 (1980).

Pollack, J. B., Toon, O. B., and Boese, R. Greenhouse models of Venus' high surface temperature, as constrained by Pioneer Venus measurements. *J. Geophys. Res.* **85**: 8223–8231 (1980).

Prinn, R. G. and Barshay, S. S. Carbon monoxide on Jupiter and implications for atmospheric convection. *Science* **198**: 1031–1034 (1977).

Prinn, R. G. and Lewis, J. S. Phosphine on Jupiter and implications for the great red spot. *Science* **190**: 274–276 (1975).

Prinn, R. G. and Lewis, J. S. Kinetic inhibition of CO and N_2 reduction in the solar nebula. *Astrophys. J.* **238**: 357–364 (1980).

Prinn, R. G. and Owen, T. Chemistry and spectroscopy of the Jovian atmosphere. In *Jupiter* (T. Gehrels, ed.). University of Arizona Press: Tucson, 1976.

Reasenberg, R. D. The moment of inertia and isostasy of Mars. *J. Geophys. Res.* **82**: 369–375 (1977).

Ree, F. Equation of state of water. *Rep. UCRL-52190,* Lawrence Livermore Lab., Livermore, Calif. (1976).

Reitsema, H. J., Hubbard, W. B., Lebofsky, L. A., and Tholen, D. J. Occultation by a possible third satellite of Neptune. *Science* **215**: 289–291 (1982).

Reynolds, R. T. and Summers, A. L. Models of Uranus and Neptune. *J. Geophys. Res.* **70**: 199–208 (1965).

Ridgway, S. T., Larson, H. P., and Fink, U. The infrared spectrum of Jupiter. In *Jupiter* (T. Gehrels, ed.). University of Arizona Press: Tucson, 1976.

Rikitake, T. *Electromagnetism and the Earth's Interior.* Amsterdam: Elsevier, 1966.

Ringwood, A. E. Phase transformations and the constitution of the mantle. *Phys. Earth and Plan. Interiors* **3**: 109–135 (1970).

Ringwood, A. E. and Anderson, D. L. Earth and Venus: a comparative study. *Icarus* **30**: 243–253 (1977).

Ringwood, A. E. and Kesson, S. E. Composition and origin of the moon. *Proc. 8th Lunar Sci. Conf.* **1**: 371–398 (1977).

Ross, M., Graboske, H. C., Jr., and Nellis, W. J. Equation of state experiments and theory relevant to planetary modelling. *Phil. Trans. R. Soc. Lond. A* **303**: 303–313 (1981).

Ross, M. and Ree, F. H. Repulsive forces of simple molecules and mixtures at high density and pressure. *J. Chem. Phys.* **73**: 6146–6152 (1980).

Russell, C. T., Elphic, R. C., and Slavin, J. A. Limits on the possible intrinsic magnetic field of Venus. *J. Geophys. Res.* **85**: 8319–8332 (1980).

Salpeter, E. E. Energy and pressure of a zero-temperature plasma. *Astrophys. J.* **134**: 669–682 (1961).

Salpeter, E. E. On convection and gravitational layering in Jupiter and in stars of low mass. *Astrophys. J.* **181**: L83–L86 (1973).

Schatz, J. F. and Simmons, G. Thermal conductivity of earth materials at high temperatures. *J. Geophys. Res.* **77**: 6966–6983 (1972).

Scherrer, P. H., Wilcox, J. M., Christensen-Dalsgaard, J., and Gough, D. Observation of low-degree 5-minute modes of solar oscillation. *Nature* **297**: 312–313 (1982).

Schubert, G., Cassen, P., and Young, R. E. Subsolidus convective cooling histories of terrestrial planets. *Icarus* **38**: 192–211 (1979).

Schubert, G., Stevenson, D. J., and Ellsworth, K. Internal structures of the Galilean satellites. *Icarus* **47**: 46–59 (1981).

Schwarzschild, M. *Structure and evolution of the stars.* Princeton: Princeton University Press, 1958.

Seiff, A., Kirk, D. B., Young, R. E., Blanchard, R. C., Findlay, J. T., Kelly, G. M., and Sommer, S. C. Measurements of thermal structural and thermal contrasts in the atmosphere of Venus and related dynamical observations: results from the four Pioneer Venus probes. *J. Geophys. Res.* **85**: 7903–7933 (1980).

Shapiro, I. I., Campbell, D. B., and DeCampli, W. M. Nonresonance rotation of Venus? *Astrophys. J.* **230**: L123–L126 (1979).

Sharma, S. K., Mao, H. K., and Bell, P. M. Raman measurements of hydrogen in the pressure range 0.2–630 kbar at room temperature. *Phys. Rev. Lett.* **44**: 886–888 (1980).

Siegfried, R. W., II and Solomon, S. C. Mercury: internal structure and thermal evolution. *Icarus* **23**: 192–205 (1974).

Sill, G. T. Sulphuric acid in the clouds of Venus. *Commun. Lunar and Plan. Lab.* No. **171**: 191–198 (1972).

Sill, G. T. and Clark, R. N. Composition of the surfaces of the Galilean satellites. In *Satellites of Jupiter* (D. Morrison, ed.). Tucson: University of Arizona Press, 1982, pp. 174–212.

Sjogren, W. L., Bills, B. G., Birkeland, P. W., Esposito, P. B., Konopliv, A. R., Mottinger, N. A., Ritke, S. J., and Phillips, R. J. Venus gravity anomalies and their correlations with topography. *J. Geophys. Res.* **85**: 8303–8313 (1983).

Slavin, J. A. and Holzer, R. E. The solar wind interaction with Mars revisited. *J. Geophys. Res.* **87**: 10285–10296 (1982).

Slavsky, D. B. and Smith, H. J. The rotation period of Neptune. *Astrophys. J* **226**: L49–L52 (1978).

Smith, B. A., *et al.* Encounter with Saturn: Voyager 1 imaging science results. *Science* **212**: 163–191 (1981).

Smith, B. A., *et al.* A new look at the Saturn system: the Voyager 2 images. *Science* **215**: 504–537 (1982).

Smith, E. J., Davis, L., Jr., and Jones, D. E. Jupiter's magnetic field and magnetosphere. In *Jupiter* (T. Gehrels, ed.). Tucson: University of Arizona Press, 1976.

Smith, E. J., Davis, L., Jr., Jones, D. E., Coleman, P. J., Jr., Colburn, D. S., Dyal, P., and Sonett, C. P. Saturn's magnetosphere and its interaction with the solar wind. *J. Geophys. Res.* **85**: 5655–5674 (1980).

Solomon, S. C. Some aspects of core formation in Mercury. *Icarus* **28**: 509–521 (1976).

Sonett, C. P. Electromagnetic induction in the moon. *Rev. Geophys. and Spa. Phys.* **20**: 411–455 (1982).

Sonett, C. P. and Duba, A. Lunar temperature and global heat flux from laboratory electrical conductivity and lunar magnetometer data. *Nature* **258**: 118–121 (1975).

Sonett, C. P., Colburn, D. S., Schwartz, K., and Keil, K. The melting of asteroidal-sized bodies by unipolar dynamo induction. *Astrophys. and Space Sci.* **7:** 446–488 (1970).

Sonett, C. P., Smith, B. F., Colburn, D. S., Schubert, G., and Schwartz, K. The induced magnetic field of the moon: conductivity profiles and inferred temperature. *Proc. Lunar Sci. Conf.* **3:** 2309–2336 (1972).

Squyres, S. W., Reynolds, R. T., Cassen, P. M., and Peale, S. J. The evolution of Enceladus. *Icarus* **53:** 319–331 (1983).

Stacy, F. D. *Physics of the earth.* New York: Wiley & Sons, 1969.

Stevenson, D. J. Miscibility gaps in fully pressure-ionized binary alloys. *Phys. Lett.* **58A:** 282–285 (1976).

Stevenson, D. J. Thermodynamics and phase separation of dense fully-ionized hydrogen-helium fluid mixtures. *Phys. Rev.* **12B:** 3999–4007 (1975).

Stevenson, D. J. Saturn's luminosity and magnetism. *Science* **208:** 746–748 (1980).

Stevenson, D. J. Anomalous bulk viscosity of two-phase fluids and implications for planetary interiors. *J. Geophys. Res.* **88:** 2445–2455 (1983).

Stevenson, D J. and Salpeter, E. E. The dynamics and helium distribution in hydrogen-helium fluid planets. *Astrophys. J. Suppl.* **35:** 239–261 (1977).

Stevenson, D. J. and Ashcroft, N. W. Conduction in fully ionized liquid metals. *Phys. Rev. A* **9:** 782–789 (1974).

Stevenson, D. J. and Salpeter, E. E. The phase diagram and transport properties for hydrogen-helium fluid planets. *Astrophys. J. Suppl.* **35:** 221–237 (1977).

Stevenson, D. J. and Salpeter, E. E. Interior models of Jupiter. In *Jupiter* (T. Gehrels, ed.). University of Arizona Press: Tucson, 1976.

Stone, E. C. and Miner, E. D. Voyager 1 encounter with the Saturnian system. *Science* **212:** 159–163 (1981).

Stone, E. C. and Miner, E. D. Voyager 2 encounter with the Saturnian system. *Science* **215:** 499–504 (1982).

Strom, R. G., Trask, N. J., and Guest, J. E. Tectonism and volcanism on Mercury. *J. Geophys. Res.* **80:** 2478–2507 (1975).

Surkov, Yu. A. Natural radioactivity of the moon and planets. *Proc. Lunar and Plant. Sci.* **12B:** 1377–1386 (1981).

Synnott, S. P., *et al.* Orbits of the small satellites of Saturn. *Science* **212:** 191–192 (1981).

Tassoul, J.-L. *Theory of Rotating Stars.* Princeton: Princeton Univ. Press, 1978.

Toksoz, M. N. and Hsui, A. T. Thermal history and evolution of Mars. *Icarus* **34;** 537–547 (1978).

Toksoz, M. N., Hsui, A. T., and Johnston, D. H. Thermal evolutions of the terrestrial planets. *The Moon and Planets* **18:** 281–320 (1978).

Torbett, M. and Smoluchowski, R. The structure and magnetic field of Uranus. *Geophys. Res. Lett.* **6:** 675–676 (1979).

Toulmin, P., III, *et al.* Geochemical and mineralogical interpretation of the Viking inorganic chemical results. *J. Geophys. Res.* **82:** 4625–4634 (1977).

Trafton, L. M. On the He-H_2 thermal opacity in planetary atmospheres. *Astrophys. J.* **179:** 971–976 (1973).

Trafton, L. and Ramsey, D. A. The D/H ratio in the atmosphere of Uranus: detection of the $R_5(1)$ line of HD. *Icarus* **41:** 423–429 (1980).

Tyler, G. L., *et al.* Radio science investigations of the Saturn system with Voyager 1: preliminary results. *Science* **212:** 201–206 (1981).

Tyler, G. L., *et al.* Radio science with Voyager 2 at Saturn: atmosphere and ionosphere and the masses of Mimas, Tethys, and Iapetus. *Science* **215:** 553–558 (1982).

Usselman, T. M. Experimental approach to the state of the core: Parts I and II. *Am. J. of Sci.* **275:** 278–303 (1975).

Veverka, J. and Burns, J. A. The moons of Mars. *Ann. Rev. Earth and Plan. Sci.* **8:** 527–558 (1980).

Wallace, L. The structure of the Uranus atmosphere. *Icarus* **43:** 231–259 (1980).

Weidenschilling, S. J. and Lewis, J. S. Atmospheric and cloud structures of the Jovian planets. *Icarus* **20:** 465–476 (1973).

Whang, Y. C. Magnetospheric magnetic field of Mercury. *J. Geophys. Res.* **82:** 1024–1030 (1977).

Wood, J. A. *The Solar System.* Englewood Cliffs, N. J.: Prentice-Hall, 1979.

Woronow, A., Strom, R. G., and Gurnis, M. Interpreting the cratering record: Mercury to Ganymede and Callisto. In *Satellites of Jupiter* (D. Morrison, ed.). Tucson: University of Arizona Press, 1982, pp. 237–276.

Yoder, C. F. and Peale, S. J. The tides of Io. *Icarus* **47:** 1–35 (1981).

Young, D. A., McMahan, A. K., and Ross, M. Equation of state and melting curve of helium to very high pressure. *Phys. Rev. B* **24:** 5119–5127 (1981).

Zapolsky, H. S. and Salpeter, E. E. The mass-radius relationship for cold bodies of low mass. *Astrophys. J.* **158:** 809–813 (1969).

Zharkov, V. N. and Trubitsyn, V. P. *Physics of Planetary Interiors.* Tucson: Pachart, 1978.

Index

Index